Telemetry Systems Engineering

For a listing of recent titles in the *Artech House Telecommunications Library,* turn to the back of this book.

Telemetry Systems Engineering

Frank Carden
Russell Jedlicka
Robert Henry

Artech House
Boston • London
www.artechhouse.com

Library of Congress Cataloging-in-Publication Data
Carden, Frank.
   Telemetry systems engineering/Frank Carden, Russell P. Jedlicka, Robert Henry.
     p. cm. — (Artech House telecommunications library)
   Includes bibliographical references and index.
   ISBN 1-58053-257-8 (alk. paper)
   1. Telecommunication systems.   I. Jedlicka, Russell P.   II. Henry, Robert.
   III. Title.   IV. Series.
TK5101 .C29824 2002
621.382—dc21   2001056564

British Library Cataloguing in Publication Data
Carden, Frank.
   Telemetry systems engineering. — (Artech House telecommunications library)
   1. Radio telemetry   2. Systems engineering
   I. Title   II. Jedlicka, Russell P.   III. Henry, Robert
621.3'84

   ISBN 1-58053-257-8

Cover design by Yekaterina Ratner

International Standard Book Number: 1-58053-257-8
Library of Congress Catalog Card Number: 2001056564

10 9 8 7 6 5 4 3 2 1

*We dedicate this book to our wives Barbara, Billie, and Sue
for their untiring and continual help in writing this book,
for their companionship during this period, and in particular,
for their almost unlimited patience during tense moments.*

Contents

Preface

This text presents introductory and advanced concepts of telemetry systems. In essence, it is written for both the beginning and the seasoned telemetering engineer and telecommunication engineer. The details of telemetry systems design are emphasized but always with a view of the overall system. Current and future systems and their design are discussed whenever possible, in the context of specifications outlined in the Inter-Range Instrumentation Group (IRIG) documents, specifically 106-00.

Digital communications is the primary thrust of *Telemetry Systems Engineering* for the simple reason that most new telemetry systems, and indeed, communication systems in general are primarily digital. The book addresses two subjects of substantial importance to the telemetry engineer: (1) M-ary communication systems, and (2) link analysis. Engineers that have studied telemetry systems from *Telemetry Systems Design* and students who have taken courses using the first book have suggested strongly that both these topics be included in the next book.

To that end, models for transmitters and receivers capable of handling QPSK, OQPSK, and Feher's patented FQPSK are discussed along with their expected bit error rates and spectral occupancy. Further, models for future telemetry systems employing more advanced modulation techniques, such as M-ary PSK, M-ary FSK, and spread spectrum are developed. New developments, with possible high spectral efficiency, such as multi-h continuous-phase modulation, are given. Chapters 6 through 10 cover modulation techniques of digital telemetry systems.

Link analysis allows the telemetry engineer to determine the carrier-to-noise ratio, and ultimately the bit error rate (BER) for digital systems, or the signal-to-noise ratio for analog signal, both of which determine the quality of the output signal for the respective systems. The link analysis section, Chapters 11 through 14, describing the characteristics of a broad class of antennas, is self-contained, comprehensive, and developed in such a way that the telemetry engineer with a limited background in electromagnetic theory can understand and use the models to perform link analysis to predict system performance.

The analysis of systems corrupted with noise is a central theme, while throughout the text efficient spectrum utilization is emphasized.

In Chapters 1 and 2, introductory concepts are developed that should help beginning engineers and managers understand bandwidth requirements for various modulation schemes.

Chapter 3, as in the first book, is devoted to the design of the preemphasis schedule for FM/FM systems in a systematic way, amenable to algorithmic code development. The process creates a unique preemphasis schedule for those anticipated operating conditions. This material remains one of the few sources available to the telemetry engineer for the algorithmic design of a preemphasis schedule.

Chapter 4 describes the difference between analog and digital signals, the parameters that characterize them, and illustrates how to determine the baud rate and the bit rate of a digital signal. The reader interested primarily in digital systems should start with this chapter. The operation, advantages, and disadvantages of serial and parallel A/D converters and the purpose of each block of a time division multiplex (TDM) system are covered. Encoder design by trading off bandwidth requirements with signal-to-noise ratio requirements is defined and illustrated. The basic digital modulation methods, ASK, FSK, and PSK are introduced.

Chapter 5 develops the protocol for the design of the data frames of the TDM system according to the IRIG requirements.

Chapter 6 is concerned with the design of systems employing PCM/FM, the past and current workhorse of the telemetry industry. For PCM/FM, a process for the determination of all parameters for the transmit and receive systems is carefully laid out, and a design procedure for a specified bit rate and attendant BER is developed.

Chapter 7 covers binary phase-shift keying, its spectrum, and predicted BER.

Chapter 8 is devoted to the development of binary digital communication system models, including the numerous forms of *frequency-shift keying*

(FSK), and *phase-shift keying* (PSK). Mathematical modeling, the generation and detection of *binary* FSK (BFSK), *binary* PSK (BPSK), and their derivates are developed. The composite and component waveforms, the spectrum, and the bandwidth efficiency are determined and illustrated. Bit error rates and power efficiency of these methods are given, and a performance comparison made.

Chapter 9 introduces M-ary modulation techniques, which are preferable to binary methods when sending data over bandwidth-constrained channels. Mathematical modeling, generation and detection, composite and component waveforms, the spectrum, bandwidth efficiency, BER, and power efficiency are determined and illustrated. The family of Feher's patented FQPSK modulation techniques is discussed, as is the enhanced FQPSK under development by JPL. Laboratory and actual flight test results of this important modulation technique are given, including spectral occupancy, defined by the 99% bandwidth, and the BER. Several important IRIG 106-00 requirements are detailed for FQPSK-B.

Chapter 10 covers the concept of *spread spectrum* and the two most common implementations, *direct sequence spread spectrum* (DSSS) and *frequency-hop spread spectrum* (FHSS). Modeling and implementation of each is presented, along with the spectral occupancy. Channel resource division according to a unique spreading code assignment known as *code-division multiple access* (CDMA) is covered. Application examples, including use in cellular phones, the Tracking and Data Relay Satellite (TDRS), and Global Positioning System (GPS) satellites are discussed.

Chapter 11 starts the section on link analysis and is devoted to basic RF components of the telemetry link, antenna gain, radiation patterns, polarization and mismatch, and the most practical form of the Friis transmission formula.

Chapter 12 continues developing the groundwork for the link analysis of telemetry systems by introducing radiating elements, the operation of microstrip patch antennas and their design, and the characteristics of conformal arrays.

Chapter 13 completes the development of the concepts and equations necessary for link analysis by introducing and defining equivalent noise temperature, receiver, and system noise figure created by the cascading of components, gain-to-noise temperature (G/T) as well as its measurement, the carrier-to-noise ratio leading to the prediction of bit error rates, and finally the link margin.

Chapter 14 summarizes the systematic steps for link analysis developed in Chapters 11 through 13; presents a process for the design of telemetry links, developing the criteria for the selection of receivers, transmitters, and antennas; and details application examples.

Chapter 15 gives an overview of the synchronization techniques employed in telemetry systems and details the demultiplexing system of the ground station.

Chapter 16 covers systems employing both analog and digital modulation.

Chapter 17 relates error-correcting convolutional coding to telemetry systems. Viterbi decoding, using both hard and soft decisions, is analyzed. Actual results of computer simulations of off-the-shelf chips that perform the encoding and Viterbi decoding in a PCM/FM system are presented, the only text available on the market today that covers this material.

Chapters 18 and 19 give an overview, as well as specific applications, of industrial telemetry applications. When telemetry is used both to monitor and control, the term *supervisory control and data acquisition* (SCADA) is often used to describe the system. Examples, industrial standards, available off-the-shelf hardware, and extensive references of both electrical and mechanical systems used for the purpose of SCADA are given in these two chapters. Process control for both centralized control using a single computer and decentralized control using *programmable logic controllers* (PLC) and the associated communications systems are discussed.

Chapter 20 outlines methods of intrusion detection for commercial applications. Available sensors and their capability are delineated.

With the continued packing of the electromagnetic spectrum with both commercial and military signals, and with the reassignment of portions of the radio frequency bands, formerly used for telemetry, to commercial endeavors, it is extremely urgent, whenever possible, that telemetry engineers employ the latest modulation techniques that exhibit a spectral efficiency greater than the 1 bit/Hz that characterizes PCM/FM. New developments are presented in this text that are the result of development efforts funded by the Advanced Range Telemetry (ARTM) program, the Telemetry Group of the Range Commanders Council, and the Consultative Committee for Space Data Systems (CCSDS).

Acknowledgments

The author would like to express his appreciation to all those engineers, scientists, and managers from the various ranges and the telemetry industry who have participated in the development of *Telemetry Systems Engineering*. A particular note of appreciation is due to those who participated as students in courses taught from the first telemetry book. Their questions, comments, and, in particular, their in-depth discussions were invaluable.

Telemetry engineer extraordinaire, Eugene Law, Naval Warfare Center Weapons Division (NAWCWD), Pt. Mugu, California, always took the time to answer my questions specifically and in detail concerning the latest advances in spectrally efficient modulation techniques, as well as adding information I was unaware of, but needed. He kept me abreast of what agency was funding what research effort. His help was invaluable.

The discussions with Tomas Chavez, director of the Test Technology Development Directorate, White Sands Missile Range, New Mexico, who had just assumed a new, challenging, and difficult position, were especially important in helping me understand what the Department of Defense (DoD) development and research programs were involved in.

I am deeply indebted to the board members of the International Foundation for Telemetry for their professional and personal encouragement and support for the past 15 years. This dynamic foundation is the heart and soul of the telemetry world.

Without the freely given and congenial support of the electrical and computer engineering (ECE) department, New Mexico State University, it would have been close to impossible to complete this work. In particular, I am indebted to Dr. Steve Horan, who now occupies the Frank Carden Telemetering and Telecommunication Chair, for his unfailing and timely logistic support. Had Dr. Steve Castillo, head of the ECE department, not set up a new computer work station for my use, this project would have been much more difficult to accomplish. The discussions with Dr. Phillip DeLeon and Dr. Jamie Ramirez made portions of this book possible.

To my two young coauthors, Russ Jedlicka and Robert Henry, I express my heartfelt thanks for their hard work, their always professional, positive, and congenial attitude, and their enthusiasm for the project.

For their professional, friendly assistance and fine writing, I would like to express my appreciation to Brian Kopp and Guillermo Rico, who worked especially hard for only a simple "Thank you."

To the staff of Artech House, I express my appreciation for their support and professional and congenial guidance. To the reviewers, who improved the clarity and accuracy of the text with their suggestions and comments, I would like them to know they did an excellent job.

On a personal note, I am indebted to my five daughters, Cathy, Lindie, Cyndi, Jan, and Patricia for their sincere interest and daily enthusiastic support. Amanda and Garret, two of my grandchildren, always pleased me with their questions about writing a book.

I will be eternally grateful to my wife, Barbara, for her support and encouragement and for her assistance in proofreading and in consistently taking care of the details necessary to have a fine book.

The interaction with all the people listed here and many others made writing this book interesting and enjoyable. Though I hope they are limited in number, any and all mistakes are mine and mine alone.

Frank Carden
Las Cruces, NM
January 2002

I am grateful to the students at UL at Lafayette who have used my class notes that are the basis for my contribution to this book. Their questions and suggestions have been invaluable feedback in improving the clarity of the manuscript. The facilities and support of the university provided the environment to develop my notes into the chapters in this book. I am equally indebted to my mentor and colleague Frank Carden, who made my collaboration in producing this book possible. I wish to acknowledge my wife, Billie, for her assistance in improving the readability, grammar, and punctuation of the manuscript. Her patience and understanding are appreciated.

Robert Henry
Lafayette, LA
January 2002

In addition to the sentiments expressed by Dr. Carden, I would like to acknowledge the support of my family, colleagues, and students. My wife, Susan Muir, has been an unending source of support and assistance. My children, Ethan, Bethany, Axel, and Isabelle have been patient throughout the development of the manuscript.

Personnel at the Physical Science Laboratory shared their telemetry and antenna design experience with me; I thank Scott Allen, Alan Baker, Bruce Blevins, Troy Gammill, David Gorman, Warren Harkey, and Stuart Head. I would also like to thank Brecken Uhl of the Radio Design Corporation for his keen review of Chapters 11, 12, and 13. The anonymous reviews provided by Artech House contained many constructive, insightful comments.

The students in the Klipsch School with whom I've worked over the years have been a constant source of inspiration. I appreciate their interest and enthusiasm. Finally, I'd like to thank Frank Carden, who offered me my first teaching assignment at New Mexico State as well as the opportunity to contribute to this text.

Russell P. Jedlicka
Las Cruces, NM
January 2002

1

Telemetry System Definition

The purpose of a telemetry system is to collect data at a place that is remote or inconvenient and to relay the data to a point where the data may be evaluated. Typically, telemetry systems are used in the testing of moving vehicles such as cars, aircraft, and missiles. Telemetry systems are a special set of communication systems. When the telemetry system is used for both control and data collection, the term *supervisory control and data acquisition* is applied. Until Chapters 18 and 19 where industrial telemetry is discussed, this text is primarily concerned with data acquisition.

1.1 Learning Objectives

This chapter is concerned with the overview of telemetry systems. Upon completing this chapter, the student should know the following:

- The names of the seven subsystems used to build a telemetry subsystem;
- The purpose of each of the seven subsystems;
- The subparts of a frequency division multiplex system;
- The definition of the frequency modulation index;
- The subparts of a time division multiplex system;
- The concept of sampling, bit, and frame rate;
- Some of the Inter-Range Instrumentation Group (IRIG) channel frequency standards.

1.2 Telemetry System Overview

An overview of a telemetering system is shown in Figure 1.1. The overall system is composed of the following:

1. Data collection system;
2. One of the following multiplex systems:
    a. A frequency division multiplexing system;
    b. A time division multiplexing system;
    c. A hybrid system, which is a combination of frequency division multiplexing and time division multiplexing;

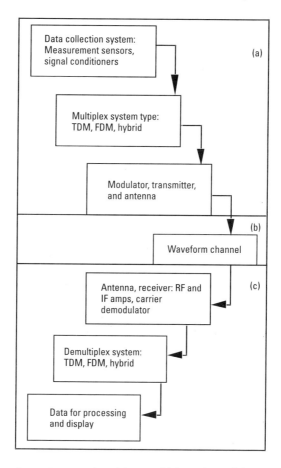

Figure 1.1 Telemetry system overview: (a) test vehicle package, (b) transmission medium, and (c) ground equipment.

3. Modulator, transmitter, and antenna;

4. Waveform or transmission channel;

5. Antenna, radio frequency receiver, intermediate frequency section, carrier demodulator;

6. Demultiplex system for time division multiplexing or frequency division multiplexing, or hybrid system;

7. Data processing.

The first subsystem, the data collection system, is composed of sensors or transducers that convert a physical variable into an electrical signal. The signal is usually very small and must be buffered or amplified before being sent to the multiplexing block. This subsystem may also contain a computer that outputs data.

The output of the data collection subsystem is fed into the multiplex system. If data from the data collection system is separated and deposited into different frequency bins for transmission, the process is referred to as *frequency division multiplexing* (FDM). If the multiplex system separates data in the time domain, it is referred to as a *time division multiplex* (TDM) system. If the system comprises both TDM and FDM subsystems, it is referred to as a hybrid system.

The third subsystem is the modulator, transmitter, and antenna. The multiplexed data that have been separated in the frequency domain or the time domain are modulated onto the carrier at the transmitter, where they are used to drive an antenna. The frequency of transmission is usually in the 1,435–1,535 MHz band or the 2,200–2,290 MHz band.

The next block is the waveform channel. The modulated carrier is radiated by the antenna over a channel described by a medium such as air.

The fifth subsystem is the receiving antenna, *radio frequency* (RF) amplifiers, *intermediate frequency* (IF) amplifiers, and carrier demodulator. The modulated carrier is received by an antenna and is sent to the receiver, where the RF signal is amplified at the RF frequency, then converted into an IF carrier and amplified again. The modulation or data is removed from the IF carrier, which is referred to as carrier demodulation or detection. This type of receiver is referred to as a *superheterodyne* receiver.

The demultiplexing subsystem is next. At this point, the data must be separated by FDM or TDM techniques, or both, to route the data from the individual sensors to the correct channels.

Once the data from the various sensors has been separated and inserted into the correct data channel, the data is now available for display, recording, and processing. Often, the signal is recorded on an analog instrumentation recorder prior to carrier demodulation.

An example of a telemetry system for the flight testing of a missile is shown in Figure 1.2. If the system is a TDM pulse code modulated system, the sensor analog data is sampled and converted into binary digital words, modulated onto the carrier, and transmitted. After traveling through space, the RF signal is received at the ground receiver, and the binary words

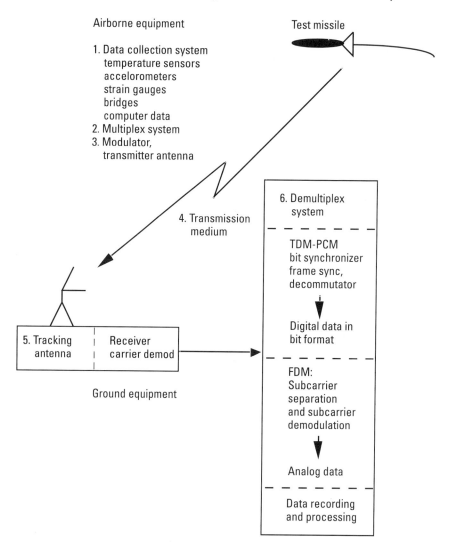

Figure 1.2 Missile test example of telemetry system overview.

are removed from the carrier and sent to the bit synchronizer and frame synchronizer. Then the separate words from the individual sensors are separated by the decommutator and are ready for processing. Typically, the data words are tagged with an identifying word and put on a bus. The composite word then goes to the correct computer peripheral device via the bus.

If the system is an FDM system, the analog data from each sensor is modulated onto each subcarrier, and the subcarriers are modulated onto the carrier and transmitted. The modulated carrier is received at the tracking antenna and sent to the receiver, which removes the subcarriers from the carrier and sends the subcarriers to the demultiplexer, which separates the subcarriers and removes the analog signal from each subcarrier. The analog signals representing sensor data are now ready for recording and processing.

1.3 Data Collection System

The data collection system comprises such devices as thermocouples, accelerometers, transducers, filters, signal conditioners, and computers. Many physical variables, such as temperatures, vibrations, pressures, force, and humidity must be measured and converted into an electrical signal. The electrical signal or message must be conditioned or amplified and mixed with other signals, attached to the carrier, and transmitted to the desired location. It is the purpose of the data collection system to obtain data from the system of sensors, condition the signals, and make them available to the multiplexing subsystem. Computer data may also be included in the system and must be handled. The computer data may be modulated onto a subcarrier.

1.4 Multiplex System

1.4.1 FDM System

In FDM, N individual signals each frequency modulate N sine waves at N different frequencies. The N sine waves, which are referred to as subcarriers, are used to form a weighted sum that frequency modulates a carrier. See Figure 1.3.

A single channel of a frequency modulation (FM)/FM multiplex system is shown in Figure 1.4. The output analog signal from a transducer or sensor is amplified in a signal conditioner whose output, m_1, is applied to an FM modulator or *voltage controlled oscillator* (VCO), whose output is the

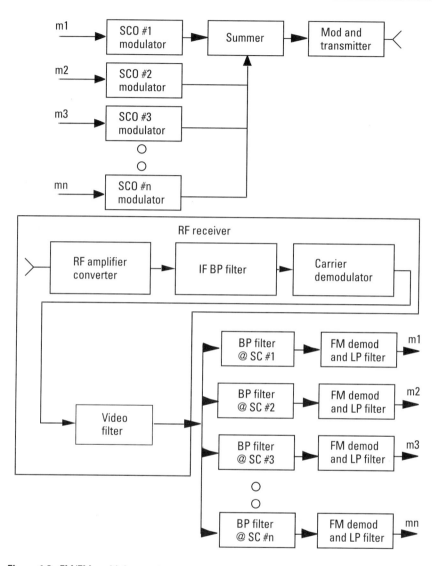

Figure 1.3 FM/FM multiplex system.

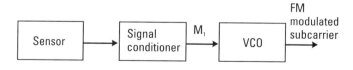

Figure 1.4 FM/FM single channel.

subcarrier. The VCO, whose output is the subcarrier, is characterized by a frequency, say, that varies between 430 and 370 Hz as the input voltage varies between +2.5V and −2.5V respectively. An input of 0V will produce an output frequency of 400 Hz, referred to as the center frequency. The difference between the center frequency and the highest or lowest frequency is referred to as the maximum peak frequency deviation and designated by f or f_d. In this case f_d = 30 Hz.

As an example of a single channel, the sensor might be a thermocouple measuring the temperature that will vary between 0° and 10.0°. The 430-Hz output from the VCO will correspond to the 10.0° reading, and the 370-Hz VCO output will correspond to the 0° temperature. Providing linearity is established, intermediate frequencies will correspond proportionally to temperatures. The sensor output voltage, in this case corresponding to temperature, will vary at some maximum rate or some maximum frequency, say, f_m. In this case, the temperature may vary at a rate equal to or less than 6 Hz. Since the thermocouple output voltage will probably vary in the millivolt range, the signal conditioner must enhance the signal such that it varies over a voltage range of, nominally, 5V. The ratio of the maximum peak frequency deviation, f_d, to the maximum signal frequency, f_m, is referred to as the modulation index or deviation ratio and is designated as β or as D. For the example here, the modulation index or deviation ratio is given by

$$D = f_d/f_m = 30/6 = 5 \tag{1.1}$$

In this book, this parameter generally will be referred to as the modulation index or mod index, although it should be remembered that the terms "modulation index" and the "deviation ratio" are used interchangeably. In some texts and occasionally in the literature, the term "modulation index" is used when the modulation signal is a pure sine wave. The mod index in this case is the ratio of the peak frequency deviation to the frequency of the sine wave. The term "deviation ratio" is used to designate the ratio of the peak frequency deviation to the 3-dB cutoff of the filter following the FM discriminator whenever a general signal with a continuous spectrum is used as the modulating signal. This is the definition of deviation ratio given by the secretariat of the Range Commanders Council [1]. This book uses the term "modulation index" for both cases.

An FM/FM multiplex system is shown in Figure 1.4 with N messages or signals (m_i) from the sensors or transducers.

1.4.1.1 IRIG Standards

The IRIG standards [1] specify the subcarrier frequencies, nominal peak frequency deviations, nominal modulation indices, maximum signal variations, and many other system parameters. IRIG standards specify two frequency modulation schemes for the subcarriers: *proportional bandwidth* (PBW) and *constant bandwidth* (CBW) channels. (See Appendix A for a complete listing.) In the PBW channels, the peak frequency deviation of the subcarrier is proportional to the subcarrier frequency, whereas in the CBW channels the deviation is a constant.

Some of the proportional bandwidth format is given in Table 1.1. In the three types of PBW systems, the peak frequency deviation is equal to 0.075 or 0.15, or 0.3 times the subcarrier frequency. The 7.5% channels are numbered 1 to 25; the 15% channels are designated A through L; and the 30% channels are designated AA through LL.

Since the information bandwidth is proportional to the deviation, the greater the deviation, the more information the channel can handle. Figure 1.5 shows the spectrum utilization by some of the 7.5% proportional channels.

1.4.1.2 IRIG Spectrum Utilization

Inspection of Figure 1.5 or Table 1.1 shows that for the 7.5% channels, a guard band between adjacent channels is built in, and hence adjacent channels may be used. When the 15% and 30% channels are used, some adjacent channels must be deleted. All of the channel IRIG standards are given in Appendix A. (See the footnotes of Appendix A concerning channel deletion.)

A nominal and a maximum frequency response are given for each channel. The impact of using one or the other may be seen in the following way. Since the deviation of the subcarrier by the message is fixed at f_d, and f_m is increased, then the mod index, D, is decreased. Because the channel output $[S/N]$ ratio varies directly as D^2 (in decibels), the output ratio will decrease. System parameters must then be selected to compensate for the

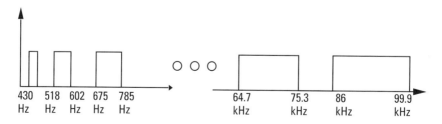

Figure 1.5 IRIG PBW channels.

decreased D. It is recommended that normally a modulation index less than 2 not be used.

For example, the lowest channel has f_d = 30 Hz, and the nominal frequency response is such that the channel will handle a signal with a maximum frequency of 6 Hz. The modulation index is given by

$$D = f_d/f_m = 30/6 = 5 \tag{1.2}$$

However, if the channel is designed to handle a signal with a maximum frequency of 30 Hz, and since f_d does not change, then

$$D = 30/30 = 1 \tag{1.3}$$

Inspection of the constant bandwidth format in Table 1.2 shows the nominal and the maximum frequencies of the channels. For example, the A channels have a nominal frequency response of 0.4 kHz, whereas the maximum frequency response is given by 2 kHz. This corresponds respectively to modulation indexes of

$$D(\text{nominal}) = 2/0.4 = 5 \tag{1.4}$$

and

$$D(\text{max freq}) = 2/2 = 1 \tag{1.5}$$

In the CBW channels, the output signal-to-noise ratio also depends on D, and it is usually recommended that a D larger than 1 be used.

1.4.2 Pulse Code Modulation TDM

In TDM using pulse code modulation (PCM), the commutator samples sequentially the analog signals from N sensors and holds each amplitude value for some pulse period. This creates a pulse amplitude modulated (PAM) pulse sequence, as shown in Figure 1.6. The PAM sequence is applied to the encoder, which quantizes the pulses and converts each pulse into a binary word, resulting in a bit sequence. The data word is nominally 8-bits long, but can be as long as 16 bits. The bit sequence is applied to a digital multiplexer, which combines data words, timing, and frame synchronization. Figure 1.7 illustrates a PCM receiver.

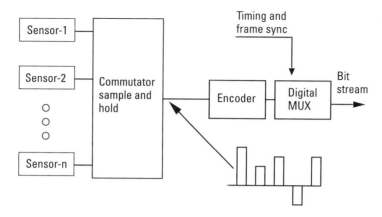

Figure 1.6 PCM system.

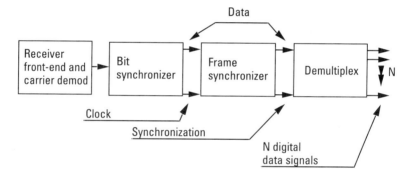

Figure 1.7 PCM receiver, bit and frame synchronizers, and demux.

1.4.2.1 Sampling Rate

For an ideal system, the sampling rate of each sensor must be equal to or greater than twice the highest frequency present in the signal spectrum. In a practical system, the signal spectrum does not cut off at a specific frequency but decays in some slower manner. To prevent distortion in the sampling and reconstruction process, in practice it is common to sample four to six times the estimated highest frequency present. Usually a signal conditioner is placed after the sensor; it is also common to have a lowpass filter in the signal conditioner to limit the upper frequency to a known value.

The sensors may be temperature sensors, that is, thermocouples with expected frequency variations of 1 Hz, or vibration sensors such as strain gauges with upper frequencies of 1 to 5 kHz. The commutator then must be programmed to sample the higher-frequency signals more often, resulting in supercommutation. The lower-frequency signals will be sampled less, resulting in subcommutation.

1.4.2.2 Bit Rate

The bit rate is equal to the sampling rate times the bit word length. For example, if the sensor had a spectrum that slowly decayed but had a maximum defined frequency of 1 kHz, then in a practical system the sampling rate would be 5 ksamples/sec. If the data word was 8-bits long, then the bit rate would be 40 Kbps. At the receiver, the bits must be detected and bit synchronization established. A bit synchronizer performs that function.

1.4.2.3 Frame Rate

After a complete cycle of sampling all the sensors, the sequence of binary words generated by the encoder is grouped together in a frame. This frame is referred to as a major frame and is composed of minor frames. To decommutate or demultiplex the data words, it is necessary to know when a minor frame starts and stops along the time axis. This is accomplished by inserting a frame synchronization word at the start of each minor frame. The synchronization word is chosen to be as distinct as possible from the data words. For system compatibility, the synchronizing word is usually chosen to be an integer multiple of the data word length. Studies performed for the IRIG committee have resulted in the specification of synchronizing words composed of pseudorandom bit patterns, which are listed in IRIG-106-00 and repeated in this book in Appendix B. Nominally, the frame synchronizing word will be at least twice as long as the data word to prevent false lock. The frame length will vary from 50 to 512 words for a class I format. The IRIG 106-00 Telemetry Standards document [1] gives two classifications of PCM formats. Class II is more general, while class I is more restrictive but in greater use. All ranges do not support class II. The formats are discussed in greater detail in Chapter 5.

1.4.3 Combination of FDM and TDM

When both FDM and TDM are used together in the same system, the system is referred to as a hybrid system.

1.5 Modulator, Transmitter, and Antenna

In this subsystem, the signals from the multiplex system are modulated onto a carrier, sent to the antenna, and transmitted. The transmitter driving the antenna is usually composed of a VCO, followed by a power amplifier that drives the antenna. The modulator is, in fact, the VCO being driven by the modulating waveform. Most FM modulators are VCOs.

1.5.1   FM/FM

In FM/FM, after the analog signals from the N sensors have been modulated onto the subcarriers, the N subcarriers are summed and modulated onto the carrier and sent to the antenna, where the modulated carrier is radiated. An example of an end-to-end system is shown in Figure 1.2.

1.5.2   PCM/FM Generation

In PCM/FM, the bit stream, shown in Figure 1.6 containing both data and frame synchronization words, is modulated directly onto the carrier. Prior to carrier modulation, the bit sequence is usually passed through a lowpass filter to decrease the fast rise time of the pulses and, hence, constrain the RF spectrum.

1.5.3   Hybrid Systems: PCM/FM + FM/FM, PCM/FM/FM

1.5.3.1   PCM/FM + FM/FM Generation

A PCM/FM + FM/FM system is shown in Figure 1.8. The TDM-PCM module is composed of the blocks shown in Figure 1.6. The M frequency modulated subcarriers are summed with the bit sequence output of the TDM-PCM module at baseband, and the resultant modulates the carrier.

1.5.3.2   PCM/FM/FM

In PCM/FM/FM, the bit sequence is modulated directly onto a subcarrier, which is summed with the other subcarriers, and the resultant is used to modulate the carrier. At the receiver the PCM modulated subcarrier channel must have a bit detector and synchronizer, frame synchronizer, and demultiplexer following the subcarrier discriminator, as shown in Figure 1.9.

1.6   Transmission or Waveform Channel

After the carrier is modulated with the information, it is transmitted through a medium. To determine if the transmitted signal arriving at the receiving antenna is of sufficient strength to be processed properly, it is necessary to

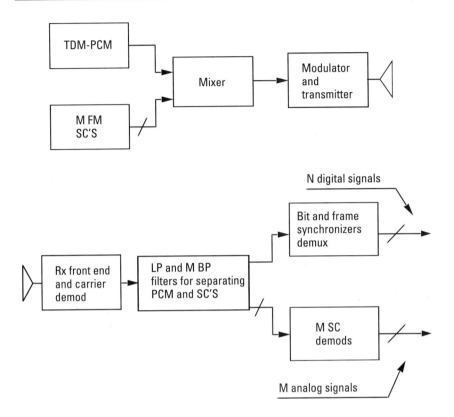

Figure 1.8 PCM/FM + FM/FM.

perform what is referred to as a link analysis. Link analysis requires knowledge of the antenna gains, attenuation of the signal as it travels though the medium, the characteristics of the receiver, and effective transmitted power. Link analysis is discussed in detail in Chapters 11 through 14. The carrier frequency chosen and the antennas must be compatible; that is, generally, the antenna dimensions and the wavelength of the carrier wave must be in the same size range.

1.7 Antennas, Receivers with RF and IF Amplifiers, and Carrier Demodulators

The incoming RF signal is intercepted by the receiving antenna, generating a signal that is fed into the RF amplifier. The amplified RF signal is converted down in frequency to an intermediate frequency and amplified again. The

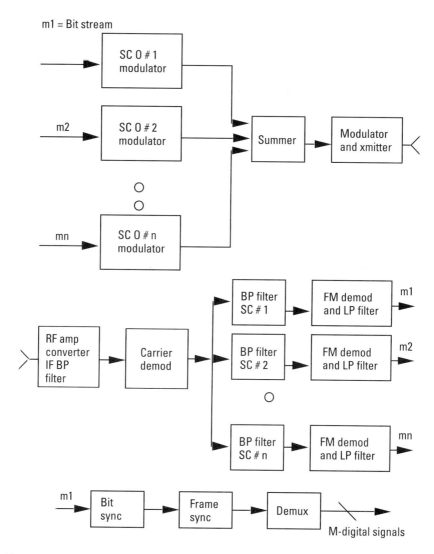

Figure 1.9 PCM/FM/FM telemetering system.

time domain signal and spectrum of the modulated IF carrier look exactly the same as the RF signal but at a lower frequency. The IF frequency is also commonly referred to as the carrier from this point on in the receiver. The process of stripping the information or modulation from the IF carrier is, in fact, referred to as carrier demodulation, and the block performing this process is referred to as the carrier demodulator. The output from the carrier demodulator is passed through a linear lowpass filter, which is referred to as the video output filter. In the receiver, two important parameters, which must be determined and used in setting up the operating receiver, are the IF bandwidth and the bandwidth or 3-dB point of the video output filter. The video output is fed into the demultiplex subsystem.

1.8 Demultiplex System

Once the information has been removed from the IF carrier and filtered by the video output filter in the receiver, it is necessary to separate the information into the proper channels. This is accomplished in the demultiplex subsystem. The separation of data into the correct channels will occur in the time domain, the frequency domain, or both.

1.8.1 Frequency Division Demultiplexing

In frequency division demultiplexing, once the IF carrier demodulation has been achieved, bandpass filters are used to separate the different signals or messages from the N different sensors or transducers. This is shown in Figure 1.4, as part of the end-to-end telemetry system.

1.8.2 Time Division Demultiplexing

1.8.2.1 PCM/FM Reception

A PCM/FM receiver is shown in Figure 1.7. The output from the carrier demodulator is a sequence of bit waveforms distorted by the channel and contaminated by noise. The function of the bit synchronizer is to make decisions between ones and zeros, generate and pass on a clean and crisp bit stream reflecting these decisions, and generate a bit clock. The purpose of the frame synchronizer is to lock on to the frame sync pattern, generate word synchronization, and pass on the data words and synchronizing clocks to the demultiplexer. The demultiplexer routes the data words into the correct channels for processing.

1.8.2.2 Time Division Demultiplexing

In the demultiplexing of TDM-PCM/FM, it is necessary to separate in the time domain the binary words representing the signals from the N sensors, once the bit sequence has been stripped from the IF carrier. This is performed with a decommutator. This type of signal transmission and modulation format is referred to as PCM/FM. Figure 1.6 also shows a block diagram of this system.

1.8.3 Hybrid Systems

Hybrid systems are combinations of FDM and TDM. It is necessary to separate the channels in both the time and frequency domain.

1.8.3.1 PCM/FM + FM/FM

In PCM/FM + FM/FM systems, after carrier demodulation at the receiver, the signal is composed of the PCM baseband signal plus M subcarriers. The subcarriers are separated with bandpass filters and sent to M subcarrier demodulators, and the resulting M analog signals are processed as before. The lowpass PCM signal is applied to a bit synchronizer, frame synchronizer, and demultiplexer, resulting in N digital data signals, which are processed as before. Figure 1.8 shows this system end-to end. Often two receivers are used to receive the two modulation packages separately.

1.8.3.2 PCM/FM/FM

In PCM/FM/FM systems, after carrier demodulation at the receiver, the signal is composed of N subcarriers, with at least one that has a PCM bit sequence modulated directly onto it. That subcarrier is applied to an FM discriminator whose output is the noisy bit sequence. The bit sequence is applied to a bit synchronizer, frame synchronizer, and decommutator, as shown in Figure 1.9.

1.9 Data Processing, Handling, and Display

Once the data has been removed from the carrier and separated into the proper channels, it is then ready to be processed. This may involve a real-time display, computer display and storage, and tape recorders. Analog instrumentation recording may be used at any point in the process. It is common practice to analog record the received signal prior to carrier demodulation.

1.10 Supporting Equipment and Operations

Two important supporting operations are diversity combining and predetection recording. The goal of both operations is to enhance the received signal.

Diversity combining, developed in the 1970s, enhances the received signal by as much as 9 dB; 3 to 6 dB is not uncommon [2]. Predetection recording allows the parameters of the receiving system to be modified to optimize the signal quality during postflight processing.

1.10.1 Diversity Combining

Diversity combining is a method of combining two independent signals in order to create a third signal that on the average is better than either of the first two. Polarization, spatial, and frequency diversity combining are methods of enhancing the received signal. Diversity combining may also be characterized as predetection and postdetection. In predetection diversity combining, the phase of the two signals must be aligned; in postdetection such alignment is not necessary.

Polarization diversity combining is the most common method employed. Conceptually, electromagnetic radiation from an antenna may be viewed as a sinusoid traveling along an axis in space. If the sinusoid has peaks and valleys falling in a single vertical plane along the axis, the wave is referred to as vertically polarized; if the plane is 90 degrees from the vertical, the wave is referred to as horizontally polarized. This class of waves is referred to as vertically or horizontally polarized. If the wave is wrapped around the axis as it travels along the axis, the wave is circularly polarized. Such waves may exhibit right-hand circular polarization (RHCP) or left-hand circular polarization (LHCP). The type of polarization is a function of the transmitting antenna. Signal degradation, due to RF signal fades caused by multipath, flame attenuation, and the fact that the antenna pattern is not perfectly omnidirectional but is composed of lobes, can be minimized by predetection or postdetection polarization diversity combining. This is possible because the two types of polarization, say, RHCP and LHCP, are of different phases and respond to RF signal fades differently.

Most telemetry receiving antennas have two independent outputs resulting from RHCP and LHCP or vertical and horizontal polarization. The theoretical best such an optimum combiner can achieve using the two signals from one class of polarization is an output signal-to-noise ratio that is the sum of the signal-to-noise ratios of the two inputs. This improvement is achieved by the combiner adding the signal voltages coherently and adding the noise powers incoherently. Predetection diversity combining takes place in the receiver prior to carrier detection. A postdetection diversity combiner accepts two signals from the video filters of two receivers, combining the signals such that the output of the video combiner has an enhanced signal-to-noise ratio. Figure 1.10 illustrates such a system.

Space diversity combining is achieved with two receiving antennas spatially separated. This method is also common and is especially effective to counteract multipath or flame attenuation.

1.10.2 Predetection Recording

Predetection recording allows the parameters of the receiving system to be modified to optimize the signal quality during postflight processing. The IF carrier with modulation is a replica of the received carrier, except it is at a lower frequency. Typically, the IF signal is taken after the second IF stage, down-converted with a receiver plug-in unit, and recorded on a wideband instrumentation tape recorder. At a later time, the recorded signal can be up-converted and reinserted into the receiver at the second IF. As the signal is being played back, parameters may be varied to obtain signal-to-noise enhancement. For example, the 3-dB point of the video filter may be varied to achieve output signal-to-noise ratio enhancement in FM/FM or a better bit error rate (BER) in PCM/FM.

1.11 IRIG Channel Standards

Some of the IRIG frequency channels are given in Table 1.1 and Table 1.2. The standards specify the center frequency of the subcarrier VCO and how

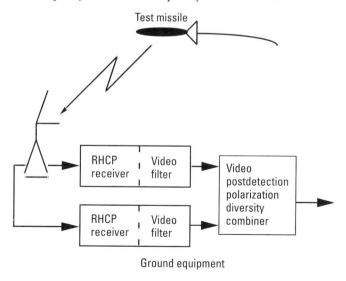

Figure 1.10 Video postdetection polarization diversity combiner.

Table 1.1
IRIG PBW Subcarrier Channels

Channels	Center Frequencies (Hz)	Lower Deviation Limit (Hz)	Upper Deviation Limit (Hz)	Nominal Frequency Response N = 5 (Hz)	Nominal Rise Time N = 5 (ms)	Maximum Frequency Response (Hz)	Minimum Rise Time (ms)
±7.5% CHANNELS							
1	400	370	430	6	58	30	11.7
2	560	518	602	8	44	42	8.33
3	730	675	785	11	32	55	6.40
4	960	888	1,032	14	25	72	4.86
5	1,300	1,202	1,398	20	18	98	3.60
6	1,700	1,572	1,828	25	14	128	2.74
7	2,300	2,127	2,473	35	10	173	2.03
8	3,000	2,775	3,225	45	7.8	225	1.56
9	3,900	3,607	4,193	59	6.0	293	1.20
10	5,400	4,995	5,805	81	4.3	405	0.864
±15% CHANNELS							
A	22,000	18,700	25,300	660	0.53	3,300	0.108
B	30,000	25,500	34,500	900	0.39	4,500	0.078
C	40,000	34,000	46,000	1,200	0.29	6,000	0.058
D	52,500	44,625	60,375	1,575	0.22	7,875	0.044
E	70,000	59,500	80,500	2,100	0.17	10,500	0.033
F	93,000	79,050	106,950	2,790	0.13	13,950	0.025
±30% CHANNELS							
AA	22,000	15,400	28,600	1,320	0.265	6,600	0.053
BB	30,000	21,000	39,000	1,800	0.194	9,000	0.038
CC	40,000	28,000	52,000	2,400	0.146	12,000	0.029

Table 1.2
IRIG CBW Subcarrier Channels

A Channels $f_d = \pm 2$ kHz Nominal frequency response = 0.4 kHz Maximum frequency response = 2 kHz		B Channels $f_d = \pm 4$ kHz Nominal frequency response = 0.8 kHz Maximum frequency response = 4 kHz		C Channels $f_d = \pm 8$ kHz Nominal frequency response = 1.6 kHz Maximum frequency response = 8 kHz		D Channels $f_d = \pm 16$ kHz Nominal frequency response = 3.2 kHz Maximum frequency response = 16 kHz		E Channels $f_d = \pm 32$ kHz Nominal frequency response = 6.4 kHz Maximum frequency response = 32 kHz	
Channel	Center Frequency (kHz)	Channel	Center Frequency (kHz)	Channel	Center Frequency (kHz)	Channel	Center Frequency (kHz)	Channel	Center Frequency (kHz)
1A	16								
2A	24								
3A	32	3B	32	3C	32				
4A	40								
5A	48	5B	48						
6A	56								
7A	64	7B	64	7C	64	7D	64		
8A	72								

far the VCO may be deviated from the center frequency by the signal from the sensor and signal conditioner. The nominal frequency response of a specific channel is given, which in fact, will specify a nominal modulation index.

Problems

Problem 1.1

List the seven subsystems of a generic telemetry system.

Problem 1.2

Describe the purpose of each subsystem.

Problem 1.3

Draw a block diagram of an FM/FM telemetering system using five subcarriers.

Problem 1.4

Discuss the purpose of the VCOs, the summer or mixer, and the composition of the signal to the transmitter for an FM/FM system.

Problem 1.5

Discuss the purpose of the subcarrier bandpass filters.

Problem 1.6

Draw a block diagram of a TDM system for five sensors.

Problem 1.7

Discuss the purpose of the sampler and commutator for a TDM system.

Problem 1.8

For channel 5, a 7.5% PBW channel, determine the modulation index for (a) the nominal frequency response, and (b) the maximum frequency response.

Problem 1.9

For channel A, a 15% PBW channel, determine the modulation index for (a) the nominal frequency response, and (b) the maximum frequency response.

Problem 1.10

For channel BB, a 30% PBW channel, determine the modulation index for (a) the nominal frequency response, and (b) the maximum frequency response.

Problem 1.11

For channels 17, 18, and 19, 7.5% PBW channels, what is the peak frequency deviation? What are the upper and lower frequencies of the three VCOs? Do they overlap?

Problem 1.12

For a modulation index or deviation ratio of 5, what is the maximum frequency response of channel 19? For a modulation index of 3, what is the maximum frequency response of channel 19?

Problem 1.13

Using a modulation index of 5, select 15 channels that can handle the following frequencies: 20, 25, 35, 45, 59, 81, 110, 160, 220, 330, 450, 600, 790, 1,050, and 1,395 Hz.

Problem 1.14

In Problem 1.13, if the modulation index is reduced to 2.5, what would be the frequencies that each of the 15 channels could handle?

Problem 1.15

The maximum frequency from a sensor is 100 Hz. If the signal is sampled at five times the highest frequency, what is the sampling rate?

Problem 1.16

For Problem 1.15, if each sample is represented by an 8-bit word, what would be the bit rate? If a 12-bit word is used, what would be the bit rate? If a 16-bit word is used, what would be the bit rate?

Problem 1.17

For the 14 constant bandwidth channels, 6A through 19A, what is the modulation index of the channels if a maximum frequency of 400 Hz is allowed? What is the modulation index if a maximum frequency of 1,000 Hz is specified?

References

[1]   Secretariat, Range Commanders Council, *Telemetry Standards*, White Sands Missile Range, NM: RCC, IRIG 106-00.

[2]   Berns, K. L., "Benefits of Polarization Diversity Reception," *ITC Proceedings*, Vol. XX, 1984, pp. 627–642.

2

Analog Frequency Modulation

In an FM/FM multiplex communication system, analog signals from N different sensors are frequency modulated onto N individual subcarriers and summed, and the resultant single waveform is frequency modulated onto a carrier and transmitted. The FM carrier is received and the carrier is demodulated, producing the resultant single waveform. That waveform is applied to N bandpass filters, which separate and recover the N subcarriers. Each individual subcarrier is applied to an FM discriminator that recovers the original signal from the sensor. A general end-to-end telemetering communication system was shown in Figure 1.1. An FM/FM system was shown in Figure 1.3, with the generalized blocks replaced with the FM/FM specific blocks. This section initially will discuss the multiplex system and the modulation subsystems shown in Figures 1.1 and 1.3. However, since noise is involved in all channels and it is essential to design the system such that the quality of each signal is maintained at an acceptable level at the output of the end-to-end telemetry (TM) system, the noise model for the receiving FM and FM/FM system will also be developed.

2.1 Learning Objectives

Upon completing this chapter, the reader should understand the following:

- Single-channel frequency modulation;
- Modulation index and FM spectrum;

25

- FM/FM;
- Functional purpose of each block in the end-to-end FM/FM system;
- Bandwidth requirements;
- Signal-to-noise model for FM and the implications of the system modulation parameters;
- Signal-to-noise model for FM/FM channels and the implications of the system modulation parameters.

2.2 Single-Channel FM

2.2.1 Time Domain FM Carrier Model

Prior to discussing an FM/FM channel, a single channel that employs frequency modulation will be examined in this section. The general expression for a carrier is given by

$$e(t) = A \sin(\omega_c t + \theta(t)) = A \sin(\Theta(t)) \tag{2.1}$$

The instantaneous frequency is the derivative of the phase and is

$$\Theta' = \omega_c + \theta'(t) \tag{2.2}$$

For frequency modulation, θ', the derivative of θ, is set proportional to the signal message, such as

$$\theta'(t) = 2\pi k m(t)$$

The units on k are hertz per volt, and k is a parameter of the VCO. Since the units on $\theta'(t)$ are radians per second, it is necessary to multiply by 2π. $\theta'(t)$ is referred to as the instantaneous radian frequency and if divided by 2π gives

$$f = k m(t) \tag{2.3}$$

Then f is referred to as the instantaneous frequency. When $m(t)$ takes on its peak value, the peak frequency deviation will occur.

To insert the phase back into the argument of the sine function, the instantaneous radian frequency must be integrated to obtain phase,

$$\Theta(t) = \int\limits_0^t \omega_c dt + 2\pi k \int\limits_0^t m(t)dt = \omega_c t + 2\pi k \int\limits_0^t m(t)dt \qquad (2.4)$$

The FM carrier is given by

$$e(t) = A \sin\left(\omega_c t + 2\pi k \int\limits_0^t m(t)dt\right) \qquad (2.5)$$

If $m(t)$ is such that its peak value is 1, then $k = f_d$ and the FM wave becomes

$$e(t) = A \sin\left(\omega_c t + 2\pi f_d \int\limits_0^t m(t)dt\right) \qquad (2.6)$$

If $m(t)$ is single tone modulation such as $m(t) = \cos \omega_m t$, the FM wave is given by

$$e(t) = A \sin(\omega_c t + (2\pi f_d / \omega_m) \sin \omega_m t) \qquad (2.7)$$

The constant $f_c = \omega_c / 2\pi$ is referred to as the center frequency. The constant $2\pi f_d / \omega_m$ is referred to as the deviation ratio or the modulation index, D, and also as β. Then,

$$e(t) = A \sin(\omega_c t + D \sin \omega_m t) \qquad (2.8)$$

2.2.2 FM Spectrum

The spectrum for the single-tone FM-modulated carrier may be obtained by expanding $e(t)$ using the trig identity $\sin(A + B) = \sin A \cos B + \sin B \cos A$, giving

$$e(t) = A[\sin \omega_c t \cos(D \sin \omega_m t) + \cos \omega_c t \sin(D \sin \omega_m t)] \qquad (2.9)$$

Relating the cosine and the sine of a function to a series representing the Bessel functions gives

$$\cos(D \sin \omega_m t) = J_o(D) + 2 \sum_{n=1}^{\infty} J_{2n}(D) \cos 2n\omega_m t$$

$$\sin(D \sin \omega_m t) = 2 \sum_{1}^{\infty} J_{2n-1}(D) \sin((2n - 1)\omega_m t)$$

Inserting these terms into (2.9) and using trigonometric identities gives

$$\text{carrier} \qquad\qquad \text{USB} \qquad\qquad\qquad \text{LSB}$$

$$e(t) = AJ_o(D) \sin \omega_c t + A \sum_{n=1}^{\infty} J_{2n}(D)[\sin(\omega_c + 2n\omega_m)t + \sin(\omega_c - 2n\omega_m)t]$$

$$\text{even order}$$

$$\text{USB} \qquad\qquad\qquad \text{LSB}$$

$$+ A \sum_{n=1}^{\infty} J_{2n-1}(D)[\sin(\omega_c + (2n - 1)\omega_m)t - \sin(\omega_c - (2n - 1)\omega_m)t]$$

$$\text{odd order}$$

$$(2.10)$$

2.2.3 FM Bandwidth Requirements

Inspection of the Bessel function coefficients as a function of their argument D, the deviation ratio, shows that all the functions have a zero value at the origin except $J_0(0)$, which has the value of 1. Hence, when $D = 0$, all the power is in the carrier and all the sidebands are zero. As D is increased, the carrier power is reduced and the sidebands start to grow, since the total power in the fm wave remains constant.

For β or $D \ll 1$, narrowband frequency modulation results and the only significant sidebands are the first two. Therefore, the approximate bandwidth requirement is given by

$$BW = B_c = 2f_m \qquad\qquad (2.11)$$

where

$$B_c = \text{bandwidth of the IF}$$

For $D \gg 1$, wideband frequency modulation results and it has been found that a good approximation for the bandwidth requirement is, by Carson's rule [1], given by

$$BW = 2(f_d + f_m) \tag{2.12}$$

Figure 2.1 shows the resulting spectrums as D is varied from less than 1 to much greater than 1.

Although this was derived for single-tone modulation, it has been found that the equations hold for a general $m(t)$ where f_m is the highest frequency present. For $D \gg 1$, BW becomes

$$BW = 2f_d(1 + 1/D) \tag{2.13}$$

The interpretation of (2.13) is that as D becomes large, the required bandwidth approaches $f_c \pm$ peak frequency deviation.

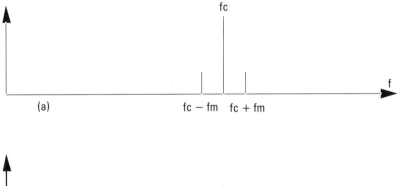

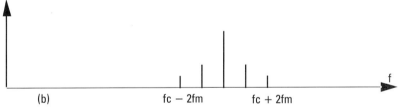

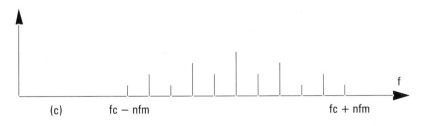

Figure 2.1 Single-tone FM spectrums as a function of β or D: (a) $D = \beta \ll 1$, (b) $D \approx 1$, and (c) $D = \beta \gg 1$.

For multitone modulation such as FM/FM, it has been found that a good prediction for the necessary carrier IF bandwidth is given by [2]

$$B_c = 2(f_{dn} + f_{sh}) \qquad (2.14)$$

where

f_{dn} = RMS of the deviations of the carrier by each subcarrier;

f_{sh} = highest frequency subcarrier.

That is,

$$f_{dn} = \sqrt{f_{dc1}^2 + f_{dc2}^2 + + + f_{dcn}^2} \qquad (2.15)$$

f_{dn} is also referred to as the norm of the carrier deviation.

Work from the literature [2] indicates this bandwidth prediction probably contains all but about 1% of the sideband power.

For FM/FM, the peak deviation of the carrier by the subcarriers is given by

$$f_{dp} = f_{dc1} + f_{dc2} + + + f_{dcn} \qquad (2.16)$$

This peak value would only occur if all the subcarrier peaks aligned, which has an extremely low probability of happening.

It is also instructive to look at the spectrum of FM/FM for the simple case of only two subcarriers. For two-tone FM, the FM wave is given by

$$= A_c \sum_{n=-\infty}^{\infty} \sum_{m=-\infty}^{\infty} J_n(D_1) J_m(D_2) \cos(\omega_c + n\omega_1 + m\omega_2)$$

where

D_1 = modulation index of the first tone;

D_2 = modulation index of the second tone.

A simple but illustrative case occurs for small modulation indexes and when $\omega_1 \gg \omega_2$. The resulting spectrum is shown in Figure 2.2. The higher frequency establishes sidebands as in single-tone modulation. The lower

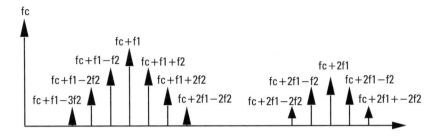

Figure 2.2 Spectrum two-tone FM.

frequency creates sidebands around the carrier and the higher frequency sidebands with the sum and difference frequency. Each higher-frequency sideband component appears to be a tone with FM modulation. The net result is that the bandwidth of the two-tone FM signal is determined primarily by the higher frequency subcarrier.

Although this is for a simple case of only two tones, it will be seen from the models for the general FM/FM case that the higher subcarriers basically set the required bandwidth.

2.2.4 FM Modulation Implementation

Frequency modulation may be achieved by applying the message voltage $m(t)$ to a VCO. The output frequency is given by

$$f = f_c + km(t)$$

f_c is the center frequency of the VCO and occurs when no signal is applied and k is a constant of the VCO with units of hertz/volt. When $|m(t)| \leq 1$, $f_d = k$. An example is shown in Figure 2.3 for this case and for a square wave-input voltage to the VCO.

2.2.5 FM Demodulation Implementation

Since the instantaneous frequency is proportional to the message, the phase of the carrier must be differentiated in order to obtain the instantaneous frequency and the message. That is,

$$\Theta(t) = \omega_c t + \theta(t) = \omega_c t + 2\pi f_d \int_0^t m(t)dt \qquad (2.17)$$

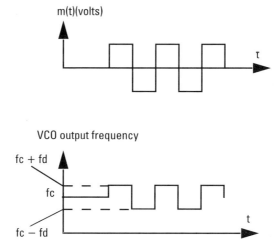

Figure 2.3 VCO output frequency for a square wave voltage input.

Then

$$\Theta' = \omega(t) = \omega_c + \theta'(t) = \omega_c + 2\pi f_d m(t)$$

This can be achieved with an FM discriminator whose output voltage is proportional to the instantaneous frequency deviation from the center frequency, f_c. That is,

$$e_o(t) = k_d f_d m(t) \qquad (2.18)$$

where k_d, a constant of the FM discriminator has units of volts/hertz. Figure 2.4 illustrates the transfer characteristic and output voltage for an FM discriminator with an input carrier whose frequency is changing as a square wave. In this case, the output voltage, $e_o(t)$ is

$$e_o(t) = \pm k_d f_d$$

Since the instantaneous frequency is proportional to the message and the number of zero crossings of the carrier is related to the instantaneous frequency, zero crossing counters are also employed as FM discriminators.

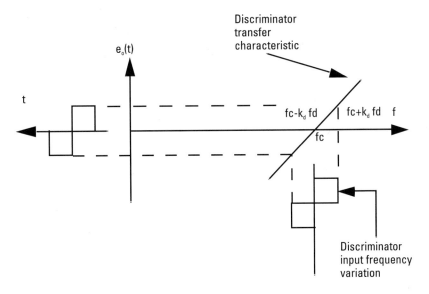

Figure 2.4 FM discriminator transfer characteristic and output voltage for a square wave frequency variation.

2.3 FM/FM

2.3.1 FM/FM Definition

In an FM/FM multiplex communication system, analog signals from N different sensors are frequency modulated onto N individual subcarriers and summed, and the resultant single waveform is frequency modulated onto a carrier and transmitted. The FM carrier is received, and the carrier is demodulated, producing the resultant single waveform. That waveform is applied to N bandpass filters, which separate and recover the N subcarriers. Each individual subcarrier is applied to an FM discriminator, which recovers the original signal from the sensor.

2.3.2 IRIG Spectrum Utilization

The modulation formats for PBW and CBW channels were given in Tables 1.1 and 1.2. It is interesting to estimate the bandwidth required by a single subcarrier channel after carrier demodulation. Two cases will be considered here, $D = 5$ and $D = 1$.

When the subcarrier signal is designed for a nominal modulation index of 5, then (2.13) applies. For example, in the first PBW channel, $f_d = 30$,

$f_m = 6$, and $D = 5$; then the bandwidth of the bandpass filter used to separate this subcarrier from all others is given by

$$B_{bpi} = 2(f_d + f_m) \qquad\qquad (2.19)$$

$$B_{bpi} = 2f_d(1 + 1/D) = 2f_d(1 + 0.2) = 2.4f_d \approx 2f_d = 2(30) = 60 \text{ Hz}$$

This bandwidth is an approximate bandwidth and contains about 95% of the power. As D becomes much larger than 1, the estimation of bandwidth contains a larger percentage of the power.

On the other hand, if the system is designed for a modulation index of 1, a maximum frequency response of 30 Hz can be handled. That is, since f_d remains fixed at 30 Hz, then $D = 1$ as indicated. Then the equation for approximating the bandwidth of narrow band FM for the single FM channel is (2.11) and is

$$B_{bpi} \approx 2f_m = 2(30) = 60 \text{ Hz}$$

or,

$$B_{bpi} = 2f_d(1 + 1/D) = 2f_d(1 + 1) = 4f_d = 120$$

Equation (2.11), which is used for predicting the approximate bandwidth for narrow band FM, is accurate only for a mod index much smaller than 1, which is not true for this second case. The second equation predicts a required bandwidth that is too large and probably contains over 99% of the power, whereas the first equation predicts a bandwidth containing only about 95% of the power. The actual bandwidth depends on the definition of power content. The bandwidth of the bandpass filter remains approximately the same in both cases, although it should be recognized that the first approximation is better than the last two. The last estimation of $4 f_d$ is probably much too large. Once again it should be pointed out that while the mod index of 1 allows the channel message to have a maximum frequency of 30 Hz, compared to 6 Hz for a mod index of 5, the lower mod index is undesirable because the channel signal-to-noise ratio will be substantially lower for the smaller mod index.

In choosing between the PBW and the CBW channels, it is necessary to know something of the spectrum generated by each of the sensors. If the spectral occupancy of each signal generated by the sensors is about the same, then the CBW channels would be selected. On the other hand, if the spectral content of each signal varies substantially, then the PBW channels would be selected. For example, if one sensor is to measure fuel that is being consumed slowly and is varying slower than, say, 6 Hz, while other sensors are measuring vibrations at, say, 14 different points where the maximum frequency at one point would not exceed 14 Hz and the highest maximum frequency at another point would not exceed 790 Hz, a PBW format would be selected.

2.4 Systems Contaminated with Noise

2.4.1 Single-Channel FM

The output signal-to-noise ratio in terms of the input carrier-to-noise ratio in the carrier IF bandwidth and in terms of signal parameters for a single FM channel with single-tone modulation will be developed later, in Section 2.7, and is given by

$$[S/N]_o = \frac{3}{2} D^2 \frac{B_{IF}}{f_m} [S/N]_i \qquad (2.20)$$

where

$[S/N]_i = [S/N]_c$ by definition;

f_m = frequency of the single tone;

f_d = peak deviation of the carrier by the single tone;

$D = f_d/f_m$ the modulation index or deviation ratio;

η = white noise power spectral density in the IF.

A single FM channel is shown in Figure 2.5. This model assumes white noise as the input to the IF carrier bandpass filter *and that the FM discriminator is followed by an ideal lowpass filter with a cutoff frequency of* f_m. Equation (2.20) shows that the output signal-to-noise will increase as the square of the modulation index, although this will not occur indefinitely. Note from Figure 2.6 that the noise power spectral density at the output of the FM discriminator is not flat but parabolic. This model also assumes that the system is operating above threshold. (Threshold will be discussed in Section 2.8.)

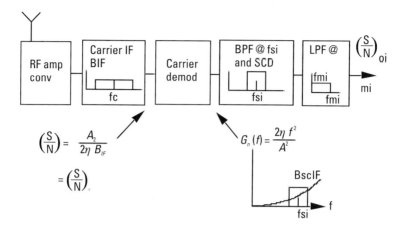

Figure 2.5 The *i*th subcarrier channel in an FM/FM multiplex.

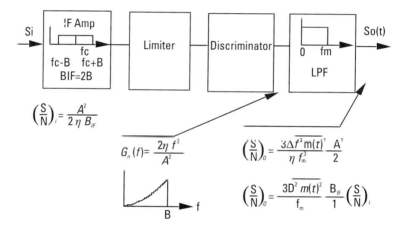

Figure 2.6 FM receiver and spectrums.

2.4.2 FM/FM Noise

Because the noise is parabolic out of the carrier discriminator, a special problem is created in the design of an FM/FM multiplex system. Since the N bandpass filters used to separate the subcarriers must contend with the noise power spectral density out of the carrier discriminator, which is increasing as f^2 (see Figure 2.5), the higher-frequency subcarrier channels must contend with more noise. That is, the higher-frequency bandpass filters must work into a greater noise power spectral density than the lower-frequency filters; hence the system and signal parameters must be designed such that the required signal quality is achieved in all the channels. Figure 2.5 depicts the *i*th channel of an FM/FM multiplex system.

Although (2.20) is for a single FM channel, it may be used for predicting the output signal-to-noise ratio for the *i*th subcarrier channel if the $[S/N]_c$ is replaced by the signal-to-noise ratio in the *i*th subcarrier IF, and if the carrier IF bandwidth is replaced by the *i*th subcarrier IF bandwidth. Further manipulations will give the signal-to-noise of the *i*th subcarrier channel in terms of the carrier-to-noise ratio.

Specifically, the equation for the output signal-to-noise ratio for an FM/FM system may be developed in the following way. There are three important cases in FM demodulation or detection in which the signal-to-noise out, in terms of the signal-to-noise in, is of importance. They are as follows:

1. The predetection filter is bandpass and the postdetection filter is lowpass.

2. The predetection filter is bandpass and the postdetection filter is bandpass.

3. A combination of the above two (FM/FM).

Block diagrams of these three cases are illustrated in Figure 2.7.

For single-tone modulation and the BP-to-LP case, the equation relating the output signal-to-noise in terms of the carrier-to-noise power in the IF is

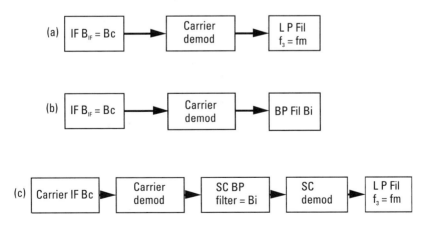

Figure 2.7 Three types of filters for FM demodulation: (a) BP-LP case, (b) BP-BP, and (c) *i*th subcarrier channel.

$$[S/N]_{o_{LP}} = \frac{3}{2}D^2\frac{B_{IF}}{f_m}[S/N]_c \qquad (2.21)$$

$B_{IF} = B_c$ = bandwidth of the bandpass filter preceding the discriminator;

f_m = highest frequency in the modulating signal and the 3-dB frequency of the lowpass filter following the discriminator;

D = FM mod index.

For the BP-to-BP case, the equation [2–4] is

$$[S/N]_{o_{BP}} = \frac{1}{4}\left[\frac{f_{dc}^2}{f_s^2}\right]\left[\frac{A^2}{\eta Bi}\right] \qquad (2.22)$$

where

f_s = frequency of a single tone modulation of the carrier; may also be thought of as the frequency of a single subcarrier;

f_{dc} = peak deviation of the carrier by the tone;

A = carrier power;

B_i = bandwidth of the postdetection bandpass filter;

η = one-sided power spectral density of the white noise applied to the IF filter.

Because the first bracket term in (2.22) is also the mod index, D_c, of the carrier with respect to the single tone or single subcarrier, (2.22) may be rewritten as

$$[S/N]_{o_{BP}} = \frac{1}{4}D_C^2\frac{A^2}{\eta B_i} \qquad (2.23)$$

The process of combining the two cases in order to predict the performance of the composite FM/FM system depicted in Figure 2.7(c) is as follows. The starting equation is (2.21) for the BP-to-LP part of the composite case. However, the IF filter of (2.21) becomes the bandpass filter of the subcarrier. Hence, B_c is replaced by B_i, and the $[S/N]_c$ must be replaced by $[S/N]_{oBP}$. Then f_m must be replaced with f_{mi}, the highest frequency in the message modulating the ith subcarrier, and the output signal-to-noise ratio of the ith channel becomes

$$[S/N]_{oi} = \frac{3}{2} D_{si}^2 \frac{B_i}{f_{mi}} [S/N]_{oBP} \qquad (2.24)$$

Converting the expression of (2.23) into an ith channel expression by adding an ith index and substituting the resultant into (2.24) gives an expression for $[S/N]_{oi}$ of

$$[S/N]_{oi} = \frac{3}{2} D_{si}^2 \frac{B_i}{f_{mi}} \frac{1}{4} D_{ci}^2 \frac{A^2}{\eta B_i} \qquad (2.25)$$

Multiplying the numerator and the denominator of (2.25) by B_c and rearranging gives

$$[S/N]_{oi} = \frac{3}{4} D_{si}^2 \frac{B_i}{f_{mi}} D_{ci}^2 B_c \frac{A^2}{2\eta B_i B_c} \qquad (2.26)$$

B_i divides out, and because the last term in (2.26) is $[S/N]_c$, (2.26) becomes

$$[S/N]_{oi} = \frac{3}{4} D_{si}^2 D_{ci}^2 \frac{B_c}{f_{mi}} [S/N]_c \qquad (2.27)$$

Note that B_{IF} is sometimes used for B_c in the literature. If the modulation index is replaced by the deviations, (2.27) becomes

$$[S/N]_{oi} = \frac{3}{4} \frac{f_{dsi}^2}{f_{mi}^2} \frac{f_{dci}^2}{f_{si}^2} \frac{B_c}{f_{mi}} [S/N]_c \qquad (2.28)$$

Therefore, the output signal-to-noise ratio for the ith channel of an FM/FM system is given in decibels if 10 log [ ] is taken of (2.28).

In terms of modulation indexes,

$$[S/N]_{oi} = \frac{3}{4} D_{si}^2 D_{ci}^2 \frac{B_c}{f_{mi}} [S/N]_c \qquad (2.29)$$

Or, in terms of voltage or numerical values gives

$$[S/N]_{oi} = \sqrt{3/4} \, \frac{f_{dsi}}{f_{mi}} \frac{f_{dci}}{f_{si}} \sqrt{\frac{B_c}{f_{mi}}} \, [S/N]_c \tag{2.30}$$

$$= \sqrt{3/4} \, D_{si} D_{ci} \sqrt{\frac{B_c}{f_{mi}}} \, [S/N]_c \tag{2.31}$$

where

f_{dsi} = deviation of the ith subcarrier by the ith message;

f_{mi} = maximum frequency of the ith message = message bandwidth or the 3-dB point of the postdetection filter if the two are not equal;

f_{dci} = deviation of the carrier by the ith subcarrier;

f_{si} = center frequency of the ith subcarrier;

B_c = carrier IF bandwidth;

D_{si} = modulation index of subcarrier and message in the ith channel;

D_{ci} = modulation index of the carrier and the ith subcarrier;

$[S/N]_c = [S/N]_i$ = carrier-to-noise ratio in the carrier IF.

Note that 20 log [ ] of the voltage equations should be taken to convert the equation to decibels.

2.5 FM/FM Multiplex Systems

2.5.1 PBW Channels

In PBW channels (PBCs), the peak frequency deviation of the subcarrier, f_{dsi}, is chosen as a fixed percentage, say P, of the subcarrier frequency. Specifically, there are three PBCs, for which P is 7.5%, 15%, and 30%. That is, in the 7.5% channels, $f_{dsi} = Pf_{si} = (0.075)f_{si}$. The modulation index, D_{si}, is nominally chosen to be 5 for all subcarriers. The message bandwidth, $f_{mi} = f_{dsi}/D_{si} = Pf_{si}/D_{si}$. Thus, the message bandwidth of the subcarrier channel is proportional to the subcarrier frequency.

2.5.2 CBW Channels

In CBW channels, f_{dsi}, is a constant for all channels, and the message bandwidth is also a constant.

2.6 Operational Filter Bandwidths

2.6.1 Receiver IF Bandwidth

Once the FM/FM package, that is, the preemphasis schedule (the deviation of the carrier by each subcarrier) has been designed, then the receiver IF bandwidth may be calculated from (2.14). This equation gives an approximate transmission bandwidth as well as the IF bandwidth. The transmission bandwidth is not necessarily the 99% bandwidth referred to in the IRIG standards [5].

2.6.2 Video Filter Bandwidth

The noise output from the carrier demodulator is also parabolic and continues in frequency until $f = B_{IF}/2$. Since the signal that is passed at this point is the subcarriers at baseband, it is necessary to pass only the frequencies up to the highest subcarrier plus half of the bandwidth occupied by the message modulating the highest frequency subcarrier. Therefore, the bandwidth of the lowpass video filter should be set to

$$B_v = f_{hsc} + BW_{bph}/2 \qquad (2.32)$$

where BW_{bph} is the bandwidth of the bandpass filter separating the highest frequency subcarrier and f_{hsc} is the highest frequency subcarrier. Because the filters are nonideal, the actual setting should be 10% larger than (2.32) indicates.

2.7 Development of the FM Noise Model and Signal-to-Noise Ratio

2.7.1 Carrier Plus Noise Model

Narrowband noise, $n(t)$, with a bandpass spectrum may be represented by

$$n(t) = n_c(t) \cos \omega_c t + n_s(t) \sin \omega_c t$$

By narrowband and bandpass, it is meant that f_c is at least twice as large as $f_c - f_{min}$, where f_{min} is the lowest frequency in the bandpass spectrum. Both $n_c(t)$ and $n_s(t)$ have equivalent lowpass spectrums, as shown in Figure

2.8(a). The result of multiplying these two lowpass processes by $\cos 2\pi f_c t$ and $\sin 2\pi f_c t$ is to move the lowpass spectrums about f_c, shown in Figure 2.8(b), creating a bandpass process. This bandpass noise is usually the result of applying Gaussian white noise with a flat two-sided power spectrum of $\eta/2$ to a bandpass filter such as the IF filter.

2.7.2 Signal-to-Noise Ratios

In the signal-to-noise region where most systems must operate, the signal is substantially larger than the noise. The signal-to-noise ratio may be deter-

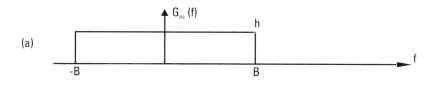

$$G_{no}(f) = \frac{\eta f^2}{A^2} \qquad \text{two-sided}$$

$$G_{no}(f) = \frac{2\eta f^2}{A^2} \qquad \text{one-sided}$$

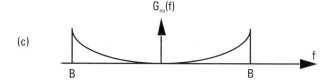

Figure 2.8 FM noise.

mined by finding the output signal when no noise is present and the output noise power when only a carrier is present with no modulation.

The model for an unmodulated carrier plus narrowband noise is

$$e(t) = A \sin \omega_c t + n(t)$$
$$= A \sin \omega_c t + n_c(t) \cos \omega_c t + n_s(t) \sin \omega_c t$$

$n(t)_c \cos \omega_c t$ is in phase quadrature with the carrier, and $n(t)_s \sin \omega_c t$ is in phase with the carrier. The limiter that precedes the discriminator removes the in-phase noise component since it adds in amplitude and gives

$$e(t) = A \sin \omega_c t + n_c(t) \cos \omega_c t \qquad (2.33)$$

Using a trigonometric identity gives

$$e(t) = \sqrt{A^2 + n_c^2} \sin(\omega_c t + \theta) \qquad (2.34)$$

where $\theta = a \tan n_c / A$.

In the region of high carrier-to-noise ratios

$$\theta \approx n_c / A \qquad (2.35)$$

and

$$\sqrt{A^2 + n_c^2} = A \qquad (2.36)$$

The baseband power spectral density of n_c is as shown in Figure 2.8(a) and is

$$G_{nc}(f) = \eta \text{ W/Hz} \qquad -B < f < B \qquad (2.37)$$

where $B = B_{IF}/2$.

Assume the discriminator constant is 1 V/Hz, and since the output is the derivative of the phase (or the difference between the instantaneous frequency and the center frequency) the discriminator output is

$$e_0(t) = f' = \theta'/2\pi = n_c'(t)/2\pi A \qquad (2.38)$$

The transfer function for a differentiator is $H(j\omega) = j\omega$. The relationship between the input and the output power spectral density is

$$G_o(f) = |H(f)|^2 G_i(f) \qquad (2.39)$$

Using (2.38) and (2.39), the two-sided power spectral density at the output of the discriminator is

$$G_o(f) = (1/2\pi A)^2 (2\pi f)^2 \eta = f^2 \eta / A^2 \quad -\frac{B_{IF}}{2} < f < \frac{B_{IF}}{2} \quad (2.40)$$

$$= 0 \qquad\qquad\qquad\qquad \text{elsewhere}$$

This spectrum is parabolic or increases as f^2 and is shown in Figure 2.8(c). The output noise power of interest is the noise power out of the lowpass filter following the FM discriminator. The lowpass filter has a cutoff frequency just high enough to pass the maximum frequency, f_m, present in $m(t)$. Assuming the postdetection LP filter is ideal *with a cutoff frequency of f_m*, the output noise power is

$$N_{out} = (\eta/A^2)2 \int_0^{f_m} f^2 df = \frac{2\eta f_m^3}{3A^2} \qquad (2.41)$$

The FM-modulated carrier without noise is

$$e(t) = A \sin\left(\omega_c t + 2\pi f_d \int_0^t m(t)dt \right) \qquad (2.42)$$

The output of the FM discriminator is

$$e_o(t) = f_d m(t) \qquad (2.43)$$

The output signal power is

$$S_o = f_d^2 \, \text{AVG}[m(t)^2] = f_d^2 \, \overline{m(t)^2} \qquad (2.44)$$

The input power is the carrier power and is

$$S_i = A^2/2 \qquad (2.45)$$

The input noise power to the discriminator is the noise power in the IF bandwidth. Assuming white noise input with a two-sided *power spectral density* (PSD) of $\eta/2$ and an idealized bandpass filter for the IF, with bandwidth $2B$, the noise power into the discriminator is

$$N_i = \eta B_{IF} = \eta 2B \qquad (2.46)$$

The signal-to-noise out of the postdetection LP filter is given by

$$[S/N]_o = \frac{S_o}{N_{out}} = \frac{f_d^2 \overline{m(t)^2}}{2\eta f_m^3 / 3A^2} = \frac{3A^2 f_d^2 \overline{m(t)^2}}{2\eta f_m^3} \qquad (2.47)$$

The signal-to-noise out in terms of signal-to-noise in, which is the carrier-to-noise in the IF bandwidth (see Figure 2.6), is obtained by multiplying the numerator and the denominator of (2.47) by B_{IF} and rearranging, giving

$$[S/N]_o = \frac{3f_d^2 \overline{m(t)^2}}{f_m^2} \frac{B_{IF}}{f_m} \frac{A^2/2}{\eta B_{IF}} \qquad (2.48)$$

$$= \frac{3f_d^2 \overline{m(t)^2}}{f_m^2} \frac{B_{IF}}{f_m} [S/N]_i$$

Equation (2.48) can be written in terms of the modulation index *if, and only if, the corner frequency of the post detection filter is equal to* f_m. Then,

$$[S/N]_o = 3D^2 \overline{m(t)^2} \frac{B_{IF}}{f_m} [S/N]_i \qquad (2.49)$$

For single-tone modulation $\overline{m(t)^2} = 1/2$, and $[S/N]_o$ becomes

$$[S/N]_o = 3D^2 \frac{B_{IF}}{2f_m} [S/N]_i \qquad (2.50)$$

2.8 Threshold

At the onset of *threshold,* the output signal-to-noise breaks sharply downward for a small decrease in signal-to-noise. This degradation is much more than the linear model predicts. "Threshold" is the term given to the point at which this occurs.

2.8.1   Threshold Description

At the threshold point, the noise out of the discriminator changes from limited variations or phase jitter to impulse type noise [6–10] as shown in Figure 2.9(d). This usually occurs at a carrier-to-noise ratio in the IF of 10 to 13 dB, depending upon the modulation and IF bandwidth.

2.8.2   Threshold Carrier Model and Description

For high signal-to-noise ratios, Figure 2.9(a) shows a phase model of carrier plus noise resulting from (2.33). The in-phase noise component has been eliminated by the limiter, and only the quadrature component remains, causing phase jitter as illustrated in Figure 2.9(a). Occasionally, depending upon the signal-to-noise ratio, the noise component becomes large enough to cause the carrier to swing 360 degrees, as shown in Figure 2.9(b), resulting in an impulse when the phase change is demodulated since the output of the discriminator is proportional to the derivative of the phase. Figure 2.9(c) shows the variations in the phase as a function of time. The noise causes

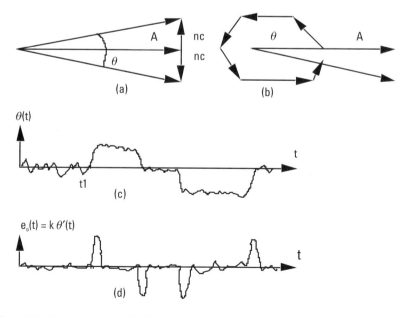

Figure 2.9 Noise out of an FM discriminator showing impulses.

only a slight variation in θ, until at $t = t_1$ the noise causes the phase of the carrier to jump by 2π degrees. Taking the derivative of the phase gives the impulsive nature of the discriminator output, as depicted in Figure 2.9(d). All of the above equations showing the output signal-to-noise ratios are only valid whenever the system is operating above threshold.

2.9 Effect of Increasing the IF Bandwidth

Increasing the IF bandwidth does not increase the noise in the output of the lowpass filter following the FM discriminator, providing threshold does not occur in the IF. Since the noise in the IF is equal to ηB_{IF}, the noise in the IF *does* increase, as shown in Figure 2.10(b) by the zigzag line; hence $[S/N]_i$ decreases. If $[S/N]_i$ falls below 12 dB, threshold will occur. Equation (2.47) indicates that the output noise will not increase as B_{IF} is increased, which is illustrated in Figure 2.10(a); however, (2.41) does show that the output noise will increase if the 3-dB point, f_m, of the postdetection lowpass filter is increased. If B_{IF} is increased and the carrier power is increased to maintain a constant $[S/N]_i$, (2.47) shows that the output signal-to-noise will increase as $A^2/2$ is increased. Equation (2.48) is obtained from (2.47) by multiplying by B_{IF}. Equation (2.48) indicates that the output signal-to-noise will increase as B_{IF} is increased, which will happen only if the carrier power is increased in order to maintain a constant $[S/N]_i$. Note that the parabolic noise increases with frequency until $f = B_{IF}/2$ is reached, and then the noise PSD goes to zero if an idealized filter is assumed; otherwise, the

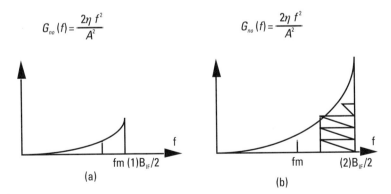

Figure 2.10 Parabolic noise spectrums out of an FM discriminator for two different IF bandwidths. In (a), the IF bandwidth is less than it is for (b). In (b), the zigzag line indicates the increase in the noise in the IF with the larger IF bandwidth.

spectrum rolls off as the filter characteristic, which is sharp for a high-order bandpass filter. Increasing the 3-dB corner frequency of the postdetection filter beyond this point does *not* increase the output noise for an ideal filter and does so only slightly for a practical filter.

2.10 Transmission Bandwidth Estimation

In this book, an estimate of the required IF bandwidth, B_c, is given by (2.11) and (2.12). For FM/FM, a generalized concept was used, which was defined by (2.14). Equation (2.12) is based upon Carson's rule supported by empirical data [5]. Equation (2.14) is based upon work by Rosen [2]. The estimate of B_c was also used as an estimate of the transmission or emission bandwidth, B_t, and usually B_c will be slightly less than B_t.

Problems

Problem 2.1

Draw a block diagram of an FM system.

Problem 2.2

Discuss the units on k, the sensitivity constant of a VCO.

Problem 2.3

If $k = 12.5$ Hz/V in (2.3), what would be the peak deviation in hertz if (a) $m(t) = 2 \sin 2\pi f_m t$, (b) $m(t) = \sin 2\pi f_m t$?

Problem 2.4

In Problems 2.2 and 2.3, if $f_m = 6$ Hz, what would be the modulation index for (a) and (b)?

Problem 2.5

What is the modulation index for channel 2 in the 7.5% PBW channels?

Problem 2.6

A typical VCO or subcarrier oscillator, SCO, as it is referred to, will deviate from the center frequency to a peak frequency when the input voltage goes from 0V to 2.5V. If the peak deviation is 30 Hz, what is k?

Problem 2.7

Repeat Problem 2.6; if the peak deviation is 225 Hz, what is k?

Problem 2.8

For f_d = 4 kHz, estimate the required bandwidth for the following values of D, (a) 0.1, (b) 1, (c) 5, and (d) 10.

Problem 2.9

If $[S/N]_i$ = 12 dB, B_{IF} = 500 kHz, and fm = 500 kHz, for a single FM channel, find $[S/N]_o$ for the following values of D: (a) 0.7, (b) 2, (c) 5, and (d) 10.

Problem 2.10

If f_{mi} = 1,395 Hz, $[S/N]_c$ = 12 dB, B_c = 500 kHz, and D_{si} = 5, find $[S/N]_o$ for the following values of D_{ci} = (a) 0.5, (b) 0.718, (c) 1, and (d) 1.5.

Problem 2.11

If f_{mi} = 1,395 Hz, $[S/N]_c$ = 12 dB, B_c = 500 kHz, D_{si} = 5, and f_{si} = 93 kHz, find f_{dci} for the following values of $[S/N]_o$: (a) 39 dB, (b) 42 dB, (c) 44 dB, (d) 46 dB, and (e) 49 dB.

Problem 2.12

For the 7.5% PBW channels, and D_{si} = 5, use (2.31) to show that the deviation of the carrier by the subcarrier, f_{dci}, increases as the subcarrier frequency, f_{si}, is raised to the 3/2 power. (Use the equation for numerical values.)

Problem 2.13

For the CBW channels, using the equation for numerical values, show that

$$f_{dci} = \frac{[S/N]_o}{[S/N]_i\sqrt{3/4}} \frac{\sqrt{f_{dsi}f_{si}}}{\sqrt{B_c}D_{si}^{1.5}}$$

References

[1]   Lathi, B. P., *Modern Digital and Analog Communication Systems*, 2nd Edition, New York: Holt, Rinehart, and Winston, 1989, p. 294.

[2]   Rosen, C. "System Transmission Parameters Design for Threshold Performance," *Proc. Int. Telemetering Conf*, Vol. XIX, 1988, pp. 145–182.

[3]   Uglow, K., "Noise and Bandwidth in FM/FM Radio Telemetering," *IRE Transactions*, May 1957, pp. 19–22.

[4]   Rechter, R. J., "Summary and Discussion of Signal-to-Noise Ratio Improvement Formulae for FM and FM/FM Links," *Proc. Int. Telemetering Conf*, Vol. III, 1967, pp. 221–255.

[5]   Secretariat, Range Commanders Council, *Telemetry Standards*, White Sands Missile Range, NM: RCC, IRIG 106-00.

[6]   Stumpers, F. L. H. M., "Theory of Frequency-Modulation Noise," *Proc. IRE*, 36, Sept. 1948.

[7]   Stumpers, F. L. H. M., "Distortion of Frequency-Modulated Signals in Electrical Networks," *Communication News*, 9, April 1948.

[8]   Rice, S. O., "Statistical Properties of a Sine-Wave Plus Random Noise," *Bell Sys. Tech. J.*, 27, Jan. 1948.

[9]   Rice, S. O., "Noise in FM Receivers," *Proc. Symp. Time Series Analysis*, M. Rosenblatt, ed., New York: Wiley, 1963.

[10] Osborne, W., and D. Whiteman, *Optimizing PCM/FM+FM/FM Systems Using IRIG Constant Bandwidth Channels,* ECE Tech Report Series No. 92-014, Las Cruces, NM: New Mexico State University, Nov. 1992.

3

Design of FM/FM Systems

In the design of communication systems, parameters must be determined such that the signal quality is equal to or greater than the design specifications. Generally, in analog systems, the output signal-to-noise ratio is one of the specifications. In FM/FM telemetry systems, the deviation of the carrier with respect to the subcarriers is a variable that will be determined in order to achieve the desired signal quality. Specifically, the preemphasis schedule is developed such that the signal quality in all the channels meets specifications. This chapter is concerned with the design of the preemphasis schedule.

3.1 Learning Objectives

Upon completing this chapter, the reader should understand the following concepts and procedures:

- The function of the modulation parameters in setting the signal-to-noise output;
- The tradeoff between bandwidth and signal-to-noise output;
- A systematic process for designing the preemphasis schedule for FM/FM systems;
- Preemphasis design for operation above threshold;
- Preemphasis design for a specified bandwidth;

- Preemphasis design for a minimum transmission bandwidth or concurrent all-channel dropout;
- Preemphasis design for different signal and subcarrier mod indexes;
- Preemphasis design for different specified output signal-to-noise ratios in the subcarrier channels;
- Setting the receiver IF and video filter bandwidths and discriminator postdetection bandwidths based upon modulation design parameters for the actual operation of the telemetry receiving system;
- Hardware implementation of the preemphasis schedule.

3.2 System Parameters

This section is concerned with what modulation parameters may be used to design the preemphasis schedule. Inspection of (2.27) and (2.28) indicates what parameters—primarily modulation (mod) indexes, or deviation ratios, as they are sometimes called—are available to vary in order to maintain equal quality of the output signals in all the channels. This problem must be addressed since the bandpass filters used to separate the subcarriers must contend with a power spectral density increasing as f^2. Further, since D_{si} is set, nominally at 5, the bandwidth of the bandpass filters is approximately given by $2f_{dsi}$, which is increasing as f_{si} increases for PBW channels and is constant for CBW channels. In an FM/FM system designed to IRIG specifications, D_{si} is nominally set to 5, and f_{mi} is specified while the carrier IF bandwidth B_c and $[S/N]_c$ are the same for all channels. The only remaining nonfixed variable is D_{ci}, the mod index of the carrier with respect to the ith subcarrier. Since D_{ci} is the ratio f_{dci}/f_{si} and the subcarrier frequency is fixed, the only parameter available to vary in order to maintain equal signal quality is, f_{dci}, the deviation of the carrier by the subcarrier. The design values of the f_{dci}'s are referred to as the preemphasis schedule.

Once the preemphasis schedule has been designed, it is possible and necessary to specify the operational parameters of the receiver such as the IF and video filter bandwidths, postdetection lowpass filter bandwidths, and filter types.

The first step in designing the preemphasis schedule is to solve (2.30) for f_{dci}, which gives

$$f_{dci} = \sqrt{\frac{4}{3} \frac{f_{si}}{f_{dsi}}} \sqrt{f_{mi}^3} \frac{1}{\sqrt{B_c}} \frac{[S/N]_{oi}}{[S/N]_c} \tag{3.1}$$

The difficulty in using this equation is that it is one equation in two unknowns, B_c and f_{dci}. To use (3.1) as is, B_c, the carrier IF bandwidth, must be estimated by a rule of thumb since it is a function of the f_{dci}'s, which are being calculated. A method to circumvent this problem will be discussed in Section 3.3. Early methods developed and used to solve this problem were presented by Downing [1], Nichols and Rauch [2], Law [3], and Rosen [4, 5].

3.3 Design Procedure

A preemphasis schedule design procedure that circumvents the problem of one equation in two unknowns will now be given. The approach is to develop an expression for B_c in terms of f_{dci} and known parameters and substitute that expression into (3.1), converting it into one equation in one unknown. The method to be developed is systematic and amenable to computer programming.

The approach is to define some intermediate variables to facilitate developing one equation in one unknown. Specifically, the equations are arranged such that an expression for B_c in terms of f_{dc1} may be found and substituted into (3.1) and f_{dc1} solved for. Then the other f_{dci}'s are solved for. The process of algebraically manipulating the equations to solve for the actual deviations will be developed next.

In the process, B_c, will be solved for in terms of the norm of the carrier deviations, f_{dn}, and f_{dc1}, to be defined. The first step in the systematic approach is to define all the variables to be used.

Repeating (2.15) gives

$$f_{dn} = \sqrt{f_{dc1}^2 + f_{dc2}^2 + + + f_{dcn}^2} \qquad (3.2)$$

Let

f_{dc1} = deviation of the carrier by the highest frequency subcarrier;

f_{dc2} = deviation of the carrier by the next highest frequency subcarrier;

$\vdots$

f_{dcn} = deviation of the carrier by the lowest frequency subcarrier.

Let

f_{s1} = highest frequency subcarrier;

f_{s2} = next highest frequency subcarrier;

$$\vdots$$

$$\vdots$$

f_{sn} = lowest frequency subcarrier.

Note that because the subcarriers are working from a parabolic noise spectrum, the highest frequency subcarrier will have the maximum noise power to contend with and hence must have the largest f_{dci} to obtain the largest signal-to-noise FM improvement. Normalizing (3.2) with respect to f_{dc1} gives

$$f_{dn} = f_{dc1} \sqrt{1 + \left(\frac{f_{dc2}}{f_{dc1}}\right)^2 + \left(\frac{f_{dc3}}{f_{dc1}}\right)^2 + + + \left(\frac{f_{dcn}}{f_{dc1}}\right)^2} \qquad (3.3)$$

Define:

$$A_1 = 1$$
$$A_2 = f_{dc2}/f_{dc1}$$
$$A_3 = f_{dc3}/f_{dc1}$$
$$\vdots$$
$$\vdots$$
$$\vdots$$
$$A_n = f_{dcn}/f_{dc1}$$

$$(3.4)$$

Using the definitions of (3.4) in (3.3) gives

$$f_{dn} = f_{dc1} \sqrt{A_1^2 + A_2^2 + A_3^2 + + + A_n^2} \qquad (3.5)$$

Defining A_p such that

$$A_p = \sqrt{A_1^2 + A_2^2 + A_3^2 + + + A_n^2} \qquad (3.6)$$

Then

$$f_{dn} = f_{dc1} A_p \qquad (3.7)$$

The A_i's, and hence A_p, are in terms of the carrier deviations, which are unknown; however, the A_i's may be expressed in terms of known parameters by formally taking the ratio of the deviations using (3.1), which gives

$$A_i = \frac{f_{dci}}{f_{dc1}} = \frac{\sqrt{4/3} \dfrac{[S/N]_{oi}}{[S/N]_c} \dfrac{f_{si}}{f_{dsi}} \sqrt{f_{mi}^3}}{\sqrt{4/3} \dfrac{[S/N]_{o1}}{[S/N]_c} \dfrac{f_{s1}}{f_{ds1}} \sqrt{f_{m1}^3}} \tag{3.8}$$

If the output signal-to-noise ratios in all the channels are to be the same, then

$$A_i = \frac{f_{si}}{f_{s1}} \frac{f_{ds1}}{f_{dsi}} \frac{\sqrt{f_{mi}^3}}{\sqrt{f_{m1}^3}} \tag{3.9}$$

Equation (3.9) is important since it allows the A_i's to be calculated in terms of known parameters. Equation (3.7) gives f_{dn}, the norm of the total carrier deviation in terms of f_{dc1} and A_p, which is known. Although f_{dc1} is not known, the equation with f_{dn} may be used to express B_c in terms f_{dc1} and substituted into (3.1), reducing this equation into one equation in one unknown. The resulting equation may be used to solve for f_{dc1}. Further, since

$$f_{dci} = f_{dc1} A_i \tag{3.10}$$

the other f_{dci}'s may be solved for. Proceeding and repeating (2.14) gives B_c in terms of f_{dn} such as

$$B_c = 2[f_{dn} + f_{s1}] = 2(f_{dc1} A_p + f_{s1}) \tag{3.11}$$

f_{dc1} must be solved for first. Repeating and rearranging (3.1), then substituting the index 1 for the ith index gives

$$f_{dc1} = \sqrt{4/3} \frac{[S/N]_{oi}}{[S/N]_c} \frac{1}{\sqrt{B_c}} \frac{f_{s1}}{f_{ds1}} \sqrt{f_{m1}^3} \tag{3.12}$$

Multiplying numerator and denominator of (3.12) by $\sqrt{f_{ds1}}$ substituting $D_{s1} = f_{ds1}/f_{m1}$ gives

$$f_{dc1} = \sqrt{4/3} \, \frac{[S/N]_{oi}}{[S/N]_c} \frac{1}{\sqrt{B_c}} \frac{f_{s1}\sqrt{f_{ds1}}}{\sqrt{D_{s1}^3}} \tag{3.13}$$

Substituting the expression for B_c, (3.11), into (3.13) results in

$$f_{dc1} = \sqrt{4/3} \, \frac{[S/N]_{oi}}{[S/N]_c} \frac{1}{\sqrt{2(f_{dc1}A_p + f_{s1})}} \frac{f_{s1}\sqrt{f_{ds1}}}{\sqrt{D_{s1}^3}} \tag{3.14}$$

If we let

$$C = \sqrt{4/3} \, \frac{[S/N]_{oi}}{[S/N]_c}$$

then (3.14) becomes

$$f_{dc1} = C \frac{1}{\sqrt{2(f_{dc1}A_p + f_{s1})}} \frac{f_{s1}\sqrt{f_{ds1}}}{\sqrt{D_{s1}^3}} \tag{3.15}$$

Inspection of (3.15) reveals that f_{dc1} may be solved for without knowing B_c explicitly. That is, if $[S/N]_c$ is specified, the A_i's are calculated, and since the other parameters are known (the D_{si}'s are nominally set to 5), f_{dc1} can be calculated. Once f_{dc1} is calculated, the remaining f_{dci}'s may be calculated from (3.10). Rearranging (3.15) results in a third order polynomial given by

$$f_{dc1}^3 + \frac{f_{s1}}{A_p}(f_{dc1}^2) - \frac{C^2 f_{s1}^2 f_{ds1}}{2A_p D_{s1}^3} = 0 \tag{3.16}$$

This third-order equation may be used to solve for f_{dc1} explicitly in terms of the known parameters. Once f_{dc1} is solved for, all the other f_{dci}'s may be solved for since the A_i's are known. That is,

$$f_{dci} = f_{dc1}A_i \tag{3.17}$$

3.4 Design Examples

3.4.1 PBW Channels

Design Example 1

To clarify the use of the FM/FM design process, an FM/FM multiplex design example will be worked for 15 subcarriers of the 7.5% PBW channel. The subcarriers used will be the 1.3 kHz through the 93 kHz.
Parameters:

$[S/N]_C$ = 12 dB (3.98 numerical);

$D_{si} = f_{dsi}/f_{mi}$ = 5 (nominal);

Desired $[S/N]_{oi}$ = 46 dB (199 numerical) minimum in all channels.

Step 1

The A_i's will be calculated. Repeating (3.9) gives

$$A_i = \frac{f_{si}}{f_{s1}} \frac{f_{ds1}}{f_{dsi}} \frac{\sqrt{f_{mi}^3}}{\sqrt{f_{m1}^3}} \qquad (3.18)$$

From the IRIG standards:

f_{s1} = 93 kHz;

f_{ds1} = 6.975 kHz;

f_{m1} = 1.395 kHz;

f_{s2} = 70 kHz;

f_{ds2} = 5.75 kHz;

f_{m2} = 5.250 kHz;

$C = \sqrt{4/3} \dfrac{[S/N]_{oi}}{[S/N]_C}$ = 57.87.

Specifically, for A_2,

$$A_2 = \frac{f_{s2}}{f_{s1}} \frac{f_{ds1}}{f_{ds2}} \frac{\sqrt{f_{mi}^3}}{\sqrt{f_{m1}^3}} = 0.653$$

Using (3.18) to calculate all the A_i's and noting that $A_1 = 1$ by definition gives the values in column 4 in Table 3.1.

Step 2

Calculate A_p. Using (3.9) to calculate A_p gives

$$A_p = \sqrt{A_1^2 + A_2^2 + A_3^2 + + + A_n^2} = 1.31$$

Step 3

Repeating (3.16) gives

$$f_{dc1}^3 + \frac{f_{s1}}{A_p}(f_{dc1}^2) - \frac{C^2 f_{s1}^2 f_{ds1}}{2 A_p D_{s1}^3} = 0 \tag{3.19}$$

Solving (3.19), a third-order polynomial, for f_{dc1} and using the parameters for this problem results in

$$f_{dc1} = 66.804 \text{ kHz} \tag{3.20}$$

Table 3.1
Design Example 1: Initial Preemphasis Schedule

f_{si} (kHz)	SCO (kHz) Upper Dev. (f_{ds})	f_{mi} (Hz)	A_i	f_{dci}
93.0	6.975	1,395	1.00	66,804.73
70.0	5.250	1,050	0.653	43,624.44
52.5	3.937	790	0.424	28,379.84
40.0	3.000	600	0.282	18,843.99
30.0	2.250	450	0.183	12,239.53
22.0	1.650	330	0.115	7,686.29
14.5	1.087	220	0.062	4,136.34
10.5	0.787	160	0.038	2,554.39
7.35	0.551	110	0.022	1,484.27
5.4	0.405	81	0.014	934.70
3.9	0.292	59	0.0086	573.69
3.0	0.225	45	0.0058	387.05
2.3	0.172	35	0.0039	259.82
1.7	0.127	25	0.0024	163.48
1.3	0.098	20	0.002	114.0

Step 4

From (3.17), the remaining f_{dci}'s may be calculated, such as

$$f_{dci} = f_{dc1} A_i \qquad (3.21)$$

and are shown in column 5 of Table 3.1.

Step 5

Calculate f_{dn}, the norm of the carrier deviations, from (3.7),

$$f_{dn} = f_{dc1} A_p = (66.804)(1.319) = 88.106 \text{ kHz} \qquad (3.22)$$

Step 6

Calculate B_c, the carrier IF bandwidth, from (3.11),

$$B_c = 2[f_{dn} + f_{s1}] = 2(88.106 + 93) = 362.2 \text{ kHz} \qquad (3.23)$$

Step 7

Calculate BW_v, the bandwidth of the output video filter following the carrier discriminator, in terms of the required bandwidth of the highest frequency subcarrier due to the modulating message, which is the bandwidth, BW_{bpfs1}, of the bandpass filter of the highest frequency subcarrier.

$$
\begin{aligned}
B_v &= f_{s1} + (BW_{bpfs1})/2 \\
&= 93 \text{ kHz} + 2(f_{ds93} + 1{,}395)/2 \\
&= 93 \text{ kHz} + (6{,}975 + 1{,}395) = 101.37 \text{ kHz}
\end{aligned}
$$

Summary of Results for Design Example 1

The only unspecified parameter available to ensure acceptable and equal signal-to-noise ratios in all the subcarrier channels in the design of an FM/FM multiplex system meeting IRIG specifications is the f_{dci}'s, the deviations of the carrier by the subcarriers. This procedure calculated the necessary deviations for a specified 46-dB signal-to-noise output in 15 PBW channels. Table 3.2 gives the resulting modulation index of the carrier with respect to the subcarrier and the calculated signal-to-noise ratio for the calculated carrier deviations.

Refinements in the Design

Inspection of column 5 in Table 3.1 indicates that the five lowest subcarriers will deviate the carrier less than 1 kHz. Variations of outputs of VCOs,

Table 3.2
D_{si} and D_{ci} for Design Example 1

f_{si} (kHz)	f_{dci} (kHz)	$(S/N)_{oi}$ (dB)	D_{si}	D_{ci}
93.0	66.80	46	5.000	0.7183
70.0	43.62	46	5.000	0.6232
52.5	28.37	46	4.984	0.5406
40.0	18.84	46	5.000	0.4711
30.0	12.23	46	5.000	0.4080
22.0	7.68	46	5.000	0.3494
14.5	4.13	46	4.943	0.2853
10.5	2.55	46	4.922	0.2433
7.35	1.48	46	5.000	0.2019
5.4	0.934	46	5.000	0.1731
3.9	0.573	46	5.000	0.1471
3.0	0.387	46	5.000	0.1290
2.3	0.259	46	5.000	0.1130
1.7	0.163	46	5.100	0.0962
1.3	0.114	46	5.100	0.088

resistors, transmitters, or intermodulation products in the system may cause larger apparent carrier deviations. Therefore, the deviations of the carrier by the lowest frequency subcarriers should be increased to a larger fixed percentage of f_{dn}. A reasonable increase is to make all the lower deviations equal to 10% of the computed f_{dn}. In this example, it would require increasing the deviations to about 8.8 kHz. However, for the purpose of illustrating the impact of this increase in the lower seven subcarriers, deviations will be increased to 2 kHz.

Increasing the f_{dci}'s changes several parameters. One change is that since $D_{ci} = f_{dci}/f_{si}$, the modulation index of the lower frequency subcarriers with respect to the carrier increases. Observation of (2.27) shows that the signal-to-noise improves as D_{ci} increases, which means that $[S/N]_{oi}$ will exceed the design requirements for the lower frequency channels. This presents an opportunity to increase the information capability of the lower channels. The modulation index of the subcarriers with respect to the message is $D_{si} = f_{dsi}/f_{mi}$ and is nominally set to 5. (Note that f_{mi} is the highest frequency in the message modulating the subcarrier.) Inspection of (2.27) shows that the signal-to-noise also varies directly as D_{si}. Since D_{ci} was increased, D_{si} may be decreased by increasing the allowed maximum frequency of the modulating signal.

Increasing the deviations of the carrier by the lower-frequency subcarriers increases f_{dn}, which makes it necessary to evaluate the increase in the required carrier IF bandwidth since B_c is a function of f_{dn}. It will be found that there is very little effect on the bandwidth. Design Example 2 will determine the effect on the system parameters of increasing the deviations of the carrier by the lower subcarriers.

Design Example 2

The effect of increasing the deviation of the carrier by the seven lower-frequency subcarriers will be evaluated for the multiplex system of Design Example 1. The deviation on all seven will be increased to 2 kHz.

Step 1

Increase f_{dci} of the seven lower subcarriers. Column 2 in Table 3.3 shows the increase in deviations.

Step 2

Compute the new f_{dn} from (3.2).

$$
\begin{aligned}
f_{dn} &= \sqrt{f_{dc1}^2 + f_{dc2}^2 + + + f_{dcn}^2} \\
&= [(66.8)^2 + (43.6)^2 + (28.4)^2 + (18.8)^2 + (12.2)^2 \qquad (3.24) \\
&\quad + (7.69)^2 + (4.12)^2 + (2.55)^2 + 7(2)^2]^{1/2} \\
&= 88.23 \text{ kHz}
\end{aligned}
$$

Table 3.3
Increased f_{dci} for Lower Subcarriers in Design Example 1

f_{si} (kHz)	f_{dci} (kHz)	$(S/N)_{oi}$ (dB)	D_{si}	D_{ci}
93.0	66.80	46.0	5.000	0.7183
70.0	43.62	46.0	5.000	0.6232
52.5	28.37	46.0	4.984	0.5406
40.0	18.84	46.0	5.000	0.4711
30.0	12.23	46.0	5.000	0.4080
22.0	7.68	46.0	5.000	0.3494
14.5	4.13	46.0	4.943	0.2853
10.5	2.55	46.0	4.922	0.2433
7.35	2.0	48.6	5.000	0.2019
5.4	2.0	52.6	5.000	0.1731
3.9	2.0	56.8	5.000	0.1471
3.0	2.0	60.2	5.000	0.1290
2.3	2.0	63.7	5.000	0.1130
1.7	2.0	67.9	5.100	0.0962
1.3	2.0	71.0	5.100	0.088

Step 3

Compute B_c from (3.11) and the new D_{ci}'s; that is,

$$B_c = 2[f_{dn} + f_{s1}] = 2(88.23 + 93) = 362.5 \text{ kHz} \qquad (3.25)$$

and

$$D_{ci} = f_{dci}/f_{si}$$

For example, D_{c15} is given by

$$D_{c15} = 2,000/1,300 = 1.54 \qquad (3.26)$$

Step 4

Compute the new $[S/N]_{oi}$ for the seven lowest subcarriers from (2.30).
Repeating (2.31) gives

$$[S/N]_{oi} = \sqrt{3/4}\, D_{si} D_{ci} \sqrt{\frac{B_c}{f_{mi}}}\, [S/N]_c \qquad (3.27)$$

The new signal-to-noise ratio for the lowest subcarrier is

$$[S/N]_{o15} = (0.866)(3.98)(5)(1.54)[(602)/4.47)] = 3,574.2 \text{ (numerical)}$$
$$[S/N]_{o15} = 20 \log 35,742 = 71 \text{ dB} \qquad (3.28)$$

Column 3 in Table 3.3 shows the resulting increase in $[S/N]_{oi}$ for the lower-frequency subcarriers. Since this is larger than the design specifications, f_{mi} will be doubled, lowering D_{mi} to approximately 2.5, which decreases $[S/N]_{oi}$.

Step 5

Double f_{mi} of the seven lowest subcarriers, and compute D_{si} and $[S/N]_{oi}$. Note that B_c remains at 362.5 kHz.

The mod index of the 15th subcarrier with respect to the message is

$$D_{s15} = f_{ds15}/f_{m15} = 98/40 = 2.45 \qquad (3.29)$$

Using equation (3.27) gives

$$[S/N]_{o15} = \sqrt{3/4} \, [S/N]_c D_{si} D_{ci} \sqrt{\frac{B_c}{f_{mi}}}$$

$$[S/N]_{o15} = (0.866)(3.98)(2.45)(1.54)[(362{,}500)/40]^{1/2}] = 1{,}263 \text{ (numerical)}$$

$$[S/N]_{o15} = 20 \log 1{,}263 = 62 \text{ dB} \tag{3.30}$$

The modified $[S/N]_{oi}$ is shown in column 3 of Table 3.4.

Summary of the Modifications

The deviations of the carrier by the seven lowest subcarriers were increased to 2 kHz, which increased the D_{ci}'s, which increased $[S/N]_{oi}$'s substantially above the design requirement. Consequently, the D_{si}'s were decreased by increasing the maximum frequency of the message, and the new $[S/N]_{oi}$'s were evaluated and found to be above the design requirement in all channels except the 7.35 kHz and the 5.4 kHz. These two channels should have their D_{si}'s increased to possible 4. The modified $B_c = 362.5$ kHz was found to be very close to the original $B_c = 362.2$ kHz.

It was determined that increasing the deviations of the seven lower frequency subcarriers to 2 kHz had very little impact on f_{dn}. In fact, increasing the seven lower subcarrier deviations to 10% of the calculated f_{dn} or to 8.8 kHz has a small effect, as will be shown next.

Table 3.4
Decreased D_{si} for Lower Subcarriers in Design Example 1

f_{si} (kHz)	f_{dci} (kHz)	$(S/N)_{oi}$ (dB)	D_{si}	D_{ci}
93.0	66.80	46.0	5.00	0.72
70.0	43.62	46.0	5.00	0.62
52.5	28.37	46.0	4.98	0.54
40.0	18.84	46.0	5.00	0.47
30.0	12.23	46.0	5.00	0.41
22.0	7.68	46.0	5.00	0.35
14.5	4.13	46.0	4.94	0.28
10.5	2.55	46.0	4.92	0.24
7.35	2.0	40.0	2.51	0.28
5.4	2.0	43.6	2.50	0.37
3.9	2.0	47.7	2.46	0.51
3.0	2.0	51.2	2.50	0.67
2.3	2.0	54.5	2.46	0.87
1.7	2.0	58.9	2.55	1.17
1.3	2.0	62.0	2.45	1.54

From (3.2)

$$
\begin{aligned}
f_{dn} &= \sqrt{f_{dc1}^2 + f_{dc2}^2 + + + f_{dcn}^2} \\
&= [(66.8)^2 + (43.6)^2 + (28.4)^2 + (18.8)^2 + (12.2)^2 \qquad (3.31) \\
&\quad + (7.69)^2 + (4.12)^2 + (2.55)^2 + 7(8.8)^2]^{1/2} \\
&= 91.1 \text{ kHz}
\end{aligned}
$$

The resulting bandwidth is

$$
B_c = 2[f_{dn} + f_{s1}] = 2(91 + 93) = 368.2 \text{ kHz}
$$

This small increase in B_c supports the model suggested in Chapter 2 and illustrated in Figure 2.4, although this case has substantially more subcarriers. Further, using only the five highest subcarrier deviations of the carrier gives

$$
\begin{aligned}
f_{dn} &= [f_{dc1}^2 + f_{dc2}^2 + + + f_{dc5}^2]^{1/2} \\
&= [(66.8)^2 + (43.6)^2 + (28.4)^2 + (18.8)^2 + (12.2)^2]^{1/2} \\
&= 76.7 \text{ kHz}
\end{aligned}
$$

The 76.7 kHz obtained using only the highest five subcarrier deviations compares favorably with the actual 88 kHz using all the deviations for a first-cut look.

Rewriting (3.7) gives

$$
f_{dn} = f_{dc1} A_p = 1.3 f_{dc1}
$$

This equation indicates that the total RMS carrier deviation or norm of the deviations is about 1.3 times the deviation of the carrier by the highest frequency subcarrier in PBW telemetry systems. This is a rule of thumb used some of the time.

3.4.2 CBW Channels

In CBW channels, f_{dsi}, the deviation of the subcarrier by the ith message, is a constant. The message bandwidth, or equivalently, the maximum frequency, f_{mi}, each message is allowed to contain is the same for all channels. Since f_{mi} and f_{dsi} are constant, then D_{si} is fixed. Inspection of (2.22) shows that

D_{ci} must remain fixed in order for all channels to have the same output signal-to-noise ratio. The parameter used in order to achieve the specified signal-to-noise is f_{dci} and is given by

$$D_{ci} = f_{dci} / f_{si} \qquad (3.32)$$

The f_{dci}'s will increase linearly with the f_{si}'s and their value will be determined by the preemphasis design procedure. The design procedure is the same as in the PBW case and it is to compute the A factors, solve for f_{dc1}, and finally solve for the numerical values of the other carrier deviations. A design example will be worked next.

Design Example 3

An example using 14 subcarriers, channels 6A through 19A, will be worked for an FM/FM CBW multiplexed system.

Parameters:

$$[S/N]_c = 12 \text{ dB (3.98 numerical)}$$

$$D_{si} = 5 \text{ nominal}$$

$$\text{Desired } [S/N]_{oi} = 46 \text{ dB (199 numerical)}$$

$$C = \sqrt{3/4} \; \frac{[S/N]_{oi}}{[S/N]_c} = 57.87$$

From the IRIG standards:

$f_{s1} = 160$ kHz;

$f_{ds1} = 2.0$ kHz;

$f_{m1} = f_{ds1} / D_{s1} = 2,000/5 = 400$ kHz;

$f_{s2} = 152$ kHz;

$f_{ds2} = 2.0$ kHz;

$f_{m2} = f_{ds2} / D_{s2} = 2,000/5 = 400$ kHz.

Step 1

The A_i's will be calculated using (3.4), which is

$$A_i = \frac{f_{si}}{f_{s1}} \; \frac{f_{ds1}}{f_{dsi}} \; \frac{\sqrt{f_{mi}^3}}{\sqrt{f_{m1}^3}} \qquad (3.33)$$

Since, in this example, all the D_{si}'s are set at 5, all the f_{dsi}'s are equal, as are the f_{mi}'s and (3.33) becomes

$$A_i = \frac{f_{si}}{f_{s1}} \qquad (3.34)$$

Specifically, the expression for A_2 is

$$A_2 = 152/160 = 0.95$$

Using (3.34) the remaining A_i's are calculated and listed in Table 3.5, column 4.

Step 2

Calculate A_p. Using (3.6) to calculate A_p gives

$$A_p = \sqrt{A_1^2 + A_2^2 + A_3^2 + + + A_n^2} = 2.64 \qquad (3.35)$$

Step 3

Solve for f_{dc1} using (3.16). Repeating (3.16),

$$f_{dc1}^3 + \frac{f_{s1}}{A_p}(f_{dc1}^2) - \frac{C^2 f_{s1}^2 f_{ds1}}{2A_p D_{s1}^3} = 0$$

Table 3.5
Design Example 3: Preemphasis Schedule

f_{si} (kHz)	f_{dsi} (kHz)	f_{mi} (kHz)	A_i	f_{dci} (kHz)
160	2	400	1.00	48.76
152	2	400	0.95	46.32
144	2	400	0.90	43.88
136	2	400	0.85	41.44
128	2	400	0.80	39.00
120	2	400	0.75	36.57
112	2	400	0.70	34.13
104	2	400	0.65	31.69
96	2	400	0.60	29.25
88	2	400	0.55	26.81
80	2	400	0.50	24.38
72	2	400	0.45	21.94
64	2	400	0.40	19.50
56	2	400	0.35	17.06

Using the parameters of this problem, solving for f_{dc1} gives

$$f_{dc1} = 48{,}759 \text{ Hz} \tag{3.36}$$

Step 4

Calculate the D_{ci}'s and the remaining f_{dci}'s:

$$D_{c1} = f_{dc1}/f_{s1} = 48{,}759/160{,}000 = 0.3047$$

and

$$D_{ci} = 0.3047 \text{ and } D_{si} = 5 \text{ for all channels}$$

Since the D_{ci}'s are fixed, the other f_{dci}'s may be calculated from D_{c1}, that is,

$$f_{dc2} = (D_{c1})(f_{s2}) = (0.3047)(152{,}000) = 46{,}321 \text{ Hz}$$

The remaining f_{dci}'s are calculated and tabulated in Table 3.5, column 5.

Step 5

Calculate f_{dn}, the norm of the carrier deviations from (3.11). Using (3.7) and the value of A_p from (3.35) gives

$$f_{dn} = f_{dc1}A_p = (48{,}759)(2.64) = 128.7 \text{ kHz} \tag{3.37}$$

Step 6

Calculating B_c, the carrier IF bandwidth gives

$$B_c = 2[f_{dn} + f_{s1}] = 2(128.7 + 160) = 577.4 \text{ kHz} \tag{3.38}$$

Step 7

Calculate the output signal-to-noise ratio in each channel. For channel 1, the highest frequency channel, utilizing (2.31) gives

$[S/N]_{o1} = \sqrt{3/4}\, D_{s1} D_{c1}\, [S/N]_c;$

$[S/N]_{o1} = (0.866)(5)(0.3047)(760/20)(3.98) = 199 \text{ (numerical)};$

$[S/N]_{o1} = 46 \text{ dB}.$

Column 3 in Table 3.6 gives all the computed signal-to-noise ratios for the channels.

Summary of Results for Design Example 3

The only unspecified parameters available to insure acceptable and equal signal-to-noise ratios in all the subcarrier channels in the design of an FM/FM multiplex system meeting IRIG specifications are the f_{dci}'s, the deviations of the carrier by the subcarriers. This example calculated the necessary deviations for a specified 46-dB signal-to-noise output in 14 CBW channels.

Since the maximum frequency allowed for each channel is constant at 400 Hz, this type of format would be chosen whenever the signals from the various sensors have about the same spectral content. This design example also fixed D_{si} at 5; however, had D_{si} been chosen to be, say 2.5, then the maximum frequency response of each channel would have been 800 Hz. In order to compensate for the decreased D_{si} in the signal-to-noise equation, D_{ci} and hence f_{dci} would have been calculated to be larger, increasing the necessary IF bandwidth.

3.5 Threshold

3.5.1 Signal-to-Noise Ratio for a Bandpass Filter

The FM/FM system may be designed such that the subcarrier signal-to-noise ratios in the bandpass filters are equal to or are larger than the carrier

Table 3.6
D_{si} and D_{ci} of Design Example 3

f_{si} (kHz)	f_{dci} (kHz)	$(S/N)_{oi}$ (dB)	D_{si}	D_{ci}
160	48.76	46	5	0.3047
152	46.32	46	5	0.3047
144	43.88	46	5	0.3047
136	41.44	46	5	0.3047
128	39.00	46	5	0.3047
120	36.57	46	5	0.3047
112	34.13	46	5	0.3047
104	31.69	46	5	0.3047
96	29.25	46	5	0.3047
88	26.81	46	5	0.3047
80	24.38	46	5	0.3047
72	21.94	46	5	0.3047
64	19.50	46	5	0.3047
56	17.06	46	5	0.3047

IF ratios in the carrier IF so that the threshold will occur later in the subcarrier bandpass filters than in the carrier IF. This is done because data will degrade less for threshold in the carrier IF than for threshold in the subcarrier bandpass filters. This will not give a minimum transmission bandwidth, and often the system is designed for concurrent thresholds.

Equation (2.22) gives the output signal-to-noise ratio for a bandpass filter following an FM discriminator. The output signal-to-noise ratio is also the signal-to-noise ratio in the output bandpass filter. It is desirable to have this greater than 12 dB in order to delay the onset of threshold in the subcarrier bandpass filter until after threshold in the carrier IF.

Multiplying (2.22) by B_c, rearranging, and inserting the subscript i for the ith subcarrier bandpass filter gives

$$\left[\frac{S}{N}\right]_{OBPi} = \frac{1}{2}\frac{f_{dci}^2}{f_{si}^2}\frac{B_c}{B_i}\left[\frac{A^2}{2\eta B_c}\right] \tag{3.39}$$

The term in the bracket is the carrier-to-noise ratio in the carrier IF. Therefore, (3.39) may be written as

$$\left[\frac{S}{N}\right]_{OBPi} = \frac{1}{2}\frac{f_{dci}^2}{f_{si}^2}\frac{B_c}{B_i}\left[\frac{S}{N}\right]_c \tag{3.40}$$

For the signal-to-noise ratio of the subcarrier bandpass filter to be greater than 12 dB, the term preceding the bracket must be greater than 1, since the carrier signal-to-noise ratio is 12 dB by design.

3.5.2 Threshold Design Margin

Specifically, after the design of an FM/FM system, since all terms in (3.40) are known and the only design variable is f_{dci}, the term preceding the bracket should be set up as an inequality and checked to see if it is equal to or greater than 1. Proceeding gives

$$\frac{1}{2}\frac{f_{dci}^2}{f_{si}^2}\frac{B_c}{B_i} \geq 1 \tag{3.41}$$

Solving for f_{dci} gives

$$f_{dci} \geq \sqrt{2}f_{si}\sqrt{B_i/B_c} \tag{3.42}$$

The inequality of (3.42) insures that the signal-to-noise ratio in the ith subcarrier bandpass filter is greater than 12 dB or 3.99 numerical. Taking 10 log of (3.40) gives

$$\left[\frac{S}{N}\right]_{OBPi} (\text{dB}) = 10 \log\left[\frac{1}{2} \frac{f_{dci}^2}{f_{si}^2} \frac{B_c}{B_i}\right] + \left[\frac{S}{N}\right]_c (\text{dB}) \qquad (3.43)$$

The first term in (3.43) gives the threshold design margin, M_{ti}, for the ith subcarrier bandpass filter.

Design Example 4

Equation (3.43) will be used to determine the design margin for the FM/FM channels of Design Example 2. This example results from the design of 15 FM/FM channels modified for larger f_{dci}'s for the lower frequency channels and with the associated D_{si}'s changed. Noting that f_{dci}/f_{si} is D_{ci}, the threshold margin for the ith subcarrier channel is given by

$$M_{ti} = 10 \log\left[\frac{1}{2} D_{ci}^2 \frac{B_c}{B_i}\right] \qquad (3.44)$$

Since the bandpass filter of the ith channel must pass the subcarrier modulated by the message and the mod index is nominally 5, the bandwidth of this filter is given by

$$B_i = 2(f_{dsi} + f_{mi}) \approx 2 f_{dsi} \qquad (3.45)$$

Using (3.45) and rearranging, (3.44) becomes

$$M_{ti} = 20 \log 0.5 \sqrt{B_c} \, \frac{D_{ci}}{\sqrt{f_{dsi}}} \qquad (3.46)$$

The threshold margin for the highest subcarrier channel is given by

$$M_{t1} = 20 \log 0.5 \sqrt{B_c} \, \frac{D_{c1}}{\sqrt{f_{ds1}}} \qquad (3.47)$$

Using the calculated IF bandwidth, B_c, of 366 kHz and Table 3.4, (3.47) becomes

$$M_{t1} = 20 \log(0.5)(366)^{1/2} \frac{0.718}{(6,975)^{1/2}} = 8.2 \text{ dB} \qquad (3.48)$$

Using (3.47) the threshold design margins for the other 14 channels are computed and are listed in Table 3.7.

Design margins calculated for the CBW channels in Design Example 3 are as follows.

The parameters are:

B_c = 577 kHz;

D_{ci} = 0.3047;

f_{dsi} = 2 kHz.

The equation for the design margin is

$$M_{ti} = 20 \log 0.5 \sqrt{B_c} + 20 \log(D_{ci}/\sqrt{f_{dsi}}) \qquad (3.49)$$

Since the parameters in this equation in all the channels are the same, the threshold design margin for all channels is

$$M_{ti} = 51.59 - 43.33 = 8.26 \text{ dB}$$

Table 3.7
Threshold Margins for Proportional Bandwidth Channels in Design Example 2

f_{si} (kHz)	M_{ti} (dB)
93	8.3
70	8.3
52.5	8.3
40	8.3
30	8.3
22	8.3
14.5	8.3
10.5	8.3
7.35	11.4
5.4	14.9
3.9	19.1
3.0	22.5
2.3	26.1
1.7	30.0
1.3	33.5

3.6 Changing the Preemphasis Schedule to Utilize Specified IF or Transmission Bandwidth

3.6.1 Bandwidth Utilization

Because the telemetry receivers have only a fixed and specified number of IF bandwidths and because the allocated transmission bandwidth is limited and fixed, it is desirable to design the modulation parameters such that the utilized bandwidth meets specifications and is approximately one of the given IF bandwidths. Once the design process is completed, the following process is used to accomplish that design.

A_i is defined by

$$A_i = \frac{f_{si}}{f_{s1}} \frac{f_{ds1}}{f_{dsi}} \frac{\sqrt{f_{mi}^3}}{\sqrt{f_{m1}^3}} \tag{3.50}$$

also,

$$f_{dci} = f_{dc1} A_i \tag{3.51}$$

and

$$B_c = 2(f_{dn} + f_{sh}) = 2(f_{dc1} A_p + f_{s1}) \tag{3.52}$$

Equation (3.50) shows that the A_i's are independent of the specified signal-to-noise ratios if they are all equal. Once the A_i's are calculated for a group of subcarriers, the A_i's and A_p will not change, even though the specified signal-to-noise may change. (Changing the carrier deviation by the lower frequency subcarriers will change the A_i's, but not appreciably.) Therefore, for a group of subcarriers whose f_{dci}'s have been determined in the design process but do not utilize the available IF bandwidth, f_{dc1} can be determined such that B_c is the required value. Then the remaining f_{dci}'s may be determined, since each is related to f_{dc1} by A_i.

3.6.2 Design Equations

For example, say A_p and f_{dc1} have been determined for a group of subcarriers and a specified output signal-to-noise ratio, but the calculated IF bandwidth does not use all the allocated bandwidth. Then, (3.52) may be used to solve for f_{dc1} in terms of the desired IF bandwidth. Such as,

$$f_{dc1} = \frac{1}{A_p}\left(\frac{B_c}{2} - f_{s1}\right) \tag{3.53}$$

Then (3.51) may be used to solve for the remaining f_{dci}'s. Specifically,

$$f_{dci}(\text{new}) = f_{dc1}(\text{new})A_i = f_{dc1}(\text{new})\frac{f_{dci}(\text{old})}{f_{dc1}(\text{old})} \tag{3.54}$$

$$f_{dci}(\text{new}) = \frac{f_{dc1}(\text{new})}{f_{dc1}(\text{old})} f_{dci}(\text{old}) = Pf_{dci}(\text{old})$$

where P is the ratio increase of f_{dc1}.

As (3.54) shows, all the deviations will be increased proportionally by the factor P. Certainly, as the new f_{dci}'s are determined and used, the $[S/N]_o$ ratios will increase. The carrier power must be increased also in order to maintain at least a 12-dB carrier-to-noise ratio in the new bandwidth of the IF bandpass filter.

Design Example 5

An example will be worked here in which the IF bandwidth is expanded to a specified value. The set of proportional bandwidth subcarriers and the design parameters determined in Design Example 1 will be used. The first set of carrier deviations will be used, since increasing the deviations of the lowest frequency subcarriers had little effect on B_c.

Say, it is desirable to utilize a 500-kHz IF or transmission bandwidth. The system parameters of Design Example 1 were:

$$A_p = 1.319;$$

$$B_c \text{ (old)} = 362 \text{ Hz};$$

$$f_{dc1} = 66.8 \text{ kHz.}$$

Substituting these values into (3.53) gives

$$f_{dc1} = \frac{1}{1,319}(500/2 - 93) = 119 \text{ kHz}$$

The proportional increase, P, is

$$P = 119/66.8 = 1.78$$

The new f_{dci}'s may be found by multiplying each of the previous carrier deviations by this factor. The resulting deviations are shown in Table 3.8. The f_{dci}'s below 8.8 kHz have been increased to 8.8 kHz for the reasons mentioned previously. Calculating f_{dn} from the sum of the squares of the deviations of the carrier by the subcarriers and calculating B_c from the generalized (3.52), gives f_{dn} = 158.8 and B_c = 502 kHz. Note that if the 10% rule, which states that all f_{dci}'s should be at least 10% of f_{dn} is implemented, then the 10 lowest subcarriers should have f_{dci}'s = 15.8 kHz in the new preemphasis schedule.

3.7 Designing to a Specified Transmission Bandwidth

3.7.1 Design Process

An example of designing to a specified transmission bandwidth will be worked next. Assume it is desired to design a preemphasis schedule using the five constant bandwidth subcarriers shown in Table 3.9 and using 3 MHz of bandwidth.

Table 3.8
Increased f_{dci}'s for Expanded B_c for Design Example 1

f_{si} (kHz)	f_{dci} (kHz) Initial	f_{dci} (kHz) New
93	66.80	119.0
70	43.60	77.7
52.5	28.37	50.5
40	18.84	33.5
30	12.24	21.78
22	7.68	13.6
14.5	4.13	8.8
10.5	2.55	8.8
7.25	1.48	8.8
5.4	0.93	8.8
3.9	0.57	8.8
3.0	0.38	8.8
2.3	0.25	8.8
1.7	0.16	8.8
1.3	0.11	8.8

Note: f_{dn} = 158 kHz; B_c = 2(f_{dn} + 93) = 2(251) = 502 kHz.

Table 3.9
Five CBW Channels

Channel	f_{si}	f_{dsi}	f_{mi}
111E	896	32	6.4
95E	768	32	6.4
79E	640	32	6.4
63E	512	32	6.4
47E	384	32	6.4

Note: Desired bandwidth = 3 MHz.

Design Example 6

The first step is to calculate the A_i's. The equation for A_i is given by

$$A_i = \frac{f_{si}}{f_{s1}} \frac{f_{ds1}}{f_{dsi}} \frac{\sqrt{f_{mi}^3}}{\sqrt{f_{m1}^3}} \tag{3.55}$$

For the CBW channels:

f_{dsi} = constant for all five channels;

f_{mi} = constant for all five channels if D_{si} is the same in all the channels.

Then

$$A_i = \frac{f_{si}}{f_{s1}} \tag{3.56}$$

and

$A_1 = 1$;

$A_2 = 768/896 = 0.85$.

The equation for A_p is given by

$$A_p = \sqrt{A_1^2 + A_2^2 + A_3^2 + + + A_n^2} \tag{3.57}$$

$$A_p = [1 + (0.857)^2 + (0.714)^2 + + (0.428)^2]^{1/2} = 1.66$$

Now f_{dc1} may be calculated,

$$f_{dc1} = \frac{1}{A_p}\left(\frac{B_c}{2} - f_{s1}\right) = 1/1.66(3/2 - 896 \text{ kHz}) = 363 \text{ kHz}$$

(3.58)

The other f_{dci}'s are given by

$$f_{dci} = A_i f_{dc1}$$

Then

$$f_{dc2} = A_2 f_{dc1} = (0.857)(363) \text{ kHz} = 311.8 \text{ kHz}$$

The signal-to-noise out must be checked and is given by

$$[S/N]_{o1} = \sqrt{3/4} \, D_{si} D_{ci} \frac{\sqrt{B_c}}{\sqrt{f_{mi}}} [S/N]_C$$

(3.59)

Numerically evaluating (3.59) gives

$$[S/N]_{o1} = 43.6 \text{ dB}$$

The specified and the computed parameters are shown in Table 3.10.
If the five lowest frequency subcarriers in the E group, 79E, 63E, 47E, 31E, and 15E had been chosen with all the same required parameters such as B_c = 3 MHz, then $[S/N]_{oi}$ = 50 dB. All things being equal, it would be

Table 3.10
Parameters for Design Example 6

Channel	f_{si} (kHz)	f_{dsi} (kHz)	f_{mi} (kHz)	A_i	f_{dcl} (kHz)	D_{ci}	$[S/N]_{oi}$ dB
111E	896	32	6.4	1.0	362	0.4	43.6
95E	768	32	6.4	0.857	311	0.4	43.6
79E	640	32	6.4	0.714	259	0.4	43.6
63E	512	32	6.4	0.571	207	0.4	43.6
47E	384	32	6.4	0.428	155	0.4	43.6

desirable to use the lower-frequency subcarriers in this group, since they use less bandwidth while handling the same signal frequency; hence, for a specified bandwidth, the output signal-to-noise will be larger (see Problem 3.42).

3.8 Designing the Preemphasis Schedule for Different Values for the D_{si}'s

3.8.1 Design Equations and Procedures

All the design procedures above have assumed that the D_{si}'s would all be the same. Typically, all the D_{si}'s were assumed to be 5. However, in those design procedures, all the D_{si}'s could have been set to any value between 1 and 5 as long as they were all the same. A design procedure will now be given for developing the preemphasis schedule for different values of the D_{si}'s. Repeating (3.55) gives

$$A_i = \frac{f_{si}}{f_{s1}} \frac{f_{ds1}}{f_{dsi}} \frac{(f_{mi})^{3/2}}{(f_{m1})^{3/2}} \tag{3.60}$$

This can be written as

$$A_i = \frac{f_{si}}{f_{s1}} \frac{f_{ds1}}{f_{dsi}} \frac{f_{mi}}{f_{m1}} \frac{\sqrt{f_{mi}}}{\sqrt{f_{m1}}} \tag{3.61}$$

$$A_i = \frac{f_{si}}{f_{s1}} \frac{D_{s1}}{D_{si}} \frac{\sqrt{f_{mi}}}{\sqrt{f_{m1}}}$$

Once the group of subcarriers have been chosen, (3.61) allows the A_i's to be computed based upon the assigned value of the D_{si}'s rather than some fixed value. This procedure will design the preemphasis schedule based upon the specified output signal-to-noise ratio with the required bandwidth falling where it will. Once the A_i's are calculated, the design procedure is the same as the first design procedure above. Any necessary modifications to the required bandwidth can be achieved as shown in the procedure in Section 3.6.

The design procedure steps are as follows:

1. Select the group of subcarriers compatible with the sensors.
2. Specify the required signal-to-noise in the output channels.
3. Specify the D_{si}'s for the individual subcarrier channels based upon the maximum f_{mi}'s the channels will be required to handle.
4. Calculate the A_i's.
5. Calculate A_p.
6. Solve for f_{dc1} using the cubic (3.16).
7. Calculate the remaining f_{dci}'s using (3.17).
8. Calculate the required bandwidth, B_c.

This completes the design of the preemphasis schedule. A design example will be worked next.

Design Example 7

A design example will be worked for specified different D_{si}'s. It will be assumed that the sensor requirement will be satisfied by the 5 sensors of Design Example 6. The required output signal-to-noise ratio is specified at 43.3 dB; thus, the required bandwidth cannot be specified and will be a function of the computed carrier deviations by the subcarriers. The specifications and frequency requirements for the subcarriers are given in Table 3.11. Note that the lowest frequency channel, 47E, must handle a signal frequency of 32 kHz and will therefore have a mod index, D_{si}, of 1. Calculating the A_i's based upon (3.61) gives the values listed in column 6, Table 3.11.

Calculating A_p as the square root of the sum of the squares of the A_i's gives

$$A_p = 5.053$$

Table 3.11
Parameters for Design Example 7

Channel	f_{si}	f_{dsi}	f_{mi}	D_{si}	A_i	f_{dci}	D_{ci}
111E	896	32	6.4	5	1	286	0.32
95E	768	32	6.4	5	0.86	245	0.32
79E	640	32	6.4	5	0.71	204	0.32
63E	512	32	6.4	5	0.57	163	0.32
47E	384	32	32	1	4.79	1,370	3.57

Frequencies are given in kilohertz.

Solving the cubic (3.16) for f_{dc1} and (3.17) for the other f_{dci}'s results in the following carrier deviations given in column 7 of Table 3.11. Solving for f_{dn} as the square root of the sum of the squares of the f_{dci}'s gives

$$f_{dn} = 1.4 \text{ MHz}$$

The resulting D_{ci}'s are listed in column 8 of Table 3.11. The required IF bandwidth is given by

$$B_c = 2(f_{dc1}A_p + f_{s1}) = 4.69 \text{ MHz}$$

This is in contrast to the required bandwidth of 3 MHz for Design Example 5 for the same required output signal-to-noise ratio of 43.3 dB. The greater required bandwidth is the result of decreasing the mod index of the signal with respect to the subcarrier and making it up by increasing D_{c5} in order to keep an output signal-to-noise of 43.3 dB.

In essence, bandwidth has been exchanged for output signal quality. This will always occur in FM/FM as the D_{si}'s are reduced in order to handle higher signal frequencies and the D_{ci}'s are increased in order to maintain the output signal-to-noise ratios.

3.8.2 Summary of the Results of Design Example 7

The required transmission bandwidth is a function of which subcarrier is assigned to handle the higher signal frequency. Listed below in Table 3.12 are the required bandwidths, B_c, where four of the five subcarriers are assigned 6.4-kHz message bandwidths, while the fifth is given the 32-kHz message. Only the subcarrier modulated by the 32-kHz message will be listed. For CBW subcarriers, it appears that assigning the highest frequency message to the lowest frequency subcarrier will require the lowest overall transmission bandwidth.

Table 3.12
B_c for Different Subcarriers Assigned the High-Frequency Message

Channel	111E	95E	79E	63E	47E
B_c	7.06	6.43	5.87	5.27	4.69

3.9 Designing the Preemphasis Schedule for the Minimum Transmission Bandwidth with Equal D_{si}'s (Concurrent All-Channel Dropout)

For a given set of subcarriers, this minimum transmission or IF bandwidth occurs when threshold happens simultaneously in both the carrier and subcarrier bandpass filters. This is referred to as concurrent all-channel dropout. The design procedure can be used to design the preemphasis schedule for a minimum transmission or IF bandwidth, B_c, resulting in concurrent all-channel dropout. Reducing the design transmission bandwidth further will endanger the data, since the receiver and subcarrier bandpass filters will be operating below threshold. All other parameters being fixed, reducing the required transmission bandwidth is achieved by reducing the required channel output signal-to-noise ratio, which will reduce the design f_{dci}'s and hence B_c. The threshold design margin, M_{ti}, will also be reduced. The minimum bandwidth achievable without threshold occurring will occur when the threshold design margin equals 0 dB. Alternatively, the minimum bandwidth will occur if the preemphasis schedule is designed such that threshold occurs simultaneously in all the bandpass filters.

3.9.1 Design Equations and Procedures

The mathematical equation for determining threshold in the subcarrier bandpass filters is (3.39), which is

$$[S/N]_{OBPi} = \left[\frac{1}{2} \frac{f_{dci}^2}{f_{si}^2} \frac{B_c}{B_i} \right] \left[\frac{A^2}{2 \eta B_c} \right] \qquad (3.62)$$

The term in the second bracket is the carrier-to-noise ratio in the carrier IF and by link design is set at a minimum of 12 dB, where threshold occurs. If the first term is made larger than 1, then the signal-to-noise in the subcarrier bandpass filter will be above 12 dB and the onset of threshold will occur later in this filter than in the carrier IF, which is desirable if there is transmission bandwidth available. However, if the system is being designed for minimum bandwidth utilization, it is necessary for the two thresholds to occur at the same time.

To insure the threshold occurred later in the subcarrier filter, the first term in (3.62) was set up as the inequality

$$\frac{1}{2} \frac{f_{dci}^2}{f_{si}^2} \frac{B_c}{B_i} \geq 1 \tag{3.63}$$

In Section 3.5, after the preemphasis design procedure was completed, this inequality was checked to determine the threshold design margin. On the other hand, if the equal sign is used in (3.68), the threshold design margin is 0 dB and the thresholds will occur simultaneously.

The procedure for designing the preemphasis schedule for a minimum bandwidth is developed next. The process is started by equating the terms in (3.63) to 1. Rearranging and noting that $B_i = 2(f_{dsi} + f_{mi})$ for large or small mod indexes gives

$$f_{dci}^2 B_c = 4 f_{si}^2 (f_{dsi} + f_{mi}) \tag{3.64}$$

The problem in using (3.64) as the starting point in a design procedure after the ith index has been converted to the 1 index is that it is one equation in two unknowns, whereas (3.63) was used after the preemphasis schedule was designed, and the parameters were known and used for checking the inequalities. The way around this dilemma is to note the relationship in (3.11) between f_{dc1} and B_c:

$$B_c = 2[f_{dn} + f_{s1}] = 2[f_{dc1} A_p + f_{s1}] \tag{3.65}$$

The design procedure is the same as the procedure used in designing for a specified output signal-to-noise ratio in that once a group of subcarriers is selected, the A_i's and A_p are calculated. Equation (3.65) is substituted into (3.64) to convert the equation into one equation in one unknown with the ith index replaced by the highest frequency index of 1. Rearranging and normalizing gives

$$f_{dc1}^3 + \frac{f_{s1}}{A_p} f_{dc1}^2 - 2 \frac{f_{s1}^2 (f_{ds1} + f_{m1})}{A_p} = 0 \tag{3.66}$$

This is a cubic equation, which must be solved for f_{dc1}. Using the A_i's determined early, the other f_{dci}'s may be calculated, giving the preemphasis schedule which insures that all thresholds will occur simultaneously and, hence, will use the minimum transmission bandwidth. This cubic equation is for a design procedure for minimum bandwidth utilization, whereas the earlier cubic equation was for a specified output signal-to-noise ratio and, in fact, had a signal-to-noise factor as a coefficient.

The design procedure for minimum bandwidth and concurrent all-channel dropout is as follows:

1. Select the group of subcarriers compatible with the sensors.
2. Specify the D_{si}'s for the individual subcarrier channels.
3. Calculate the A_i's.
4. Calculate A_p.
5. Solve for f_{dc1} using the cubic (3.66).
6. Calculate the remaining f_{dci}'s using (3.17).
7. Calculate the signal-to-noise in the output channels.
8. Calculate the required bandwidth, B_c, from (3.65).

Design Example 8

The eight steps in Section 3.9.1 will be used to design for concurrent thresholds and, hence, a minimum transmission bandwidth.

1. Step 1 will be carried out by selecting the five constant bandwidth channels of Design Example 6 to demonstrate the minimum bandwidth design procedure.
2. All the D_{si}'s will be set to 5.
3. The A_i's are the same as in Design Example 6.
4. From Design Example 6, $A_p = 1.66$.
5. Substituting the specific values of these parameters into the cubic (3.66) and solving gives

$$f_{dc1} = 221 \text{ kHz}.$$

6. Using the relationship $f_{dci} = A_i f_{dc1}$, gives the following values in kilohertz in Table 3.13.
7. The output signal-to-noise in all the channels is computed to be

$$[S/N]_{oi} = 38.53 \text{ dB}.$$

8. The required bandwidth, B_c and f_{dn}, is

$$f_{dn} = 367 \text{ kHz};$$
$$B_c = 2.5 \text{ MHz}.$$

Table 3.13
Parameters for Design Example 8

Channel	f_{si} (kHz)	f_{dsi} (kHz)	f_{mi} (kHz)	A_i	f_{dci} (kHz)
111E	896	32	6.4	1.0	221
95E	768	32	6.4	0.857	189
79E	640	32	6.4	0.714	158
63E	512	32	6.4	0.571	126
47E	384	32	6.4	0.482	95

All the D_{ci}'s = 0.246.

B_c cannot be reduced further without having the receiver operate in threshold in the subcarrier bandpass filters if the carrier IF is operating at 12 dB.

In designing for all-channel dropout, regardless of the set of subcarriers chosen, whether using CBW or PBW subcarriers, the signal-to-noise output will be 38.5 dB if the D_{si}'s equal 5 and the carrier-to-noise ratio is 12 dB. This can be shown by setting B_i equal to $2(f_{dsi} + f_{mi})$ in (3.64) and solving for B_c which gives

$$B_c = \frac{4(f_{dsi} + f_{mi})}{D_{ci}^2} \qquad (3.67)$$

Substituting (3.67) into (2.29) results in

$$[S/N]_{oi} = \frac{3}{4} D_{si}^2 D_{ci}^2 \frac{4(f_{dsi} + f_{mi})}{D_{ci}^2} \frac{1}{f_{mi}} [S/N]_c \qquad (3.68)$$

$$[S/N]_{oi} = 3D_{si}^2 (D_{si} + 1) [S/N]_c$$

Setting $D_{si} = 5$ and taking 10 log of the result will give 38.5 dB. Equation (3.68) shows that whenever the design margin is set equal to 0 dB for the concurrent threshold design procedure, the output signal-to-noise is a function of only D_{si} and the input carrier-to-noise ratio. For example, setting $D_{si} = 1$ gives a $[S/N]_c = 16.77$ dB, which is an unsatisfactory result. *Prudence must be exercised whenever an all-channel dropout design is initiated with the D_{si}'s less than 5.*

3.10 Summary of Design Examples 6 and 8

To put the results of these two examples into context and show the results of accepting a lower output signal-to-noise ratio for a smaller transmission bandwidth, the results of a preemphasis design example for a specified output signal-to-noise ratio, 46 dB, will also be given. The required bandwidth and design threshold margin resulting from this design will be given first:

1. $[S/N]_{oi} = 46$ dB;
    $B_c = 3.3$ MHz;
    $M_{ti} = 8.3$ dB.

2. $[S/N]_{oi} = 43.3$ dB;
    $B_c = 3.0$ MHz;
    $M_{ti} = 5.6$ dB.

3. $[S/N]_{oi} = 38.7$ dB;
    $B_c = 2.4$ MHz;
    $M_{ti} = 0$ dB.

From an observation of 1, 2, and 3, it can be seen that the higher the required output signal-to-noise ratio, the larger the transmission bandwidth. An implicit assumption in these examples is that the transmitted power may be increased whenever necessary in order to keep the carrier-to-noise ratio at 12 dB as the IF bandwidth is increased.

3.11 Designing the Preemphasis Schedule for All-Channel Dropout and Unequal D_{si}'s

In designing the preemphasis schedule for all-channel dropout, but with unequal D_{si}'s, (3.8) must be used before the signal-to-noise ratios are divided out because this design will create unequal output signal-to-noise ratios. Under these conditions, (3.8) for A_i becomes

$$A_i = \frac{f_{dci}}{f_{dc1}} = \frac{\sqrt{4/3} \dfrac{[S/N]_{oi}}{[S/N]_c} \dfrac{f_{si}}{f_{dsi}} \sqrt{f_{mi}^3}}{\sqrt{4/3} \dfrac{[S/N]_{o1}}{[S/N]_c} \dfrac{f_{s1}}{f_{ds1}} \sqrt{f_{m1}^3}} \tag{3.69}$$

Rearranging (2.31) gives

$$\frac{[S/N]_{oi}}{[S/N]_C} = \sqrt{3/4}\, D_{si}\, D_{ci}\sqrt{\frac{B_c}{f_{mi}}} \tag{3.70}$$

Substituting the value of B_c from (3.67) into (3.70) gives

$$\frac{[S/N]_{oi}}{[S/N]_C} = \sqrt{3}\, D_{si}\sqrt{1 + D_{si}} \tag{3.71}$$

Solving for the signal-to-noise ratios for the first channel in the same manner gives

$$\frac{[S/N]_{oi}}{[S/N]_C} = \sqrt{3}\, D_{s1}\sqrt{1 + D_{s1}} \tag{3.72}$$

Substituting (3.71) and (3.72) into (3.69) results in

$$A_i = \frac{D_{si}}{D_{s1}}\sqrt{\frac{1 + D_{si}}{1 + D_{s1}}}\,\frac{f_{si}}{f_{s1}}\,\frac{f_{ds1}}{f_{dsi}}\,\frac{\sqrt{f_{mi}^3}}{\sqrt{f_{m1}^3}} \tag{3.73}$$

The A_i's can be computed, then A_p solved for f_{dc1} can be solved for from the cubic (3.66). The other f_{dci}'s may be solved for using the computed values of the A_i's.

Design Example 9

An example will be worked for an all-channel dropout or concurrent threshold design with different specified D_{si}'s. The first three subcarriers of Design Example 8 will be used, given that we want channel 111E to handle a 32-kHz signal and the other two channels to handle 6.4-kHz signals. This information is summarized in Table 3.14.

The first step is to compute the A_i's from (3.73). These are listed in column 6 of Table 3.14. Computing A_p gives

$$A_p = 1.32$$

Solving (3.66) for f_{dc1} gives

$$f_{dc1} = 284 \text{ kHz}$$

Table 3.14
Parameters for Design Example 9

Channel	f_{si}	f_{dsi}	f_{mi}	D_{si}	A_i	f_{dci}	D_{ci}	M_{ti}	S/N_o
111E	896	32	32	1	1.0	284	0.32	0	19.7
95E	768	32	6.4	5	0.66	189	0.24	0	38.5
79E	640	32	6.4	5	0.55	157	0.24	0	38.5

Note: B_c = 2.5 mHz.

Because $f_{dci} = A_i f_{dc1}$, these deviations may be solved for and are listed in column 7 of Table 3.14, and the D_{ci}'s may then be computed. The threshold margin in each channel is 0 dB. Notice that the output signal-to-noise ratios in the channels vary, and the ratio is probably unacceptably low in the 111E channel. This is a result of requiring D_{s1} to be 1. Again it should be noted that requiring small D_{si}'s in channels for the all-channel dropout design will result in small output signal-to-noise ratios.

3.12 Designing the Preemphasis Schedule for Different Specified Signal-to-Noise Ratios in the Channels

The equations and the process for designing the preemphasis schedule for different specified signal-to-noise ratios will be given. Since different signal-to-noise ratios are to be specified, (3.69) must be used to calculate the A_i's. Rearranging gives

$$A_i = \frac{[S/N]_{oi}}{[S/N]_c} \frac{f_{si}}{f_{s1}} \frac{f_{ds1}}{f_{dsi}} \frac{\sqrt{f_{mi}^3}}{\sqrt{f_{m1}^3}} \tag{3.74}$$

Once the A_i's are calculated using (3.74), which has the ith channel specified signal-to-noise ratio as a factor, all the steps given in Section 3.2 are appropriate. These steps will be repeated in this section in working an example.

Design Example 10

Design a preemphasis schedule for the three CBW subcarriers of Design Example 9 with the same D_{si}'s all equal to 5, but with different specified signal-to-noise ratios. The steps are as follows:

1. Select the set of subcarriers to be used.
2. Specify the different $[S/N]_{oi}$'s.
3. Calculate the A_i's using (3.74).
4. Calculate A_p.
5. Solve (3.16) for f_{dc1}.
6. Complete the preemphasis schedule by solving for the remaining f_{dci}'s using $f_{dci} = f_{dc1} A_i$.
7. Compute B_c.

The specified $[S/N]_{oi}$'s are given in column 4 of Table 3.15. The computed A_i's are given in column 6 and the preemphasis schedule in column 7. The required B_c is 3.6 MHz.

The design was also done for the same subcarrier channels, but with all the $[S/N]_{oi}$'s specified at 40 dB. The resulting f_{dci}'s and D_{ci}'s are given in columns 9 and 10 respectively. The required B_c is 2.56 MHz. Note that requiring channel 111E to have an output signal-to-noise of 49 dB adds an additional 1 MHz of required bandwidth.

Note that the additional bandwidth requires additional transmitted power in order to maintain an acceptable carrier-to-noise ratio in the IF. This example was not worked with fixed transmitter power.

If the three lowest frequencies subcarriers in the E-channels are used, and the D_{si}'s = 5 and $[S/N]_{oi}$ = 40 dB, the required bandwidth will be less. For this case, B_c = 1.17 MHz. All things being equal, and since the lowest frequency subcarriers will handle the same signal frequencies as the higher subcarriers, the lowest frequency subcarriers would be the subcarriers of choice.

3.13 Hardware Implementation of the Preemphasis Schedule

Once the deviation schedule of the carrier with respect to the subcarriers is determined, the next step is to implement the schedule in hardware. An S-band transmitter might have a deviation sensitivity such as 500 kHz/V RMS.

A typical maximum voltage output of the subcarrier modulator, referred to as the subcarrier oscillator (SCO), would be 0.325V RMS operating into 10K ohms. Off-the-shelf SCOs with output amplitude control can have the output signal voltage set without affecting the SCO's output impedance. This special feature allows the user to precisely set the outputs of each SCO for the desired system preemphasis. With the amplitude pin open-circuited, the signal output amplitude is a maximum, and with the amplitude pin connected to ground, the signal output amplitude is zero. The desired signal output amplitude is obtained by connecting a selected resistor between the amplitude pin and ground. The resistor size may be obtained from Figure 3.1.

Design Example 11

Say we want to implement the preemphasis schedule given in Table 3.8 for the new f_{dci}'s. The maximum output voltage from the SCO will cause the following deviation

Table 3.15
Parameters for Design Example 10

Channel	f_{si}	f_{mi}	S/N_o	D_{si}	A_i	f_{dci}	D_{ci}	f_{dci}	D_{ci}
111E	896	6.4	49dB	5	1.0	615	0.68	259	0.29
95E	768	6.4	40dB	5	0.85	527	0.68	222	0.29
79E	640	6.4	40dB	5	0.71	439	0.68	185	0.29

Note: $B_c = 3.6$ MHz (different $[S/N]_{oi}$'s). $B_c = 2.56$ MHz (all $[S/N]_{oi}$'s = 40 dB).

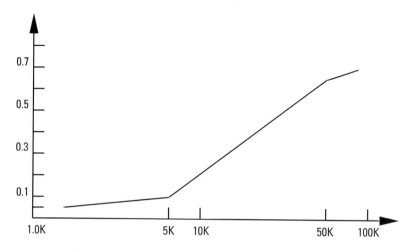

Figure 3.1 Relative amplitude of output for values of amplitude resistor.

$$f_{dc\max} = (500 \text{ kHz})\left(\frac{0.325}{1}\right) = 162.5 \text{ kHz}$$

Now say we want to set the deviation of the 93-kHz subcarrier at 119 kHz.

Therefore, the output voltage should be set to

$$v_{o93} = \left(\frac{119}{162.5}\right)(0.325) = (0.73)(0.325) = 0.238$$

The ratio of the desired voltage to the max is

$$R_{93} = \frac{0.238}{0.325} = 0.73$$

From the graph in Figure 3.1, it can be seen that the 0.73 ratio requires about a 100-k resistor to be connected between the amplitude pin and the ground. Inspection of the equations shows that the ratio to be determined in order to go to the graph is the ratio of the desired deviation to the maximum obtainable with the maximum output voltage for the particular SCO. The process is repeated until all resistor sizes are determined.

Specifically, for the 8.8-kHz deviation of the lower frequency subcarriers, say the 1.3-kHz subcarrier, the ratio is given by

$$R(N)_{1.3} = \frac{8.8}{162.5} = 0.054$$

Then, from the graph of Figure 3.1, the resistor size would be approximately $R_{1.3} = 1.2K$. From Figure 3.1 it can be seen that this is about the smallest resistor size that can be read accurately from the figure.

3.14 Summary of Design Procedures

The design of the preemphasis schedule for various options was studied. Systematic design procedures were developed for the various options. Design procedures were developed for three main options: (1) specified output signal-to-noise ratios, (2) a specified transmission bandwidth, and (3) concurrent-all-channel dropout. There were a number of modifications to these three options, such as different D_{si}'s for the channels and different specified output signal-to-noise ratios.

The transmission bandwidth used was based upon the root mean square of the carrier deviations by the subcarriers. This is neither the 99% bandwidth nor the −60-dB bandwidth discussed in the *Telemetry Standards* document IRIG 106-00 [6]. However, extensive investigations by [4] indicate that there is less than 1% of the sideband power outside this bandwidth. It seems to be a good estimate of the 99% bandwidth, but this has not been verified in an IRIG document.

The equation for the FM/FM channel output signal-to-noise ratio used is the one from the document IRIG 106-00. This equation does not include the degradation due to bandedge effects, nonidealized filters, intermodulation products, or harmonics from one channel falling in another. For this type of analysis, the reader is referred to [3, 7].

3.15 IRIG B_{IF} Specifications

Some receiver bandwidths available at the ranges are shown in Table 3.16. Statements from IRIG 106-00 concerning channel spacing for transmission are as follows.

Narrowband telemetry channel spacing is in increments of 1 MHz, beginning with the frequencies 1,435.5; 2,200.5; and 2,310.5 MHz. These numbers will be used as the base from which all frequency assignments are to be made. Medium bandwidth channels, utilizing 1 to 3 MHz and wideband channels using 3 to 10 MHz are permitted and will be centered on the center frequency of the narrowband channels.

Table 3.16
IRIG IF Bandwidths at the Ranges

B_{IF}
300 kHz
500
750
1.0 MHz
1.5
2.4
3.3
4.0
6.0
10.0

Preliminary Problems

Preliminary Problem 3.1

Using the three CBW channels, 111E, 95E, and 79E, design the preemphasis schedule with specifications as follows:

$$[S/N]_c = 12 \text{ dB} \ (3.98 \text{ numerical});$$
$$D_{si} = f_{dsi}/f_{mi} = 5 \text{ (nominal)};$$
$$[S/N]_{oi} = 40.$$

Answer: $A_p = 1.49$, $f_{dc1} = 259$ kHz, $f_{dn} = 388$ kHz, and $B_c = 2.57$ MHz.

Preliminary Problem 3.2

For Preliminary Problem 3.1, compute D_{ci} for all the channels.
Answer: $D_{ci} = 0.289$ in all channels.

Preliminary Problem 3.3

Compute M_{ti} in decibels for all the channels for Preliminary Problem 3.1.
Answer: 1.46 dB in all channels.

Preliminary Problem 3.4

Using the five PBW channels, 15, 16, 17, 18, and 19, design the preemphasis schedule with specifications as follows:

$[S/N]_c$ = 12 dB (3.98 numerical);

$D_{si} = f_{dsi}/f_{mi}$ = 5 (nominal);

$[S/N]_{oi}$ = 43.5.

Answer: A_p =1.31, f_{dc1} = 66.8 kHz, f_{dn} = 87.6 kHz, B_c = 361 kHz.

Preliminary Problem 3.5

Using the five CBW channels, 111E, 95E, 79E, 63E, and 47E, design the preemphasis schedule with specifications as follows:

$[S/N]_c$ = 12 dB (3.98 numerical);

$D_{si} = f_{dsi}/f_{mi}$ = 5 (nominal);

$[S/N]_{oi}$ = 49.

Answer: A_p = 1.66, f_{dc1} = 602 kHz, f_{dn} = 999 kHz, B_c = 3.7 MHz.

Preliminary Problem 3.6

Using the five lowest frequency subcarrier CBW E-channels, 79E, 63E, 47E, 31E, and 15E, design the preemphasis schedule with specifications as follows:

$[S/N]_c$ = 12 dB (3.98 numerical);

$D_{si} = f_{dsi}/f_{mi}$ = 5 (nominal);

$[S/N]_{oi}$ = 49.

Compare with Preliminary Problem 3.5.
Answer: A_p = 1.48, f_{dci} = 502, 402, 302, 201, 100 kHz, f_{dn} = 745 kHz, and B_c = 2.7 MHz.

Regular Design Problems

Problem 3.1

Using the 15 PBW 7.5% channels, 5 through 19, and these specifications: $[S/N]_c$ = 12 dB (3.98 numerical) and $D_{si} = f_{dsi}/f_{mi}$ = 5 (nominal), (a) design the system for $[S/N]_{oi}$ = 49 dB (281 numerical) minimum in all channels and complete Tables 3.1 and 3.2, and (b) compute f_{dn} and B_c.

Problem 3.2

Using the 15 PBW 7.5% channels, 5 through 19, and these specifications: $[S/N]_c$ = 12 dB (3.98 numerical) and $D_{si} = f_{dsi}/f_{mi}$ = 5 (nominal), (a) design the system for $[S/N]_{oi}$ = 43dB (141 numerical) minimum in all channels and complete Tables 3.1 and 3.2, and (b) compute f_{dn} and B_c.

Problem 3.3

Using the 15 PBW 7.5% channels, 5 through 19, and these specifications: $[S/N]_c$ = 12 dB (3.98 numerical) and $D_{si} = f_{dsi}/f_{mi}$ = 5 (nominal), (a) design the system for $[S/N]_{oi}$ = 40 dB (100 numerical) minimum in all channels and complete Tables 3.1 and 3.2, and (b) compute f_{dn} and B_c.

Problem 3.4

Using the 15 PBW 7.5% channels, 5 through 19, and these specifications: $[S/N]_c$ = 12 dB (3.98 numerical) and $D_{si} = f_{dsi}/f_{mi}$ = 5 (nominal), (a) design the system for $[S/N]_{oi}$ = 38 dB (79 numerical) minimum in all channels and complete Tables 3.1 and 3.2, and (b) compute f_{dn} and B_c.

Problem 3.5

Using the 15 PBW 7.5% channels, 5 through 19, and these specifications: $[S/N]_c$ = 12 dB (3.98 numerical) and $D_{si} = f_{dsi}/f_{mi}$ = 5 (nominal), (a) design the system for $[S/N]_{oi}$ = 36 dB (63 numerical) minimum in all channels and complete Tables 3.1 and 3.2, and (b) compute f_{dn} and B_c.

Problem 3.6

For Problem 3.1, increase the deviation of the carrier by all subcarriers to be at least 10% of f_{dn} and (a) compute the new $[S/N]_{oi}$ for the channels with the changed deviation of the carrier by the subcarriers; then (b) decrease the D_{si} to 2.5 on the channels with the increased deviation and compute the new $[S/N]_{oi}$.

Problem 3.7

Compute the new B_c, for Problem 3.6.

Problem 3.8

(a) For Problem 3.1 and Problem 3.6(b), double the frequency of the message modulating the subcarriers such that the new D_{si} = 2.5 on all channels and calculate the new $[S/N]_{oi}$; then (b) determine a new preemphasis schedule for Problem 3.1 by doing the complete design starting with a required $D_{si} = 2.5$. Compute the new B_c.

Problem 3.9

For Problems 3.1 through 3.5, check the design parameters to determine if the inequality of (3.43) is satisfied.

Problem 3.10

Determine the threshold margin for all channels for Problems 3.1, 3.2, 3.3, 3.4, and 3.5.

Problem 3.11

Using the 14 CBW channels, 6A through 19A, and these specifications: $[S/N]_c = 12$ dB (3.98 numerical) and $D_{mi} = f_{dsi}/f_{mi} = 5$ (nominal), (a) design the system for $[S/N]_{oi} = 49$ dB (281 numerical) minimum in all channels and complete Tables 3.1 and 3.2, and (b) compute f_{dn} and B_c.

Problem 3.12

Using the 14 CBW channels, 6A through 19A, and these specifications $[S/N]_c = 12$ dB (3.98 numerical) and $D_{mi} = f_{dsi}/f_{mi} = 5$ (nominal), (a) design the system for $[S/N]_{oi} = 43$ dB (141 numerical) minimum in all channels and complete Tables 3.1 and 3.2, and (b) compute f_{dn} and B_c.

Problem 3.13

Using the 14 CBW channels, 6A through 19A and these specifications: $[S/N]_c = 12$ dB (3.98 numerical) and $D_{mi} = f_{dsi}/f_{mi} = 5$ (nominal), (a) design the system for $[S/N]_{oi} = 40$ dB (100 numerical) minimum in all channels and complete Tables 3.1 and 3.2, and (b) compute f_{dn} and B_c.

Problem 3.14

Using the 14 CBW channels, 6A through 19A, and these specifications: $[S/N]_c = 12$ dB (3.98 numerical) and $D_{mi} = f_{dsi}/f_{mi} = 5$ (nominal), (a) design the system for $[S/N]_{oi} = 38$ dB (79 numerical) minimum in all channels and complete Tables 3.1 and 3.2, and (b) compute f_{dn} and B_c.

Problem 3.15

Using the 14 CBW channels, 6A through 19A and these specifications: $[S/N]_c$ = 12 dB (3.98 numerical) and $D_{mi} = f_{dsi}/f_{mi}$ = 5 (nominal), (a) design the system for $[S/N]_{oi}$ = 36 dB (63 numerical) minimum in all channels and complete Tables 3.1 and 3.2, and (b) compute f_{dn} and B_c.

Problem 3.16

Using Figure 3.1, determine the size of the resistors on the amplitude pin of the SCOs necessary to implement the deviations of the carrier by the subcarriers as indicated in Table 3.5. Assume the sensitivity of the transmitter is k = 500 kHz/V RMS and that the maximum voltage out from the VCOs is 0.325 RMS.

Problem 3.17

Using Figure 3.1, determine the size of the resistors on the amplitude pins of the SCOs necessary to implement the preemphasis schedule of Problem 3.11. Assume the sensitivity of the transmitter is k = 500 kHz/V RMS and that the maximum voltage out from the VCOs is 0.325 RMS.

Problem 3.18

Determine the threshold design margins for Problems 3.11, 3.12, 3.13, 3.14, and 3.15.

Problem 3.19

For Problem 3.11, design the system such that B_c = 500 kHz; then (a) list the new f_{dci}'s and (b) compute the new $[S/N]_{oi}$'s.

Problem 3.20

Design a preemphasis schedule for eight of the 15% channels using alternate channels starting with channel A, given the design specifications: $[S/N]_c$ = 12 dB and $[S/N]_{oi}$ = 46 dB. Start with a nominal D_{si} = 5. (a) Find f_{dn} and B_c and (b) redesign the preemphasis package such that B_c is designed to the nearest integer MHz bandwidth from part (b).

Problem 3.21

(a) Design a preemphasis schedule for three of the 30% channels using AA, FF, and KK. $[S/N]_c$ = 12 dB, $[S/N]_{oi}$ = 46 dB, and D_{si} = 5. (b) Find f_{dn} and B_c. (c) Redesign the preemphasis schedule such that B_c is designed to the nearest integer MHz bandwidth from part (b).

Problem 3.22

(a) Design a preemphasis schedule for the 11 CBW B channels. $[S/N]_c$ = 12 dB, $[S/N]_{oi}$ = 46 dB, and D_{si} = 5. (b) Find f_{dn} and B_c. (c) Design the preemphasis schedule such that B_c is designed to the nearest integer MHz.

Problem 3.23

Design a preemphasis schedule for six of the CBW B channels using 13B through 23B. $[S/N]_c$ = 12 dB, $[S/N]_{oi}$ = 46 dB, and D_{si} = 5. Find f_{dn} and B_c.

Problem 3.24

Design a preemphasis schedule for 15 of the 7.5% PBW channels 5 through 19, and 6 of the B CBW channels using 13B through 23B. $[S/N]_c$ = 12 dB, $[S/N]_{oi}$ = 46 dB, and D_{si} = 5. (a) Find f_{dn} and B_c and compare the B_c of this schedule with that of Problem 3.23 before the 15 PBW channels were added.

Problem 3.25

For Problem 3.24, modify the preemphasis schedule of the lower-frequency subcarriers such that all deviations are at least 10% of f_{dn}. Compute the new f_{dn}, and new B_c. Compute the new $[S/N]_{oi}$ for the subcarriers with the changed f_{dcs}.

Problem 3.26

For Problem 3.25, change the D_{si} of the modified subcarriers to 2.5 and calculate the new $[S/N]_{oi}$'s.

Problem 3.27

For Problem 3.26, calculate all the threshold margins.

Problem 3.28

Show that the all-channel dropout design procedure will give a $[S/N]_{oi} = 38.5$ dB by solving (3.69) for B_c and substituting the resulting equation into (2.27). Set $B_i = 2(f_{dsi} + f_{mi})$ and $D_{si} = 5$ and $[S/N]_c = 12$ dB.

Problem 3.29

Design a preemphasis schedule for the five channels of Preliminary Problem 3.5, except let $[S/N]_{oi} = 43.5$ and $D_{s1} = 1$. Let all the other D_{si}'s = 5. Answer: $A_p = 1$, $f_{dc1} = 2.6$ MHz, and $B_c = 7.06$ MHz.

Problem 3.30

Repeat Problem 3.29, except set $D_{s2} = 1$. Let all the other D_{si}'s = 5. Answer: $A_p = 9.6$, $f_{dc1} = 240$ kHz, $f_{dc2} = 2.3$ MHz, and $B_c = 6.4$ MHz.

Problem 3.31

Repeat Problem 3.29, except set $D_{s3} = 1$. Let all the other D_{si}'s = 5. Answer: $A_p = 8.1$, $f_{dc1} = 0.251$ MHz, $f_{dc2} = 0.21$ MHz, $f_{dc3} = 2$ MHz, and $B_c = 5.8$ MHz.

Problem 3.32

Repeat Problem 3.29, except set $D_{s5} = 1$. Let all the other D_{si}'s = 5. Answer: $A_p = 5.0$, $f_{dc1} = 0.287$ MHz, $f_{dc2} = 0.25$ MHz, $f_{dc3} = 0.2$ MHz, $f_{dc5} = 1.37$ MHz, and $B_c = 4.7$ MHz.

Problem 3.33

Repeat Problem 3.29, except set $D_{s5} = 2$. Let all the other D_{si}'s = 5. Answer: $A_p = 2.33$, $f_{dc1} = 338$ kHz, $f_{dc2} = 0.290$ MHz, $f_{dc5} = 0.57$ MHz, and $B_c = 3.37$ MHz.

Problem 3.34

Repeat Problem 3.33, except set $D_{s1} = 2$. Let all the other D_{si}'s = 5. Answer: $A_p = 1$, $f_{dc1} = 1.18$ MHz, $f_{dc2} = 0.25$ MHz, $f_{dc5} = 0.12$ MHz, and $B_c = 4.29$ MHz.

Problem 3.35

Using the three CBW channels, 111E, 95E, and 79E, design the preemphasis schedule for all-channel dropout, that is, set M_{ti} = 1 or 0 dB. This will also give the minimum transmission bandwidth. Specifications are as follows: $[S/N]_c$ = 12 dB (3.98 numerical) and $D_{si} = f_{dsi}/f_{mi}$ = 5 (nominal). Answer: A_p = 1.49, f_{dc1} = 0.224 MHz, f_{dn} = 335 kHz, B_c = 2.46MHz, and D_{ci} = 0.25.

Problem 3.36

For Problem 3.35, compute $[S/N]_{oi}$ for all the channels.
Answer: $[S/N]_{oi}$ = 38.5 dB.

Problem 3.37

Repeat Problem 3.35, except set D_{si} = 1 in all the channels.
Answer: A_p =1.49, f_{dc1} = 0.279 MHz, f_{dn} = 418 kHz, B_c = 2.6 MHz, and D_{ci} = 0.31.

Problem 3.38

For Problem 3.37, compute $[S/N]_{oi}$ for the channels.
Answer: $[S/N]_{oi}$ = 19.7 dB. Note that requiring the channels to handle 32 kHz of message bandwidth, effectively setting D_{si} = 1, gives an unsatisfactory $[S/N]_{oi}$ when designing for all-channel dropout.

Problem 3.39

For the three CBW channels 111E, 95E, and 79E, and required $[S/N]_{oi}$ ratios equal to 49 dB, 40 dB, and 40 dB, respectively, and with these specifications: $[S/N]_c$ = 12 dB (3.98 numerical) and $D_{si} = f_{dsi}/f_{mi}$ = 5, (a) design the preemphasis schedule, (b) determine B_c, (c) find D_{ci}, and (d) compute M_{ti}.
Answer: A_p = 1.5, f_{dc1} = 0.614 MHz, f_{dc2} = 0.527 MHz, f_{dc3} = 0.439 MHz, B_c = 3.6 MHz, D_{ci} = 0.686, and M_{ti} = 10.46 dB.

Problem 3.40

In Design Example 10, B_c = 2.56 MHz was given for the three subcarriers when a specified $[S/N]_{oi}$ = 40 was used. Show this to be true. Also find the f_{dci}'s.
Answer: f_{dci}'s = 0.289 MHz.

Problem 3.41

At the end of Design Example 10, it was pointed out that if the three lowest frequency subcarriers in the E-channels were used, $B_c = 1.17$ would be required. Show this to be true. Also find the f_{dci}'s.
Answer: 164, 109, and 54.7 kHz.

Problem 3.42

Work Design Example 6 using the five lowest frequency subcarriers in the E group and show $[S/N]_{oi} = 50.8$. Explain. Find D_{ci}.
Answer: 0.9.

Problem 3.43

Using the five CBW channels, 111E, 95E, 79E, 63E, and 47E, design the preemphasis schedule with specifications as follows:

$[S/N]_c = 12$ dB (3.98 numerical);

$D_{si} = f_{dsi}/f_{mi} = 5$ (nominal) in the 3 highest frequency channels;

$D_{si} = 1$ in the other two;

$[S/N]_{oi} = 46$ dB.

Answer: $A_p = 8.12$. $f_{dci} = 315$; 270; 225; 2,000; and 1,500 kHz. $f_{dn} = 2.5$ MHz. $B_c = 6.9$ MHz.

Problem 3.44

Observing the new nomenclature and using the five CBW F-channels, 3,840F; 3,584F; 3,328F; 3,584F; and 3,840F, design the preemphasis schedule with specifications as follows:

$[S/N]_c = 12$ dB (3.98 numerical);

$D_{si} = f_{dsi}/f_{mi} = 2.5$ (nominal);

$[S/N]_{oi} = 46$ dB.

Answer: $A_p = 1.95$. $f_{dci} = 3,175$; 2,963; 2,752; 2,540; and 2,328 kHz. $f_{dn} = 6.190$ MHz. $B_c = 20$ MHz.

Problem 3.45

Using the channels and specified parameters in Problem 3.44, except $[S/N]_{oi}$ = 46 dB, design the preemphasis schedule for a specified B_c = 12,000 kHz.
Answer: A_p = 1.95. f_{dci} = 1,108; 1,034; 960; 886; and 813 kHz. D_{ci} = 0.287. $[S/N]_{oi}$ = 34.6 dB.

Problem 3.46

Using the channels and specified parameters of Problem 3.44, except $[S/N]_{oi}$ = 46 dB, design the preemphasis schedule for all channel dropout or minimum transmission bandwidth.
Answer: A_p = 1.95. f_{dci} = 711, 663, 616, 568, and 521 kHz. f_{dn} = 1.38 MHz. D_{ci} = 0.185. $[S/N]_{oi}$ = 30.2 dB.

Problem 3.47

Observing the new nomenclature and using the lowest five constant bandwidth F-channels, 256F; 512F; 768F; 1,024F; and 1,280F, design the preemphasis schedule with specifications as follows:

$[S/N]_c$ = 12 dB (3.98 numerical);

$D_{si} = f_{dsi}/f_{mi}$ = 2.5 (nominal);

$[S/N]_{oi}$ = 46 dB.

Problem 3.48

Using the channels and specified parameters of Problem 3.47, except $[S/N]_{oi}$ = 46 dB, design the preemphasis schedule for a specified B_c = 12,000 kHz.

Problem 3.49

Using the channels and specified parameters of Problem 3.47, except $[S/N]_{oi}$ = 46 dB, design the preemphasis schedule for all-channel dropout or minimum transmission bandwidth.

Problem 3.50

Observing the new nomenclature and using the five CBW G-channels, 3,584G; 3,072G; 2,560G; 2,048G; and 1,536G, design the preemphasis schedule with specifications as follows:

$[S/N]_c = 12$ dB (3.98 numerical);

$D_{si} = f_{dsi}/f_{mi} = 5$ (nominal);

$[S/N]_{oi} = 46$ dB.

Answer: $A_p = 1.66$. $f_{dci} = 1,821$; $1,565$; $1,304$; $1,043$; and 282 kHz. $f_{dn} = 3.02$ MHz, $D_{ci} = 0.5$. $B_c = 13.2$ MHz.

Problem 3.51

Using the channels and specified parameters of Problem 3.50, except $[S/N]_{oi} = 46$ dB, design the preemphasis schedule for a specified $B_c = 12,000$ kHz.
Answer: $A_p = 1.66$. $f_{dci} = 1,459$; $1,254$; $1,045$; 836; and 627 kHz. $D_{ci} = 0.408$. $[S/N]_{oi} = 43.6$ dB.

Problem 3.52

Using the channels and specified parameters of Problem 3.50, except $[S/N]_{oi} = 46$ dB, design the preemphasis schedule for all-channel dropout or minimum transmission bandwidth.
Answer: $A_p = 1.66$. $f_{dci} = 882, 758, 632, 505$, and 379 kHz. $f_{dn} = 1.47$ MHz. $D_{ci} = 0.247$. $B_c = 10$ MHz. $[S/N]_{oi} = 38.5$ dB.

References

[1] Downing, J. J., *Modulation Systems and Noise,* Englewood Cliffs, NJ: Prentice-Hall, 1964, pp. 121–131.

[2] Nichols, M. H., and L. L. Rauch, *Radio Telemetry,* New York: Wiley, 1956.

[3] Law, E. L., and D. R. Hust, *Telemetry Applications Handbook,* Tech. Pub. TP000044, Point Mugu, CA: Weapons Instrumentation Division, Pacific Missile Test Center, Sept. 1987, pp. 1–21 (sec. 3–7).

[4] Rosen, C., "System Transmission Parameters Design for Threshold Performance," *Proc. Int. Telemetering Conf.,* Vol. XIX, 1988, pp. 145–182.

[5] Rosen, C., "Systems Transmissions Parameters Design for Threshold Performance," *Proc. Int. Telemetering Conf.* Vol. III, 1983, pp. 221–255.

[6] Secretariat, Range Commanders Council, *Telemetry Standards,* White Sands Missile Range, NM: RCC, IRIG 106-00.

[7] *Telemetry FM/FM Baseband Structure Study,* Vol. I and II, Defense Technical Information Center, access numbers AD-621139 and AD-621140, June 14, 1965.

4

Digital Communication Systems

This chapter introduces the fundamental components of a digital communication system. The basic function and operation of each component is presented.

4.1 Learning Objectives

Upon completion of this chapter, the reader should understand the following:

- The difference between analog and digital signals and the parameters that characterize them;
- The determination of the baud rate and the bit rate of a digital signal;
- Analog-to-digital quantization noise;
- Analog-to-digital signal-to-quantization noise ratio as a function of word length, voltage levels, and dynamic range;
- The operation, advantages, and disadvantages of serial and parallel analog-to-digital converters;
- The rate of an encoder and input/output signal relationships;
- Encoder design and trading bandwidth requirements with signal-to-noise ratio requirements;
- The purpose of each block of a TDM system;

- How PCM is generated;
- Basic digital modulation methods amplitude-shift keying, frequency-shift keying, and phase-shift keying.

4.2　Digital Communication System Overview

A block diagram overview of a point-to-point digital communication system is shown in Figure 4.1(a). The signal from an analog source, located at the transmitting site, is converted into a digital signal by the *analog-to-digital* (A/D) converter. If the source supplies a digital signal, then the Λ/D converter

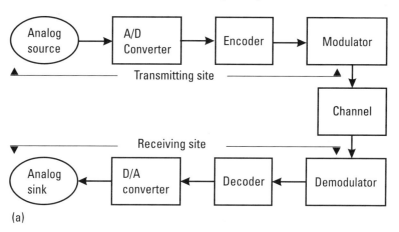

(a)

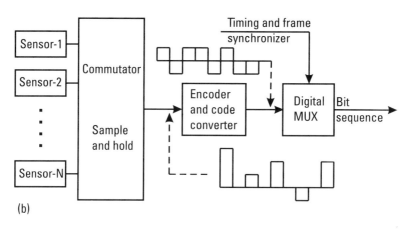

(b)

Figure 4.1　(a) Digital communication system overview, and (b) TDM telemetry system.

is not necessary. The encoder encodes the input digital signal into another digital signal, which in turn is used by the modulator to generate a modulated carrier output. The modulated signal is transmitted through the communication channel to the remotely located receiving site, where the corresponding inverse functions are performed by the demodulator, decoder, and *digital-to-analog* (D/A) converter.

For a TDM telemetry system, each sensor channel may be considered a single channel of the point-to-point digital communication system described. The specific blocks of a TDM telemetry system are shown in Figure 4.1(b). In such a system, the output bit sequence is usually the result of sampling, digitizing, and combining the outputs from a number of different sensors. The commutators's purpose is to sequentially sample the output of the sensors and produce an amplitude-modulated pulse for each sample. The encoder quantizes the samples and converts each into an n-bit binary word. Hence, the commutator and the encoder behave as the single A/D converter and encoder of Figure 4.1(a). The purpose of the *multiplexer* (MUX) is to insert the minor and major frame synchronization words into the bit stream, and to perhaps add computer data words from an onboard computer. The MUX basically builds the frame structure along with the frame synchronization timing, all to be discussed in Chapter 5. A lowpass filter is inserted between the digital MUX and the transmitter to limit the transmission bandwidth.

The individual blocks of the system depicted in Figure 4.1 are now described in detail.

4.3 Communication System Signals

4.3.1 Analog Signals

An analog signal may be described as a *continuous-amplitude* (CA) and *continuous-time* (CT) signal. That is, the signal amplitude is continuous between the maximum and minimum values, and is continuously changing in time. A typical analog signal is shown in Figure 4.2(a). Two metrics describe this signal—an amplitude metric and a time metric. Typical amplitude metrics are peak-to-peak voltage and RMS voltage. The time metric is usually given in terms of the maximum frequency or bandwidth. For the signal shown, the amplitude is 20V peak-to-peak, and the maximum frequency is 2,000 Hz.

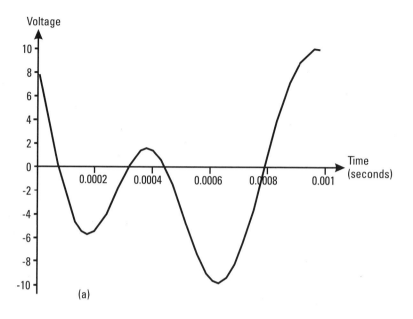

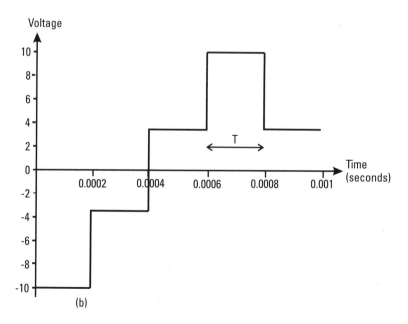

Figure 4.2 (a) Typical analog signal, and (b) typical digital signal.

4.3.2 Digital Signals

A digital signal is a *discrete amplitude* (DA) and *discrete time* (DT) signal. The signal amplitude may assume one of a finite number of discrete levels between the maximum and minimum values, and may only change at discrete times. A typical digital signal is shown in Figure 4.2(b). The amplitude metric is the number of discrete levels the signal may assume, with each level corresponding to a unique symbol. The time metric is usually given in terms of the number of symbols per second, which is also known as the baud rate. Thus, we define the following parameters:

$E \equiv$ the number of levels the signal amplitude may assume;

$B \equiv$ baud rate = the maximum number of signal symbols per second.

A numeric representation of the E levels requires a binary number of n bits, where

$$n = \log_2 E \qquad (4.1)$$

Since n is the number of bits of information required to represent one change (or symbol) of the signal, and there are baud rate symbols per second, we have: $I \equiv$ information rate = baud rate $(\log_2 E)$ bps, or

$$I = Bn = B \log_2 E \qquad (4.2)$$

Example 1

In Figure 4.2(b), the signal may assume one of four levels specifying that $E = 4$. The time between changes is $T = 0.0002$ seconds, and $B = 1/T = 5,000$ symbols/second. Thus we have $I = B(\log_2 E) = (5,000)\ (\log_2 4) = 10,000$ bps.

4.4 Quantization and A/D Conversion

The quantization process changes the continuously varying pulse amplitudes created by sampling an analog signal into a finite number of levels. Each level is assigned an n-bit binary word by the A/D converter.

4.4.1 Quantization Errors

From (4.1), the number of levels E, is given by

$$E = 2^n \qquad (4.3)$$

where n is the length in bits of the binary word. In telemetering, n is nominally 8.

If V_{pp} is the peak-to-peak voltage and the peak voltage V_p is $V_{pp}/2$ of the signal being quantized, the voltage separation between each level is given by

$$q = V_{pp}/E = 2V_p/E \qquad (4.4)$$

The quantization error is the difference between the quantized signal and the continuous signal. The error is random, has a maximum value of $q/2$ and is uniformly distributed such that

$$p(e) = 1/q \qquad -q/2 < e < q/2 \qquad (4.5)$$

The variance or noise power is given by

$$\sigma_q^2 = \frac{1}{q} \int_{-q/2}^{q/2} e^2 de = \frac{q^2}{12} \qquad (4.6)$$

In terms of the number of levels, the quantization noise power is

$$\sigma_q^2 = \frac{V_p^2}{3E^2} \qquad (4.7)$$

If $\overline{m(t)^2}$ is the RMS power of the message, $m(t)$, being quantized, the signal-to-noise power ratio out of the quantizer is given by

$$\frac{S}{N} = \frac{\overline{m(t)^2}}{\sigma_q^2} = \frac{\overline{m(t)^2}}{q^2/12} \qquad (4.8)$$

Using (4.7) and (4.8) gives

$$\frac{S}{N} = \frac{3E^2 \overline{m(t)^2}}{V_p^2} \qquad (4.9)$$

4.4.2 Hardware Implementation of Quantizers and Analog Converters

Two types of quantizers and A/D converters often used are the serial and the parallel or flash converter. These will be discussed next, along with the delta-sigma oversampled converter.

4.4.2.1 Serial Converters

The serial conversion process for $n = 3$ is illustrated by the flow chart shown in Figure 4.3.

In the three-bit serial quantizer, the input sample $m(NT)$ is shifted and scaled to have a range between 0 and V_p. When $m(NT)$ is applied to the first comparator, a decision is made as to whether $m(NT)$ is above or below $V_p/2$. If it is above, the most significant bit, b_2, is set to 1 and $-1/2\ V_p$ is subtracted from $m(NT)$, reducing it to fall between 0 and $1/2\ V_p$. If $m(NT)$ is less than $1/2\ V_p$, b_2 is set to 0, and $m(NT)$ is sent to the next comparator, which now must decide if the input pulse is above or below $1/4\ V_p$. The process is repeated and the results applied to the next comparator. Although the A/D converter illustrated is a 3-bit encoder, it is relatively simple to add another stage.

A simpler version of the serial encoder, known as a *successive approximation* A/D is shown in Figure 4.4. In this case $m(NT)$ is scaled by dividing by V_p, so that the input will fall between 0 and 1.

Example 2

It is desired to use the successive approximation A/D shown in Figure 4.4 to generate a 3-bit word. This will quantize the signal into eight levels. The threshold of each will be at $1/8(0.125)$, $2/8(0.25)$, $3/8(0.375)$, $4/8(0.5)$, $5/8(0.625)$, $6/8(0.75)$, and $7/8(0.875)$. Say the input signal $m(NT)/V_p$ is 0.7. This value falls between 5/8 and 6/8 and would be quantized to 101 in binary. In the first comparator, $m(NT)/V_p$ triggers the $b = b_2 = 1$ set.

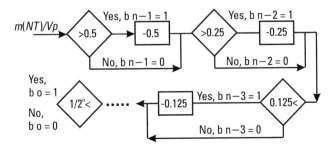

Figure 4.3 Eight-level serial quantizer and 3-bit encoder.

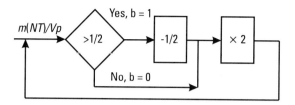

Figure 4.4 Successive approximation quantizer and encoder.

The sample goes into the subtractor and becomes 0.2 and is multiplied by 2 to become 0.4. This is now fed into the comparator, and since it is less than 0.5, $b = b_1 = 0$ is set. The 0.4 value is inserted into the multiply-by-2 module, becoming 0.8, and this is fed into the comparator. Since this value is greater than 0.5, the comparator sets $b = b_0 = 1$. The process is completed and the A/D is ready for the next sample.

4.4.2.2 Parallel A/D Converters and Oversampled Delta-Sigma Converters

A parallel converter is shown in Figure 4.5. The input $m(NT)$ is scaled to be between 0 and 1V. The logic zeros and ones from the n comparators are fed into the coder which uses combinatorial logic to set the n-bit word output.

The oversampled delta-sigma A/D converter, shown in Figure 4.6, allows the use of less expensive, less stringent, analog lowpass filters for both

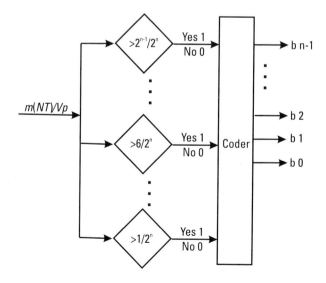

Figure 4.5 N-bit parallel converter.

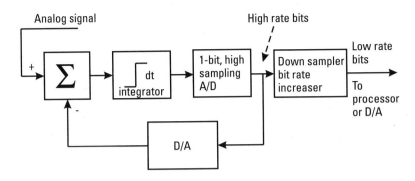

Figure 4.6 Oversampled delta-sigma converter.

the antialiasing filter following the sensor, and the anti-image filter following the D/A converter, when and if the signal is to be converted back to an analog signal. Internally, the delta-sigma converter uses an increased sampling rate and, commonly, a simple, inexpensive 1-bit A/D. In essence, this converter makes two tradeoffs: (1) It exchanges complex analog filters on either end of the system for an internal DSP chip, which does the processing, and (2) it exchanges bits in the operation of the A/D for a higher sampling rate, while maintaining the same signal quality, which allows a simple A/D.

The advantage of oversampling is that the problem of aliasing is reduced. Figure 4.7(a) shows the spectrum of a lowpass signal sampled at the minimum rate f_L, while the spectrum of the same signal oversampled at f_H, three times the minimum rate, is shown in Figure 4.7(b). Clearly a simple first-order lowpass filter will eliminate aliasing in the oversampled signal, while a more sophisticated higher-order lowpass filter would be needed in the minimally sampled signal.

In the delta-sigma converter shown in Figure 4.6, the A/D converter is a simple 1-bit converter, but operating in a high sampling mode. The output of the A/D converter goes to the feedback loop D/A converter, whose analog output signal goes to the summer, where it is subtracted from the incoming signal. The difference signal is smoothed by the integrator, basically acting as a lowpass filter, and sent to the A/D converter, continuing the process.

The purpose of the down-sampler and bit-rate increaser, referred to in the literature as a decimator filter, is to restore the bit-rate to the desired rate, which is accomplished with a DSP chip that adds bits by linearly interpolating between the rough low-bit samples. This chip must also partially eliminate quantization noise by pushing it out of the lowpass spectrum of

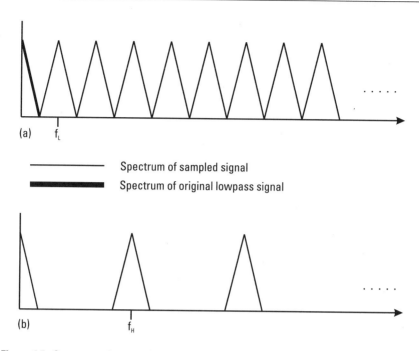

Figure 4.7 Spectrum of a sampled lowpass signal.

interest (a process referred to as noise shaping) as well as perform highpass filtering to remove other noise components added by the process.

An example of a delta-sigma converter is the one used by Phillips in their first CD player, which enabled them to use a 14-bit word instead of a 16-bit word, while maintaining the same signal quality. For more on delta-sigma converters, and the concept of trading bits for oversampling see [1, 2].

4.4.2.3 Converter Comparison

Typically, the flash converter is used extensively in digital real time video. For other applications in which the speed of the serial converter is adequate, it is preferred because of its compatibility.

Parallel or flash converter characteristics:

- Very fast, an 8-bit word in 50 ns, 8-bit words.
- Small working voltage range, 1 to 3V.
- Small input impedance, variable. Input is working into E comparators.

Serial or successive approximation characteristics:

- Comparatively slow, an 8-bit word every 0.1 ms, up to 16-bit words.
- Good voltage range, ±10V, compatible with op-amps.
- Large input impedance.

Delta-sigma oversampled converter:

- Fairly slow, a 16-bit word every 50 ms.
- Very simple structure.
- Internally uses simple low-bit A/D, typically a 1-bit A/D.
- Low voltage and power operation.
- Compatible with complementary metal-oxide semiconductor (CMOS) and may be integrated onto a chip with a microprocessor.

4.5 Encoding

Encoding methods are often classified into one of two categories: source encoding and channel encoding. Source encoding techniques attempt to remove redundant information from the source signal, and thus reduce the information rate of the signal. Channel encoding techniques transform the source digital signal into another digital signal that better matches the bandwidth and signal-to-noise characteristics of the channel. This section introduces the concept of channel encoding, assuming that any required source encoding has been implemented.

The encoder input digital signal $e(t)$ is encoded into a digital output signal $m(t)$ as shown in Figure 4.8. For each input symbol (change), the encoder generates N output symbols, (changes), each representing one of M possibilities. Thus we have

$$B_m = NB_e \tag{4.10}$$

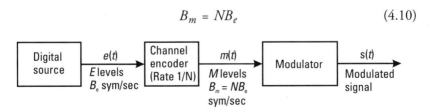

Figure 4.8 The channel encoder and modulator.

where the encoder rate is defined as $1/N$. At the encoder input the information rate is

$$I_e = B_e \log_2 E \qquad (4.11)$$

where B_e is the input baud rate, and E is the number of levels of the input digital signal. In a similar way, at the encoder output, the information rate is

$$I_m = B_m \log_2 M \qquad (4.12)$$

where B_m is the output baud rate, and M is the number of levels of the output digital signal.

If there is a one-to-one correspondence between the encoder input combinations and output combinations, we have

$$E = M^N \qquad (4.13)$$

or taking the log base 2 of each side

$$\log_2 E = N \log_2 M \qquad (4.14)$$

Substituting (4.10) into (4.12) and using (4.14) and (4.11), gives

$$I_m = B_e N \log_2 M = B_e \log_2 E = I_e \qquad (4.15)$$

Encoders that map the input signal to the output signal in accordance with (4.13) have the property that $I_m = I_e$. This is desirable in the sense that the encoder preserves the original information and adds no redundant information to the original signal. From (4.10) and (4.13), it is evident that as N is increased, the output baud rate B_m increases, while the number of output signal levels M decreases. The inverse is true as N is decreased. Furthermore, an increase (decrease) in B_m requires an increase (decrease) in channel bandwidth; and an increase (decrease) in M requires an increase (decrease) in channel signal-to-noise ratio at the receiver. Therefore, we have the following principle: The channel encoder allows the designer to "match" the digital signal to the channel characteristics by trading off channel bandwidth requirements with receiver signal-to-noise ratio requirements.

4.5.1 Design Example

Design an encoder to transform a four-level digital signal into a two-level digital signal.

Here $E = 4$, and $M = 2$. From (4.14), $N = (\log_2 E)/(\log_2 M) = (\log_2 4)/(\log_2 2) = 2/1 = 2$. Thus, there are 2 two-level output symbols for every four-level input symbol. The encoder mapping may be specified by Table 4.1:

Table 4.1
Encoder Input/Output Mapping

Encoder Input	Encoder Output
0	0 0
1	0 1
2	1 0
3	1 1

A representative encoder input signal $e(t)$, and the corresponding encoder output signal $m(t)$, for this mapping is shown in Figure 4.9. The output baud rate is twice the input baud rate, while the input and output information rates are equal.

4.6 PCM

An overview of a general telemetry system was shown in Figure 1.1, along with the subsystems to be discussed in this section. The subsystems of the

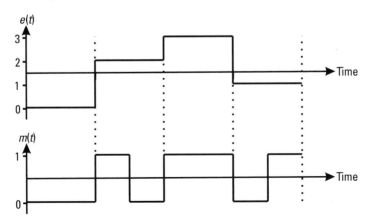

Figure 4.9 Encoder input and output signals.

general system are expanded into specific blocks for a TDM system in Figure 4.1(b). The sensor blocks are also shown for cohesiveness, although they comprise a different subsystem.

In a telemetering system, the binary sequence is usually the result of sampling, digitizing, and combining the outputs from a number of different sensors resulting in a TDM system as shown in Figure 4.1(b). The purpose of the commutator is to sample the outputs of the sensors and produce an amplitude-modulated pulse for each sample. The encoder quantizes the samples and converts them into an n-bit binary word, a sequence referred to as PCM. The bit sequence could also be a computer output.

In Figure 4.1(b), the encoder also behaves as a code converter and changes the binary logic into a voltage waveform suitable for driving the transmitter. A lowpass filter is inserted between the digital MUX and the transmitter, in order to limit the transmission bandwidth. The purpose of the MUX is to insert the minor and major frame synchronization words and perhaps add some computer data words from an onboard computer. The MUX basically builds the frame structure along with the frame sync timing.

4.7 Modulation

Three basic digital carrier modulation methods are described in this section. Details of these and advanced digital modulation techniques are presented in Chapters 8, 9, and 10.

For each input symbol, the carrier output from the modulator is a sinusoidal time signal $s(t)$ given by

$$s(t) = A \cos(2\pi ft + \phi) \tag{4.16}$$

where A is the amplitude of the transmitted signal, f is the frequency of the transmitted signal, and ϕ is the phase of the transmitted signal.

Basic digital modulation methods will vary one of the above parameters in proportion to the input signal amplitude, while holding the other two parameters constant. The modulator input signal $m(t)$ comes from the encoder output as shown in Figure 4.8.

4.7.1 Amplitude-Shift Keying

In amplitude-shift keying (ASK), the amplitude A of the carrier is shifted between two levels, while f and ϕ are held constant. This corresponds to the encoder output $m(t)$ being a binary signal ($M = 2$). A representative modulated signal for a binary input signal is shown in Figure 4.10(a). If the number of amplitude levels E is greater than 2, this process is known as M-ary ASK.

4.7.2 Frequency-Shift Keying

In frequency-shift keying (FSK), the frequency f of the carrier is shifted between two levels, while A and ϕ remain constant. The encoder output $m(t)$ is a binary signal ($M = 2$). A representative modulated signal for a binary input signal is shown in Figure 4.10(b). If the number of amplitude levels E is greater than 2, this process is known as M-ary FSK.

4.7.3 Phase-Shift Keying

In phase-shift keying (PSK), the phase ϕ of the carrier is shifted between two levels, while A and f are held constant. The encoder output $m(t)$ is a

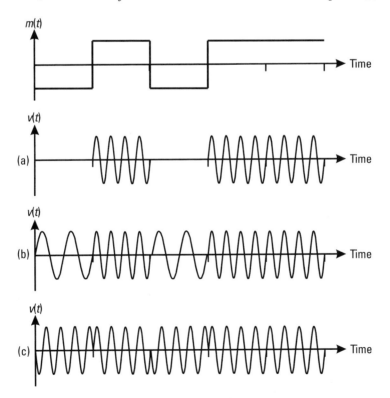

Figure 4.10 Digital modulation waveforms: (a) ASK, (b) FSK, and (c) PSK.

binary signal with $M = 2$. A representative modulated signal for a binary input signal is shown in Figure 4.10(c). If the number of amplitude levels E is greater than 2, this process is known as M-ary PSK.

Problems

Problem 4.1

Sketch an overview of an analog communication system using blocks similar to Figure 4.1(a).

Problem 4.2

What is the purpose of a commutator?

Problem 4.3

Sketch a digital signal waveform with eight levels and a baud rate of 2,400 changes/second.

Problem 4.4

What is the bit rate of the signal specified in Problem 4.3?

Problem 4.5

What is the bit rate of the signal specified in Problem 4.3 if the number of levels is changed to 64?

Problem 4.6

What is the purpose of an A/D converter?

Problem 4.7

Given that the dynamic range of a voltage is -5 to $+5$V, find the quantization noise power when the word length, n equals (a) 12 bits, (b) 8 bits, and (c) 6 bits.

Problem 4.8

If $m(t) = \cos \omega t$, find S/N after the quantizer for the word lengths used in Problem 4.7.

Problem 4.9

It is desired to use the successive approximation A/D shown in Figure 4.4 to generate a 2-bit word. (a) Determine the comparison thresholds used by the converter, and (b) What is the A/D output word?

Problem 4.10

Discuss when a parallel converter would be used and when a serial converter would be used.

Problem 4.11

What is the purpose of an MUX?

Problem 4.12

The output signal $m(t)$ from an encoder produces a new binary symbol every 0.01 seconds. The output sequence is $1,-1,1,1,-1$. Sketch the modulator output signal for (a) ASK, (b) FSK, and (c) PSK.

References

[1] Gray, R. M., "Oversampled Sigma-Delta Modulation" *IEEE Transaction on Communication, Com-35,* 1978, p. 481.

[2] Orfanidis, S., *Introduction to Signal Processing,* Englewood Cliffs, NJ: Prentice-Hall, 1996.

5

TM Channel Format Design

An overview of a telemetry system was shown in Figure 1.1, while in Figure 4.1 the general blocks were expanded into specific parts. Chapter 5 is concerned with the output of the encoder, which is the binary line code that will represent each digital word, and the purpose of the MUX, which is the creation of synchronized frames.

5.1 Learning Objectives

Upon completion of this chapter, the reader should understand the following:

- A deeper understanding of the purpose of the encoder and the MUX;
- Line coding;
- Advantages and disadvantages of various line coding formats;
- Frame construction using major frames, minor frames, subframes, and synchronization words;
- Commutation using supercommutation and subcommutation;
- IRIG PCM formats.

5.2 Line Coding or Transmission Format

Once the analog signal has been sampled, quantized, and converted into an n-bit digital word, it is necessary to convert the logic word into voltage

121

levels for transmission. Digital data can be transmitted by various transmission formats or line coding. There are advantages and disadvantages to each of the line codes.

5.2.1 Line Code Formats

The various line codes are shown in Figure 5.1(a) and (b) for a given logic sequence. Conceptually, an L indicates the voltage level is determined by the current bit, a mark of M indicates the level changes when a mark or a 1 occurs and a space or S indicates that a level change occurs whenever a space or zero occurs. Three code types will be discussed: (1) nonreturn to zero (NRZ); (2) biphase, BiΦ; and (3) delay modulation, (DM) (Miller).

5.2.1.1 NRZ and BiΦ

- NRZ-L represents a 1 or 0 by the correct voltage level for the entire duration of the bit period.

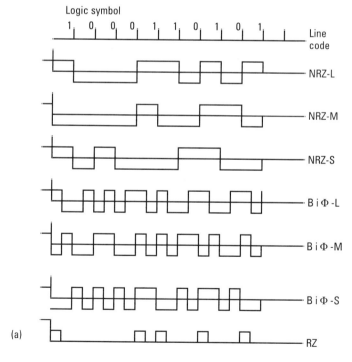

Figure 5.1 (a) Waveforms for NRZ and BiΦ line codes, and (b), waveforms for DM line codes.

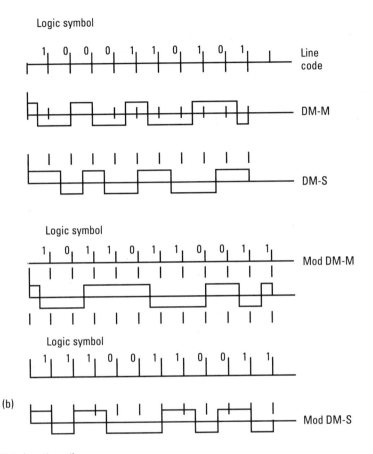

Figure 5.1 (continued).

- NRZ-M changes the voltage level whenever a mark or a 1 occurs and the voltage level remains the same for the bit period.

- NRZ-S changes the voltage level whenever a space or a 0 occurs, and the voltage level remains the same during the bit period.

- BiΦ-L represents a 1 by the 1 voltage level during the first half of the bit period and by the 0 voltage level during the last half. The 0 is represented by the negative of the voltage waveform representing the 1.

- BiΦ-M changes the level whenever a 1 occurs and again in the middle of the bit period. The level changes whenever a 0 occurs but not in the middle of the bit period.

- BiΦ-S changes the level whenever a 0 occurs and again in the middle of the bit period. The level changes whenever a 1 occurs but not in the middle of the bit period.
- RZ. A 1 is represented by a voltage level at the start of a bit period and returns to 0 at the middle of a bit period. The 0 is represented by a zero voltage level.

5.2.1.2   DM

- DM-M. A 1 is represented by a level change at mid-bit. A 0 followed by a 0 is represented by a level change at the end of the first 0 bit. No level change occurs when a 0 is preceded by a 1.
- DM-S. A 0 is represented by a level change at mid-bit. A 1 followed by a 1 is represented by a level change at the end of the first 1 bit. No level change occurs when a 1 is preceded by a 0.
- Modified DM-M. Same as DM-M, except the level change is omitted for the last 1 in an even number sequence of 1s only when preceded by an odd number of 0s since the last omission of a level change.
- Modified DM-S. Same as DM-S except the level change is omitted for the last 0 in an even number sequence of 0s only when preceded by an odd number of 1s since the last omission of a level change.

The line code should have the following properties:

- The transmission bandwidth should be as small as possible.
- The code should be transparent to long strings of either 0s or 1s.
- There should be adequate timing content.
- The spectrum of the line code should match the channel frequency response, for example, if AC coupling is used, the spectrum should have no DC content.

5.2.2   Characteristics of the NRZ and BiΦ Line Codes

5.2.2.1   NRZ Codes

NRZ codes use the smallest transmission bandwidth.
   NRZ-L codes have the following properties:

- They are polarity sensitive.
- Spectrum contains DC, and hence, DC coupling is usually required.
- Long strings of 1s or 0s create a timing problem.

NRZ-M or S:

- Since bit detection is a function of level changes, these codes are not polarity sensitive.
- If long strings of 1s or 0s are expected, then the NRZ-M or NRZ-S would be used, respectively, to help the synchronization problem. If the opposite happened, a timing problem would occur.

5.2.2.2 BiΦ Codes

BiΦ codes use twice the bandwidth of the NRZ codes. The built-in level changes enhance synchronization. BiΦ-L codes are polarity sensitive, whereas BiΦ-M and BiΦ-S are not.

BiΦ codes can be generated by using an exclusive-or logic module with the appropriate NRZ and clock as the input.

5.2.3 Delay Modulation Code Characteristics

The DM characteristics make the code especially desirable for recording and are as follows:

- The spectrum is concentrated around the normalized $f = 0.4 \ T_b$ with little energy at $f = 0$.
- The code is insensitive to 180° phase ambiguity.

Because of the spectrum concentration, lower tape speed may be used, which allows higher packing density. Since recorders have a poor DC response, no energy around DC also makes this code appealing.

5.2.4 Power Spectrum

The power spectrum for the NRZ codes is given by

$$S_{NRZ}(f) = A^2 T_b \, \text{sinc}^2(fT_b) \qquad (5.1)$$

T_b = bit period;
A = amplitude of the bit.

The power spectrum for the BiΦ codes is given by [1],

$$S_{BP}(f) = A^2 T_b \operatorname{sinc}^2\left(\frac{fT_b}{2}\right) \sin^2\left(\frac{\pi f T_b}{2}\right) \qquad (5.2)$$

The power spectrum for the DM codes is given by [1],

$$S_{MD}(f) = \frac{23 - 2\cos a - B - C + D + E + F - G + H}{2a^2(17 + 4H)} \qquad (5.3)$$

where

$a = \pi f T_b$;

$B = 22 \cos 2a$;

$C = 12 \cos 3a$;

$D = 5 \cos 4a$;

$E = 12 \cos 5a$;

$F = 2 \cos 6a$;

$G = 8 \cos 7a$;

$H = 2 \cos 8a$.

Equations (5.1), (5.2), and (5.3) are plotted in Figure 5.2. The plots show that the BiΦ codes take twice the bandwidth of the NRZ codes, since the first 0 of the BiΦ spectrum occurs at $2R_b$, while the first 0 of the NRZ spectrum occurs at R_b. In actual practice, the BiΦ bandwidth used is only $1.8R_b$.

5.3 Frame Design and Creation

In TDM, it is necessary to establish a structure to the data stream in order to route the data from the various sensors to the proper channels at the ground station. The structure is established by breaking the data stream into minor and major frames. To separate the data, it is necessary to know when a major and minor frame start. That is, frame synchronization must be achieved at the ground station.

5.3.1 Frame Design

To achieve synchronization, a synchronization word (Sc) is normally inserted at the start of each minor frame. A simple major frame containing M minor

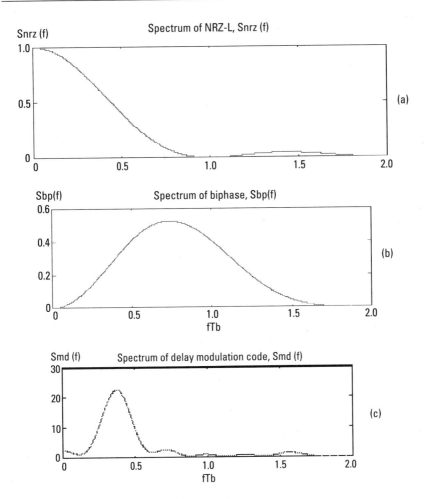

Figure 5.2 (a) NRZ-L, (b) BiΦ, and (c) DM code spectrums.

frames is shown in Figure 5.3(a). From left to right, the rows represent a minor frame. When *M* minor frames have been completed, another major frame will start.

5.3.1.1 Minor Frame

The numbers across the minor frame row represent the sensor data. A minor frame synchronization word has been inserted at the start of each minor frame with the designation of sc. The next word in the minor frame is the data word from the first sensor. Since there are *N* slots in the minor frame and one has been used for the synchronization word, there is room for

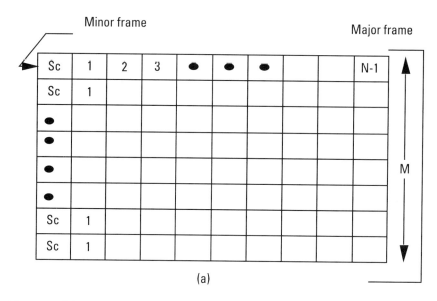

Minor frame

Major frame

(a)

Figure 5.3 (a) A simple major frame containing *M* minor frames; (b) a simple generalized commutator for *N* − 1 sensors.

$N - 1$ data words from the $N - 1$ sensors. When the first row has been completed, another synchronization word is inserted and another minor frame started.

5.3.1.2   Major Frame

A major frame is composed of a number of minor frames. In a practical system, all sensors are not sampled at the same rate; therefore, every minor frame may not have data from a specific sensor. Whenever every sensor has been sampled at least once, a major frame has been completed. The bit sequence should be thought of as laying the minor frames end-to-end along the time axis.

5.3.2   Frame Creation

The frame is created by quantizing and converting samples from analog signals into digital words in a sequential and periodic process. Marking the periodicity of this sequence with synchronization words creates the frames. The purpose of a commutator is to perform the sequential sampling. The encoder and A/D converter convert the sampled voltage into a digital word.

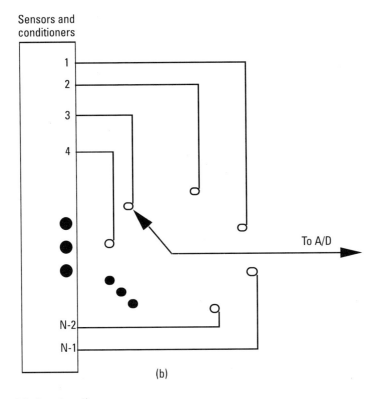

Sensors and conditioners

To A/D

(b)

Figure 5.3 (continued).

The digital MUX combines the digital data words with the timing or synchronizing words. This entire process is usually achieved with one piece of off-the-shelf hardware.

5.3.2.1 Commutator

The purpose of the commutator is to cycle through and sample each sensor. At a particular sampling moment, the output of the commutator, represented by a pulse whose amplitude is the voltage value of the sampled sensor, is fed into the encoder. A simple generalized commutator for $N - 1$ sensors is shown in Figure 5.3(b). The output from the number one sensor is encoded as an n-bit word and inserted into the number one slot in the minor frame. The process is continued until the value of the $N - 1$ sensor is sampled, converted into a digital word and inserted into the $N - 1$ slot. Another minor frame is then started. A simple generalized major frame composed of

M minor frames is shown in Figure 5.3(a). This frame would result from the commutator shown in Figure 5.3(b).

5.3.2.2 Encoder and Analog-to-Digital Converter

The encoder and analog-to-digital converter (ADC) convert the sampled voltage value into a digital word. This digital word is sent to the digital MUX.

5.3.2.3 Digital MUX

The digital MUX combines the data words with the synchronization words and perhaps digital words from a computer and creates the desired frame format.

5.3.2.4 Example of Simple Frame Creation

Assume a simple system is composed of eight sensors, all of which are sampled at the same rate and each sample is converted into an 8-bit word. A commutator that would perform this sampling is shown in Figure 5.4. The resulting major frame, composed of M minor frames, is shown in Figure

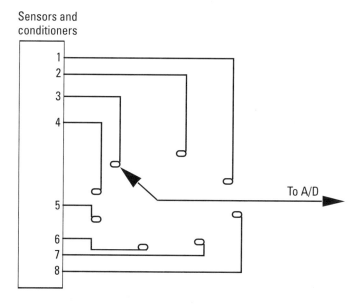

Figure 5.4 Eight-sensor commutator.

5.5. In this simple case, only one minor frame per major frame is needed, since all sensors are sampled once per minor frame.

5.3.2.5 Supercommutation

In practice, the maximum frequency present in the process being converted into a voltage by the sensor is seldom the same for all the sensors. The result is that some of the sensors must be sampled more often and the resulting data words from this sensor will appear more than once in a single minor frame. For the example given in Section 5.4.2.4, assume the number one sensor must be sampled a little less than twice as often as the other sensors. The commutator of Figure 5.6 will perform this operation. The resulting frame is shown in Figure 5.7. From a practical standpoint, the sampling rate for supercommutation must be a multiple of the minor frame rate.

5.3.2.6 Subcommutation

In actual systems, there are times when a number of the sensors will produce signals whose maximum frequency will be substantially below the maximum frequency of a group of other sensor signals. One sensor of this group of low-frequency signals may be sampled only once a minor frame. The commutator and subcommutator are shown in Figure 5.8. The resulting major frame and the subframes are shown in Figure 5.9. The commutator will cycle through the fast sensors four times before it completes cycling through the slower signals. The signal from the 5-1 sensor will be inserted into the number five slot in the first minor frame. The 5-2 sensor will be inserted into the number five slot in the second minor frame, and so on. From a practical standpoint, the sampling rate for subcommutation must be a submultiple of the minor frame rate.

SC	1	2	3	4	5	6	7	8
SC	1							
SC	1							
SC	1							
SC	1							
SC	1							
SC	1							
SC	1							

Figure 5.5 Eight-sensor commutator frame array.

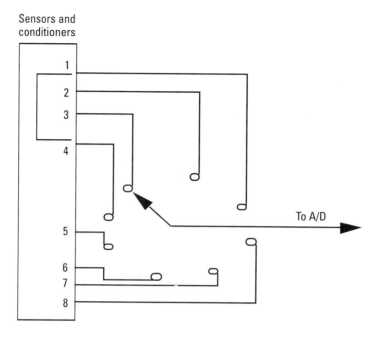

Figure 5.6 Supercommutation.

SC	1	2	3	1	5	6	7	8
SC	1			1				
SC	1			1				
SC	1			1				
SC	1			1				
SC	1			1				
SC	1			1				

Figure 5.7 Supercommutation frame.

5.4 Frame Synchronization

It is necessary to insert a minor frame synchronization word at the start or the end of each minor frame. In this work, it will be assumed that the minor frame synchronization word is inserted at the start of each minor frame. The synchronization word should look as little like a data word as possible. Past work adopted by the IRIG standards committee has resulted in a table

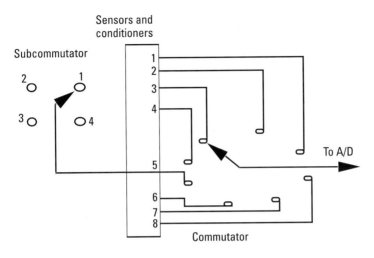

Figure 5.8 Subcommutation.

Sc	1	2	3	4	5-1	6	7	8
Sc	1	2	3	4	5-2	6	7	8
Sc	1	2	3	4	5-3	6	7	8
Sc	1	2	3	4	5-4	6	7	8

Figure 5.9 Subcommutation frame.

of synchronization words. These words vary in length from 7 bits to 33 bits. (A list of these words is given in Appendix B.) Nominally, the sync word is twice the length of the data word. If the sync word is too long, too much overhead is required, although synchronization may occur faster. For a statistical analysis of lock, see [2].

5.4.1 False Lock

If the sync word is 16-bits long and the data words are 8-bits long, it is possible for two data words to look like the sync word and for the synchronizing process to cause a false lock on the two data words. Since the process searches, checks, and then locks, for false lock to occur, the two data words must be static over a period of time. To circumvent false lock due to data words, the complement of the sync word may be used every other time.

5.4.2 Subframe Synchronization

It is necessary to know where in the major frame the system locked up, and for decommutation it is necessary to know what major frame, minor frame, and row or 8-bit word is being processed. To do this, it is necessary to know what subframe row is being processed.

Three of the techniques used for subframe synchronization are:

- Minor frame sync word complement;
- Recycle subframe synchronization;
- Identification subframe synchronization.

5.4.3 Minor Frame Sync Word Complement

At the start of the subframe, the complement of the minor frame synchronizing word is inserted in the column devoted to minor frame sync. This precludes using sync word complement to prevent false data word lock. Figure 5.10 shows such a system.

5.4.4 Recycle Subframe Synchronization

A subframe synchronizing word is inserted in a fixed row and column. That is, a single data cell is dedicated to a subframe synchronizing word. The same cell is used in every major frame. Figure 5.11 shows such a synchronizing format. This technique requires more time to obtain complete synchronization than the ID sync method, since the subframe sync process starts only after minor frame sync has occurred. The subframe sync process must also go through the search, check, and lock procedure, and the output data is

SC	1	2	3	4-1	5	6	7	8
SC	1			4-2				
SC	1			4-3				
SC	1			4-4				
SC	1			4-5				
SC	1			4-6				
SC	1			4-7				

Figure 5.10 Subframe sync by minor frame sync word complement.

SC	1	2	sfs	4-1	5	6	7	8
SC	1			4-2				
SC	1			4-3				
SC	1			4-4				
SC	1			4-5				
SC	1			4-6				
SC	1			4-7				

Figure 5.11 Subframe sync by recycle.

not considered valid until lock occurs. However, if the data has been recorded, playback techniques may be used to recover the data before sync was established. This type of subframe sync uses fewer data spaces than the ID sync method.

5.4.5 Identification Subframe Synchronization

In ID subframe synchronization, a counting sequence is inserted into a data column, as shown in Figure 5.12. As soon as minor frame sync is accomplished, the decommutator knows immediately which row the system is working on. With respect to the subframe word sync method, the ID method is much faster, but uses a data column.

SC	1	2	000	4-1	5	6	7	8
SC	1		001	4-2				
SC	1		010	4-3				
SC	1		011	4-4				
SC	1		100	4-5				
SC	1		101	4-6				
SC	1		111	4-7				

Figure 5.12 Subframe sync by identification word.

5.5 IRIG Specification Overview

Two classes of PCM formats are covered in IRIG 106-00 [3]: The basic and simpler type is type I, while the more complex is type II. Type I uses a fixed format. Use of type II requires concurrence of the range involved. Specifications of the two types are given in Table 5.1.

5.5.1 IRIG Fixed Format, Type I

Bit, word, and frame formats for type I given in IRIG 106-00 will be given next.

5.1.1.1 Bit Formats

The transmitted bit stream shall be continuous and shall contain adequate transitions to ensure bit acquisition and continued bit synchronization.

During any period of desired data, the bit rate shall not differ from the specified nominal bit rate by more than 0.1%.

The bit jitter shall not exceed ±0.1 bit interval referenced to the expected transition time with no jitter. The expected transition time shall be based on the measured average bit period as determined during the immediately preceding 1,000 bits.

The most significant bit shall be numbered 1.

5.5.1.2 Word Requirements

Words may vary from 4 bits to 16 bits. Words of different lengths may be multiplexed in a single minor frame, although the word length shall be

Table 5.1
PCM IRIG Formats

Type I	Range	Nominal
Word length	4–16 bits	8 bits
Maximum minor frame length	8,192 bits/512 words	1,600/200
Maximum major frame length	256 minor frames	200
Maximum bit rate	5 Mbps	1–2 Mb
Minimum bit rate	10 bps	
Type II		
Word length	4–64 bits	Fragmented words variable
Maximum minor frame length	16,384 bits/> 512	(variable)
Maximum bit rate	>5 Mbps	

constant. Fragmented words are not allowed by the type I format, and the word length in any position shall be constant.

The first word after the synchronization word shall be the number "one."

5.5.1.3 Frame Structure

The synchronization word shall be fixed and at least 16-bits long but not over 33 consecutive bits.

A major frame is defined as the number of minor frames necessary in order to include one sample of every parameter.

A subframe is one cycle of the parameters from a commutator whose rate is a submultiple of the minor frame rate.

The subframe counter or identifier is the standard method used.

The frame counter provides a natural binary count corresponding to the minor frame number in which the frame count word appears. It is recommended that such a counter be included in all minor frames whether type I, or type II and is especially desirable in a type II format to assist with data processing. In type I formats where subcommutation is present, the subframe counter can serve as the frame counter.

5.5.2 IRIG Type II

The type II format is characterized by changes with regard to frame structure, word length or location, commutation sequence, sample interval, and fragmented words. Format changes are inherently disruptive to test data processing, and fixed format methods are preferred. In cases in which there is necessity to vary the format, the methods shall conform to the characteristics described in IRIG 106-00.

Problems

Problem 5.1

Complete a drawing like Figure 5.1 for the coding formats shown for the bit sequence 11010010. Assume the first preceding bit is a 1.

Problem 5.2

Convert the FM/FM package of Problem 3.1 into a PCM system. Sample each channel at $5f_m$ and use 8-bit words. What is the bit rate if the synchronization overhead is 3% of the data bit rate?

Problem 5.3

Convert the FM/FM package of Problem 3.11 into a PCM system. Sample each channel at $5f_m$ and use 8-bit words. What is the bit rate if the synchronization overhead is 3% of the data bit rate?

Problem 5.4

Convert the FM/FM package of Problem 3.20 into a PCM system. Sample each channel at $5f_m$ and use 8-bit words. What is the bit rate if the synchronization overhead is 3% of the data bit rate?

Problem 5.5

Convert the FM/FM package of Problem 3.21 into a PCM system. Sample each channel at $5f_m$ and use 8-bit words. What is the bit rate if the synchronization overhead is 3% of the data bit rate?

Problem 5.6

Convert the FM/FM package of Problem 3.22 into a PCM system. Sample each channel at $5f_m$ and use 8-bit words. What is the bit rate if the synchronization overhead is 3% of the data bit rate?

Problem 5.7

Convert the FM/FM package of Problem 3.23 into a PCM system. Sample each channel at $5f_m$ and use 8-bit words. What is the bit rate if the synchronization overhead is 3% of the data bit rate?

References

[1]   Lindsay, W. C., and M. K. Simon, *Telecommunication Systems Engineering*, Englewood Cliffs, NJ: Prentice-Hall, 1973, pp. 20–22.

[2]   Roden, M. S., *Digital Communication System Design*, Englewood Cliffs, NJ: Prentice-Hall, 1988, pp. 262–270.

[3]   Secretariat, Range Commanders Council, *Telemetry Standards*, White Sands Missile Range, NM: RCC, IRIG 106-00.

6

PCM/FM

This section will discuss TDM-PCM/FM telemetry systems from end to end. The transmission end contains a data acquisition block, the multiplexer, the modulator, transmitter, and antenna. In digital telemetering systems, PCM/FM is one of the modulation formats used to convey a binary sequence. The transmitter is typically a VCO followed by a power amplifier and antenna.

Systems combining PCM/FM and FM/FM, referred to as hybrid systems, and the design process for combining the two will be examined in Chapter 16.

6.1 Learning Objectives

Upon completion of this chapter, the reader should understand the following:

- PCM/FM waveform and spectrum;
- PCM/FM bandwidth as a function of the modulation parameters;
- PCM/FM detection;
- Design of PCM/FM systems;
- Bit error rates for PCM/FM;
- Setting receiver parameters based upon modulation design parameters;
- Predicting the $[S/N]_O$ for the analog signal after the D/A conversion in a PCM/FM channel.

6.2 PCM/FM

In a telemetering system, the binary sequence is probably the result of sampling, digitizing, and combining the outputs from a number of different sensors resulting in a TDM, as shown in Figure 6.1. Output from a computer may also go into the bit sequence. The purpose of the commutator is to sample the output of the sensors and produce an amplitude-modulated pulse for each sample. The samples are applied to the encoder, which converts each sample into a digital word creating the binary sequence. The box designated as the overall encoder, which consists of the commutator, the sample-and-hold, and the encoder, is usually one piece of off-the-shelf hardware, which will handle up to 256 inputs and is referred to simply as the encoder. The binary sequence is applied to the transmitter, which consists of a VCO and power amplifier driving an antenna. Each binary level of the bit sequence causes a frequency deviation of either $f_c + \Delta f$ or $f_c - \Delta f$. Typically, in range PCM/FM, 8-bit words are used.

6.2.1 Time Waveform

In PCM/FM, in response to the PCM code, the carrier is switched between two frequencies by driving the VCO with two voltage levels representing a logic one and zero. Equations (6.1) and (6.2) represent the carrier, which remains at one of the frequencies for the duration of the bit period, T.

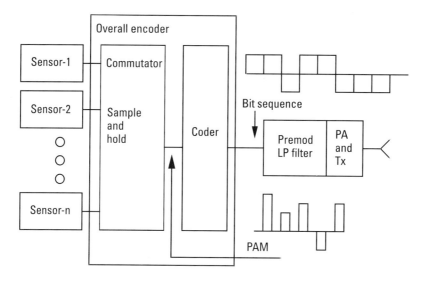

Figure 6.1 TDM PCM/FM system at the transmission end.

$$s_1(t) = A\cos(\omega_c t + \Delta\omega) \tag{6.1}$$

$$s_2(t) = A\cos(\omega_c t - \Delta\omega) \tag{6.2}$$

Conceptually, this modulation sequence may be thought of as two carriers at two different frequencies ASK, during different intervals, as depicted in Figure 6.2. Note that when the signals are generated by driving a VCO, a continuous waveform results, which is probably not the case in ASK.

6.2.2 PCM/FM Spectrum

Inasmuch as PCM/FM may be approximately represented as two ASK signals in the time domain, the spectrum of PCM/FM is similar to the two ASK spectrums under certain conditions. Specifically, and conceptually, whenever $(2\Delta f)T \gg 1$, the PCM/FM spectrum will tend toward two $\mathrm{sinc}^2 x$ distributions, each centered at $f_c + \Delta f$ and $f_c - \Delta f$, as shown in Figure 6.3(c). As this condition is relaxed, the two sinc^2 curves move toward each other, and the spectrum becomes just one pronounced lobe, as in Figure 6.3(a).

6.2.3 PCM/FM Bandwidth

The power spectral density for PCM/FM is difficult to derive for a truly random binary sequence of waveforms; however, a rough approximation of the required transmission bandwidth, B_t, is possible. For an exact expression of the power spectrum density of a PCM/FM sequence see [1] and Section 6.9 for a discussion of the actual spectrum. Bandwidth efficiency given in bits per second per hertz is an important parameter in digital communication systems and may also be estimated for PCM/FM. Carson's rule, which predicts the transmission bandwidth and hence the approximate IF bandwidth for FM analog systems, is

$$B_t = 2(\Delta f + f_m)$$

In digital systems, letting $f_m = R_b = 1/T$, where the first null of the $\mathrm{sinc}\, x$ spectrum occurs (where R_b is the bit rate and T is the bit period), and if

$$2\Delta f T > 1$$

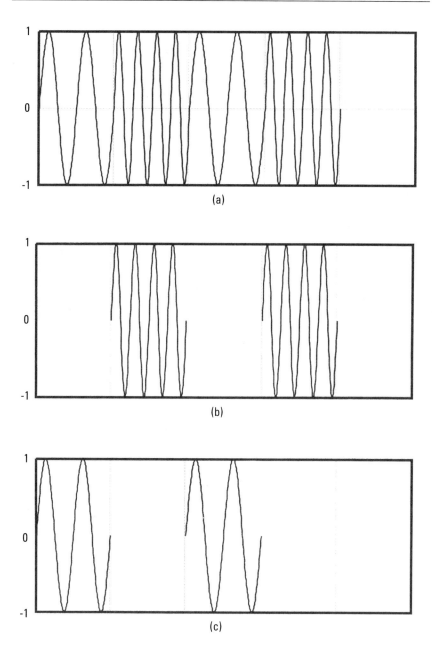

Figure 6.2 Waves showing the decomposition of (a) a PCM/FM wave into two ASK waveforms, (b) $s_1(t)$, and (c) $s_2(t)$.

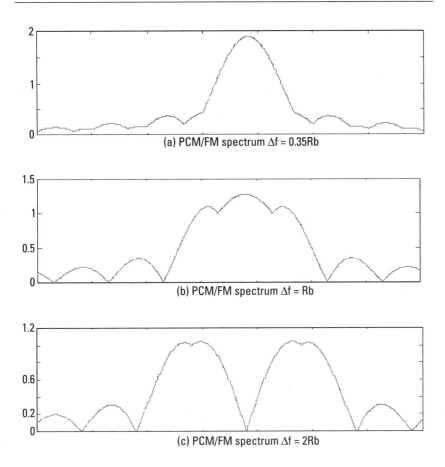

Figure 6.3 PCM/FM conceptual and approximate spectrum: (a) $0.35R_b$, (b) R_b, and (c) $2R_b$.

then,

$$B_t = 2(\Delta f + R_b) = 2(\Delta f + 1/T) \tag{6.3}$$

As is usually the case in PCM/FM, however

$$2\Delta f T < 1 \tag{6.4}$$

For this case, B_t will be approximately constrained by

$$2\Delta f < B_t < 2B_m$$

where B_m is the 3-dB frequency of the one-sided PSD of the modulating signal and is $R_b/2$ for PCM/FM. Hence,

$$2\Delta f < B_t < R_b \qquad (6.5)$$

For coherent FSK, it can be shown [2] that for a minimum BER, $\Delta f = 0.35R_b$. Further, it has also been determined experimentally and by simulations [3–8] that for a minimum BER for noncoherent detection of PCM/FM and for $B_{IF} = R_b$, that

$$\Delta f = 0.35R_b \qquad (6.6)$$

This value of R_b reduces (6.5) to

$$0.7R_b < B_t < R_b$$

It has also been determined experimentally at the receiver that $B_{IF} = R_b$ for a minimum BER for a given carrier-to-noise ratio [3–8]. Therefore, an approximation for PCM/FM bandwidth is

$$B_t = B_{IF} = R_b \qquad (6.7)$$

This estimation of bandwidth usually contains about 95% of the power [2]. It is common practice [9], to limit the spectrum of PCM/FM systems by passing the bit sequence (prior to modulation of the transmitter) through a filter with a 3-dB point between $R_b/2$ and R_b. This makes (6.7) an even better approximation for the transmission bandwidth of PCM/FM.

The transmission bandwidth, B_t, given by (6.7) is not the 99% bandwidth, B_{99}, but a good estimate. According to [9], the relationship between the two bandwidths when employing a premodulation fourth-order filter with a 3-dB corner frequency of $0.7R_b$, B_{99} is given by

$$B_{99} = 1.16R_b$$

Work by [10] and illustrated in Section 6.9 gives the relationship between B_t and B_{99} as a function of $\Delta f/R_b$ and for several premodulation filters. Usually on the ranges, the required transmission bandwidth is taken to be B_9, giving a spectral efficiency of 0.86 for PCM/FM. However, in this text for ease of mathematical representation, the transmission bandwidth is assumed to be R_b, while the spectral efficiency is taken to be 1.

6.3 PCM/FM Overview

An end-to-end block diagram of a PCM/FM system, starting with the generation of a bit sequence in the airborne equipment and ending with a detected bit sequence in the ground equipment, is shown in Figure 6.4.

6.3.1 PCM/FM Transmitter System

The bit sequence is usually the output from a TDM system. Prior to modulation of the transmitter, a linear phase premodulation filter is employed to filter the bit sequence. The purpose of the premodulation filter is to limit the RF transmission bandwidth. Since this lowpass filter causes an increased BER, its bandwidth should only be as small as necessary to meet the specifications for RF bandwidth. As the spectrum becomes more crowded and adjacent-channel interference more of a problem, there is a trend toward using three- to six-pole filters instead of first-order filters. The transmitter is usually operating at L-band or S-band.

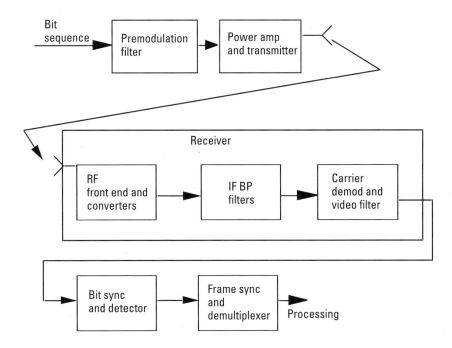

Figure 6.4 PCM/FM system.

6.3.2 PCM/FM Receiving System

The receiving system is composed of a receiver, bit detector and synchronizer, and frame synchronizer, as shown in Figure 6.4. At the ground station, the receiver must extract the bit sequence from the modulated carrier. An FM discriminator is normally used to noncoherently demodulate PCM/FM.

6.3.2.1 Receiver

The receiver comprises an RF amplifier in the front end and converters that convert the carrier down to an intermediate frequency, which usually occurs in two stages. The IF carrier, FM-modulated by the bit sequence, is demodulated by the carrier demodulator. It is common practice to refer to the demodulator as a carrier demodulator, although it actually demodulates the IF carrier. The demodulated output bit sequence is passed through a lowpass filter, which is referred to as the video output filter. The 3-dB point, B_v, of the video filter is normally set to a value between R_b and $2R_b$ [7]. That is,

$$B_v = R_b \text{ to } 2R_b \qquad (6.8)$$

For PCM/FM noncoherent demodulation, the hardware is much simpler than that required for coherent demodulation. In coherent demodulation, it is necessary to employ carrier-tracking loops, such as those used in PSK, which may lose lock, causing data dropout. For that reason, practically all RF PCM/FM systems in operation on government missile ranges are noncoherent PCM/FM. As the new receivers capable of supporting coherent systems are deployed on the ranges, there will certainly be more telemetry systems employing phase coherence. A noncoherent PCM/FM discriminator is shown in Figure 6.5 for demodulating a carrier whose instantaneous frequency is varying as a rectangle with no noise.

6.3.2.2 Bit Synchronizer

The bit sequence output from the receiver video filter is noisy and distorted as shown in Figure 6.6, and is the input to the bit synchronizer. The purpose of the bit synchronizer is to establish bit synchronization and during each bit period determine if a one or a zero was transmitted. The output of the bit synchronizer is a bit clock and a sequence of clean crisp ones and zeros as illustrated in Figure 6.6. Bit synchronizers currently employed will support bit rates up to 35 mbps.

Data-tracking loops are employed to establish and keep bit synchronization. The loop bandwidth is usually set to a value between $0.01R_b$ and

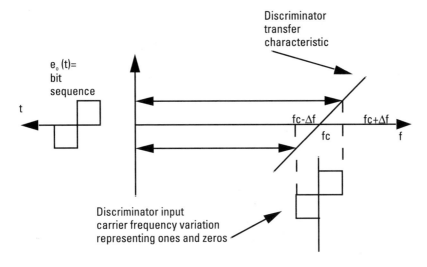

Figure 6.5 FM discriminator transfer characteristic showing the input and output when no noise is present.

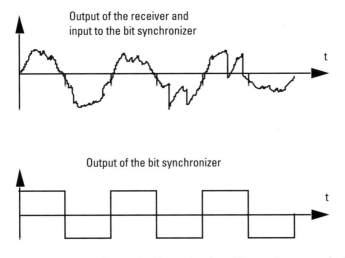

Figure 6.6 Input-output waveforms of a bit synchronizer. The receiver output is the input to the bit synchronizer.

$0.3R_b$. The ability to establish bit synchronism is a function of signal-to-noise ratios in the IF and the loop and frequency jitter and the difference between the actual bit rate and the set bit rate.

BERs for the commercial off-the-shelf bit synchronizers are advertised to be within 1 dB of theoretical. Signal-to-noise ratios, filter bandwidths, line coding and modulation indices are factors that determine the BER.

6.3.2.3 Frame Synchronizer

The purpose of the frame synchronizer is to lock onto the frame synchronizing word and establish frame sync. For an analysis of the probability of frame synchronization, see [11].

6.4 BER in a PCM/FM System

In a binary digital system, the quality of the system is usually judged by the BER. This is true for the PCM/FM system. The overall system is usually designed for a BER of 10^{-5} to 10^{-6}. For a BER of 10^{-6}, there is only one error, on the average, in a million bits. P_b and the term BER are used interchangeably in the literature; however, it should be understood that this is true if P_b is averaged over a large number of bit periods.

6.4.1 BER in Terms of System Parameters

Research has shown that a bit error is usually caused by the occurrence of an impulse [5, 6, 12, 13], as discussed in Chapter 2. When noise causes the carrier phase to jump by 360° in the IF, the FM discriminator demodulates this as an impulse. If it is a positive going impulse, and the bit is represented by a negative pulse, there is a high probability that the detecting circuit of the bit synchronizer will make a mistake. A number of research papers [5, 6, 12, 13] have equated the impulse rate to the BER with varying degrees of success. The relationship is extremely complex and depends on many factors, including the modulation, IF bandwidth, bit rate, type of IF filter, and type of demodulator. Regardless, whenever the receiver is operating several decibels above threshold, the BER is usually extremely small and is in the neighborhood of 10^{-6}. Since threshold usually occurs in the receiver whenever the carrier-to-noise drops below 12 dB, the link is normally designed such that the receiver is always operating above 12 dB.

For PCM/FM in general, a study of experimental or simulation results will show that a plot of the BER or the probability of a bit error as a function of E_b/N_o will follow an exponential curve [3, 5, 6, 7]. Such a fit was suggested by [14]. A general exponential curve is given by

$$p_b = 0.5e^{-kE_b/N_o} \tag{6.9}$$

where

E_b = bit energy;

N_o = one-sided power spectral density of the white noise going into the IF;

k varies between 0.5 and 0.9.

E_b is a variable that occurs in the theoretical analysis of digital systems, but it is carrier power, C, at the receiver that is measured or determined by link analysis and is a function of antenna gains, path distance, transmitted power, and losses. Further, it is noise power in the receiver, not N_o, that is determined from the noise figure of the receiving system. However, these parameters can be related as follows:

$$E_b = CT_b \tag{6.10}$$

and

$$N_p = \text{noise power in the IF} = N_o B_{IF} \tag{6.11}$$

where

C = carrier power at the receiver;

T_b = bit period = $1/R_b$;

B_{IF} = bandwidth of the IF.

Using these relationships, (6.9) may be written as

$$p_b = 0.5e^{-kCT_b/N_o} \tag{6.12}$$

$$p_b = 0.5e^{-kC/R_b N_o} \tag{6.13}$$

Since usually $B_{IF} = R_b$, (6.13) may be expressed as

$$p_b = 0.5e^{-kC/N_p} \tag{6.14}$$

Since C/N_p is the carrier-to-noise ratio in the IF, which is determined by link analysis, the BER may be predicted based on this engineering parameter.

Shifting the carrier between two frequencies may be achieved in a number of ways. Two ways will be discussed here. The shifting between two frequencies may be achieved by switching between two unsynchronized oscillators such that there is a discontinuity between the waveforms. The second method is to drive a VCO with a bit sequence represented by positive and negative rectangular pulses. The pulses are usually filtered prior to application to the VCO. This last method creates a continuous waveform. Almost always, PCM/FM is generated by the latter method.

The term PCM/FM in this text will always refer to the continuous waveform type of generation.

Theoretically, the bit error rate for the *noncontinuous* and noncoherently detected PCM/FM (k = 0.5) can be shown [15] to be

$$p_b = 0.5e^{-0.5E_b/N_o} \tag{6.15}$$

It has been determined by simulations and testing [3–8] that the BER for PCM/FM improves over that of (5.15). Curve fitting (k = 0.7) to the curves generated by simulations [4–7], gives an equation predicting the BER as

$$p_b = 0.5e^{-0.7E_b/N_o} \tag{6.16}$$

Noting that if $B_{IF} = R_b$, then $E_b/N_o = [S/N]_c$ in the IF. For the two cases and for a required P_b of 10^{-5}, these two equations give the needed carrier-to-noise ratios of

$$[S/N]_C(NCPCM/FM) = 13.3 \text{ dB}$$
$$[S/N]_C(PCM/FM) = 11.9 \text{ dB}$$

Equations (6.15) and (6.16) have been plotted in Figure 6.7 for the case $B_{IF} = R_b$ and with the parameter $[S/N]_c$ replaced by E_b/N_o, since the two are numerically equivalent. These curves show that for PCM/FM at E_b/N_o = 12 dB, a 1-dB increase in the carrier-to-noise ratio in the IF will decrease the BER to 10^{-6}. Further, inspection of Figure 6.7 shows that in the BER area of interest PCM/FM requires about 1.4 dB less transmitted power than discontinuous PCM/FM. In practice, as indicated earlier, almost all PCM/FM is generated by driving a VCO with a bit sequence; hence, the benefit of continuous PCM/FM with respect to the BER is inherent in almost all PCM/FM systems, a concept to be discussed in Chapter 8 as continuous-phase FSK.

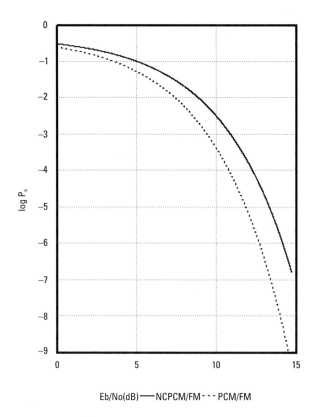

Eb/No(dB) ——NCPCM/FM - - - PCM/FM

Figure 6.7 BER curves for PCM/FM and noncontinuous PCM/FM.

6.4.2 Comparison of PCM/FM and Noncontinuous PCM/FM

PCM/FM, produced by driving a VCO with a bit sequence, requires approximately 1.4 dB less transmitted power for the same BER as noncontinuous PCM/FM generated by switching between two unsynchronized oscillators. It is interesting to look into the cause of this difference.

The IF filter, the FM discriminator, and the detection module operate as a single entity, a discriminating block, in establishing the BER along with the incoming waveform. PCM/FM has a continuous waveform, while the latter waveform is discontinuous. In the PCM/FM or continuous case, the waveform in the current bit period is a function of what occurred in a number of past bit periods. Since the discriminating block is composed of capacitors and inductors, the block has memory and inherently uses the fact of correlation in the waveforms over several bit periods to improve on the decision making process. Since in noncontinuous PCM/FM there is no correlation between bit periods, the memory of the discriminating block provides no advantage.

A discriminating block operating explicitly over a fixed number of bit periods has been suggested [16] and shown to give a reduced BER. In order to compare the three cases, the BER as a function of E_b/N_o, for noncontinuous PCM/FM, PCM/FM, and a 3-bit-wide decision module operating on PCM/FM has been plotted [4]. PCM/FM is approximately 1.5 dB better than the noncontinuous case, while the discriminator block operating with a 3-bit-wide window, is about 1.5 dB better than PCM/FM.

6.4.3 BER As a Function of Bit Rate

Inspection of (6.14) shows that the BER is determined primarily by the exponential term given by

$$\text{system parameters of the exponential term} = CT_b/N_o \qquad (6.17)$$

The larger this ratio, the smaller the BER. If, after link analysis, this term is too small, and if the antenna size, transmitted power, and so on cannot be increased and the receiver cannot be changed, then the only other parameter available to be changed is the bit rate. Since T_b is the reciprocal of R_b, decreasing the bit rate increases T_b. Therefore, for a given received carrier power and receiver and a specified BER, there is a maximum bit rate the system will support.

6.5 Bit Synchronizer Performance

Two tasks of the bit synchronizer are to (1) establish bit synchronizer and output a bit clock and (2) during each bit period make a decision as to whether a one or a zero was transmitted and output a clean bit sequence. The detection module is usually a matched filter followed by a sample and hold (S/H). The output voltage point of the S/H is applied to a voltage comparator, which makes the decision as to whether the voltage is less than or greater than a set value, thereby deciding if a one or a zero was transmitted.

The signal-to-noise input to the bit synchronizer determines the performance of the synchronizer in that it is an indicator of where the receiver is operating with respect to threshold. The signal-to-noise output from the receiver video output filter is the input signal-to-noise of the bit synchronizer.

The process of determining this input signal-to-noise will be discussed next. Although (2.20) is the signal-to-noise improvement of single-tone FM, it may be modified for PCM/FM and used to determine the signal-to-noise out of the video filter. This equation can be used since in PCM/FM the bit sequence that is smoothed by the premodulation filter drives a VCO, creating a continuous waveform. The highest frequency of the modulating signal is taken to be the 3-dB corner frequency of this filter. This equation is also appropriate since the receiver is composed of the carrier IF followed by an FM discriminator, which is followed by a lowpass filter, the video filter. Repeating (2.20)

$$[S/N]_o = 3D^2 \frac{B_{IF}}{2f_m} [S/N]_i = 3 \left[\frac{\Delta f}{f_m} \right]^2 \frac{B_{IF}}{2f_m} [S/N]_i \qquad (6.18)$$

Equation (6.18) was developed assuming that the 3-dB point of the postdetection filter, that is, the video filter, was set at the 3-dB frequency of the modulation, f_m. This may or may not be the case. The f_m term in (6.18) should always be set to the 3-dB corner frequency of the video filter of the receiver or the cutoff frequency of the parabolic noise, which is $B_{IF}/2$, whichever is smaller. [See Section 2.6.2, (2.40) and Figure 2.10(c).] Normally, $B_{IF}/2$ is equal to $R_b/2$ and hence will be less than the 3-dB point of the video lowpass filter. Making this substitution gives $[S/N]_o$ of the video filter as

$$[S/N]_o = 3 \left[\frac{\Delta f}{R_b/2} \right]^2 \frac{B_{IF}}{R_b} [S/N]_i \qquad (6.19)$$

where

Δf = peak frequency deviation;

R_b = bit rate;

$R_b/2$ = cutoff of the parabolic noise;

$[S/N]_i = [S/N]_c$ = carrier-to-noise ratio in the IF;

$[S/N]_o$ = signal-to-noise ratio in the output lowpass video filter and in PCM/FM the signal-to-noise going into the bit synchronizer.

Equation (6.19) may be written in numerical or voltage values as

$$[S/N]_o = \frac{\Delta f}{R_b/2} \sqrt{\frac{3B_{IF}}{R_b}} [S/N]_i \qquad (6.20)$$

For a minimum BER, the system design parameters in terms of the bit rate are from (6.6), (6.7), and (6.8)

$$\Delta f = 0.35R_b \qquad (6.21)$$

$$B_v = R_b \qquad (6.22)$$

$$B_{IF} = R_b \qquad (6.23)$$

A link is designed such that the carrier-to-noise ratio, $[S/N]_i$, in the IF is equal to or greater than 12 dB or 3.98 numerically. Substituting the parameter values of (6.21) and (6.23) and $[S/N]_i$ of 12 dB into (6.20) gives

$$[S/N]_o = \frac{0.35R_b}{0.5R_b} \frac{\sqrt{3R_b}}{\sqrt{2R_b}} [S/N]_i \qquad (6.24)$$

$$= (0.7)(1.73)(3.98) = 4.82 \ (13.6 \ dB)$$

Therefore, for a minimum value of 12 dB for the carrier-to-noise ratio in the IF, the output signal-to-noise ratio of the video filter or the input signal-to-noise ratio of the bit synchronizer will be 13.6 dB. For a slight design margin, it is not uncommon to require this ratio to be 15 dB [17]. If this is required and since there are no other parameters to vary, the link must be designed for a $[S/N]_i$ of 13.3 dB (4.64 numerically), which will give the required $[S/N]_o$ of 15 dB out of the video filter and into the bit synchronizer.

PCM/FM Example 1:

$$R_b = 1 \ mb/s = 1 \times 10^6 \ mb/s$$

The system parameters should be selected as follows:
Transmitter:

$f_3 = 0.7$ mHz = premodulation lowpass filter 3-dB frequency;
$\Delta f = 0.35R_b = 0.35$ mHz = peak carrier frequency deviation;

Receiver:

$B_{IF} = R_b = 1$ MHz;

$B_v = R_b = $ video filter 3-dB frequency.

For example, for an S-band transmitter at, say, a center frequency of 2,250.5 MHz, the upper frequency would be 2,250.85 MHz and the lower 2,250.15 MHz.

In setting up the overall link, the carrier power, antenna gains, and distance should be such that the $[S/N]_i = 13.3$ dB if it is desired to have a signal-to-noise ratio of 15 dB into the bit synchronizer.

6.6 PCM/FM System Design

In PCM/FM, system design is concerned with using the minimum transmission bandwidth and obtaining a minimum BER. Based upon a number of published works [3–8], the system parameters normally used to achieve these ends will be given next.

6.6.1 Setting the PCM/FM Transmission and Receiver System Operating Parameters in Terms of the Bit Rate for NRZ

6.6.1.1 Setting the Peak Transmitter Deviation

For an optimum BER, the peak deviation of the transmitter is chosen to be $0.35R_b$, that is,

$$\Delta f = 0.35R_b \qquad (6.25)$$

6.1.1.2 Setting the Premodulation Filter Bandwidth

Typically, the 3-dB bandwidth of this filter will be between 0.5 and 1.2 of the bit rate. The corner frequency is usually set close to $0.7R_b$. Further, for this type of modulation format, the filter is nominally a three-pole to six-pole linear filter; that is, the premodulation filter corner frequency, f_{3dB} (pm), is

$$f_m = f_{3dB} = 0.5 \text{ to } 1.2R_b \qquad (6.26)$$

This equation gives the bonds on the corner frequency but is usually close to $0.7R_b$.

6.1.1.3 Setting the Receiver IF Bandpass Filters

The bandwidth of the IF bandpass filter is usually chosen to be equal to the bit rate, R_b. However, parameter variations and the impact of Doppler may require a slightly larger IF, that is,

$$B_{IF} = R_b \text{ to } 1.5R_b \qquad (6.27)$$

6.1.1.4 Setting the Receiver Video Output Filter

The 3-dB point of the video filter, B_v, is set to a value between R_b and $2R_b$; that is,

$$B_v = f_{3dB} \text{ (video out)} = 1R_b \text{ to } 2R_b \qquad (6.28)$$

Extensive laboratory test results at the Pacific Missile Test Center [7, 10] illustrate specific empirical relationships between modulation parameters, transmission bandwidth, and the BER for PCM/FM.

6.7 Design of PCM/FM for BiΦ-L (Manchester)

Figure 5.2 showed the spectrum of NRZ-L and BiΦ-L, and gave the advantage of NRZ-L with respect to bandwidth and the advantage of BiΦ with respect to coupling and synchronization. The advantages of BiΦ are that it is self-clocking with transitions every bit period, and that because there is not DC, AC coupling may be used.

The disadvantages of BiΦ-L are that it requires twice the bandwidth of NRZ and 3 dB more received power than NRZ, since the IF bandwidth will be at least twice as large.

6.7.1 BER

The BER of BiΦ is closely related to the impulse rate in the IF; hence, the signal-to-noise ratio in the IF must be large enough to keep the receiver out of threshold. Threshold usually occurs in the neighborhood of 12 dB. To compare the probability of a bit error of BiΦ and NRZ, the probability will be given in terms of system parameters.

In general, the probability of a bit error is given by

$$p_b = 0.5e^{-kE_b/N_o} \qquad (6.29)$$

Expressing E_b/N_o in terms of carrier power gives

$$\frac{E_b}{N_o} = \frac{CT_b}{N_o} = \frac{C}{N_o R_b} \qquad (6.30)$$

To convert this expression into carrier-to-noise ratio, it is necessary to express R_b in terms of the IF bandwidth. Since the bandwidth requirement for BiΦ is at least $2R_b$, the IF bandwidth is given by

$$B_{IF} = 2R_b$$

The noise power in the IF is given by

$$N_p = N_o B_{IF} = N_o 2R_b \qquad (6.31)$$

Solving (6.31) for R_b and substituting into (6.30) gives

$$\frac{E_b}{N_o} = 2\frac{C}{N_p} \qquad (6.32)$$

Equation (6.31) shows that for BiΦ the noise power in the IF will be twice the noise power for NRZ for the same bit rate since B_{IF} will be twice as large. Therefore, to achieve the same carrier-to-noise ratio, the carrier power must be 3 dB greater. Since in NRZ the IF bandwidth is equal to R_b, E_b/N_o is numerically the same as the carrier-to-noise ratio; whereas in BiΦ, E_b/N_o will be 3 dB greater, as shown by (6.32). That is, if the carrier-to-noise ratio is kept at 12 dB for both NRZ and BiΦ, E_b/N_o will be 12 dB and 15 dB, respectively. Published results of simulations and analysis [8, 18] indicate that if the carrier-to-noise ratio is kept the same for the two cases, the BER will be approximately the same. It should be noted again that since the IF will be twice as wide in BiΦ the received carrier power must be 3 dB greater than for NRZ.

6.7.2 Setting System Parameters for a Minimum BER for BiΦ

A number of research papers have been published on setting system parameters for a minimum BER [8, 18, 19]. These results will now be given.

6.7.2.1 Setting the Peak Transmitter Deviation

For a minimum bit error rate [8, 18], the peak deviation of the transmitter is

$$\Delta f = 0.65 R_b \qquad (6.33)$$

6.7.2.2 Setting the Premodulation Filter Bandwidth

Typically, the 3-dB bandwidth of this filter will be between

$$f_m = f_{3dB} \text{ (pm)} = 1.4 \text{ to } 2.4R_b \qquad (6.34)$$

6.7.2.3 Setting the Receiver IF Bandpass Filters

The bandwidth of the IF bandpass filter is usually chosen to be

$$B_{IF} = 1.5 \text{ to } 4R_b \qquad (6.35)$$

6.7.2.4 Setting the Receiver Video Output Filter

The 3-dB point of the video filter, B_v, is set to a value of

$$B_v = f_{3dB} \text{ (video out)} = 1.8R_b \text{ to } 3R_b \qquad (6.36)$$

6.8 Signal-to-Noise Ratio from PCM, Including Quantization and Bit Error Noise

From Chapter 4, the variance due to the quantization error is given by

$$\overline{q(t)^2} = \frac{m_p^2}{3(2^{2n})} \qquad (6.37)$$

The variance due to bit errors is given by [19];

$$\overline{b(t)^2} = \frac{4P_e(2^{2n} - 1)m_p^2}{3(2^{2n})} \qquad (6.38)$$

where

P_e = probability of a bit error;

n = bit word size;

m_p = peak voltage of message encoded.

Since the two errors are independent, the total error power is the sum of the two,

$$\overline{e(t)^2} = \overline{q(t)^2} + \overline{b(t)^2} \tag{6.39}$$

The signal-to-noise ratio is

$$[S/N]_o = \frac{\overline{m(t)^2}}{\overline{e(t)^2}} \tag{6.40}$$

$$[S/N]_o = \frac{\overline{m(t)^2}}{\dfrac{m_p^2}{3(2^{2n})} + \dfrac{4m_p^2 P_e(2^{2n} - 1)}{3(2^{2n})}} \tag{6.41}$$

$$[S/N]_o = \frac{3(2^{2n})}{1 + 4P_e(2^{2n} - 1)} \frac{\overline{m(t)^2}}{m_p^2} \tag{6.42}$$

Note that $\overline{m(t)^2}/m_p^2$ is the RMS power to peak value squared of the message and is 1/2 for a sine wave.

For noncontinuous PCM/FM, the probability of a bit error is given by

$$P_e = 0.5e^{-0.5E_b/N_o} \tag{6.43}$$

But, $E_b = CT_b = C/R_b$, and $N_p = N_o R_b$. Then

$$P_e = 0.5e^{-0.5C/N_p} \tag{6.44}$$

Substituting (6.44) into (6.42) gives

$$[S/N]_o = \frac{3(2^{2n})}{1 + 2e^{-C/2N_p}(2^{2n} - 1)} \frac{\overline{m(t)^2}}{m_p^2} \tag{6.45}$$

Equation (6.45) is shown plotted in Figure 6.8 for $m(t)$ equal to a sine wave, and hence the signal power to peak value squared is 1/2. The interpretation of Figure 6.8 is that in the low signal-to-noise region $[S/N]_o$ is determined primarily by the BER; however, as this component becomes small, the quantization noise, which is not a function of the carrier-to-noise power, saturates the equation and $[S/N]_o$ becomes a constant. Increasing the carrier power will not increase the $[S/N]_o$ once this point is reached.

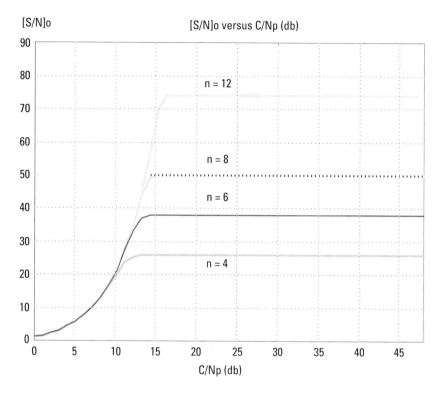

Figure 6.8 $[S/N]_o$ for PCM/FM as a function of both quantization noise and bit errors.

Note that while the equation used was for noncontinuous PCM/FM, which will indicate a slightly smaller signal-to-noise output initially, the best signal-to-noise that may be achieved is determined by the quantization noise, and this limit will be the same for both types of PCM/FM.

6.9 Actual PCM/FM Spectrum

The actual spectrum of PCM/FM resulting from modulating a NRZ-L bit sequence onto a carrier using a VCO and creating a continuous waveform is given by [20]

$$S(f) = \frac{4}{R_b} \left[\frac{D}{\pi(D^2 - X^2)} \right]^2 \left[\frac{(\cos \pi D - \cos \pi X)^2}{(1 + \cos^2 \pi D) - 2 \cos \pi D \cos \pi X} \right]$$

$$(6.46)$$

where

$S(f)$ = PCM/FM spectrum;

R_b = bit rate;

$D = 2\Delta f/R_b$;

$X = 2(f - f_c)/R_b$;

Δf = peak deviation;

f_c = carrier frequency.

Taking 10 log of (6.46) and plotting gives Figure 6.9. This figure illustrates the PCM/FM spectrum for $\Delta f/R_b$ = 0.25, 0.35, and 0.45.

It is important to know functionally in terms of modulation parameters at what frequency these nulls occur, since they are an indicator of required bandwidth. The nulls in the plot result from the numerator of the third term going to zero. The first zero of this term cancels out the zero in the denominator of the second term, so that the second zero of the third term causes the first null in the spectral plot of Figure 6.9. The denominator of the third term has no zeros, because for parameters of interest the term in parentheses is always larger than $2 \cos \pi D$.

The zero of the denominator of the second term occurs when

$$X = \pm D \qquad (6.47)$$

In terms of frequencies and deviation this is

$$\frac{2(f - f_c)}{R_b} = \pm \frac{2\Delta f}{R_b} \qquad (6.48)$$

Solving for the frequency at which this zero occurs gives

$$f = f_c \pm \Delta f \qquad (6.49)$$

The first zero occurs in the numerator of the third term whenever

$$\cos \pi D = \cos \pi X \qquad (6.50)$$

or

$$\pi X = \pi D \qquad (6.51)$$

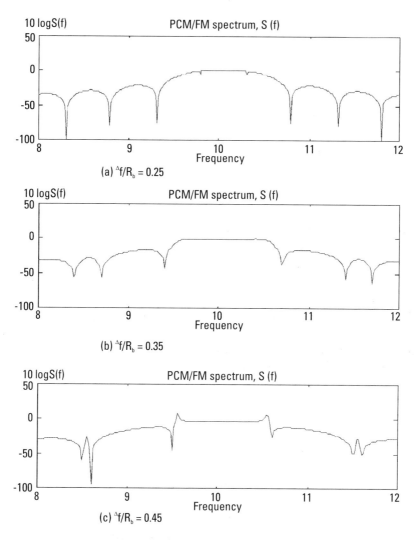

Figure 6.9 PCM/FM spectrum for a $\Delta f/R_b$ of (a) 0.25, (b) 0.35, and (c) 0.45R_b = 1bps and f_c = 10Hz.

This occurs at the same frequency as indicated in (6.50); hence, the pole of the second term cancels the zero of the third term and there is no null. There are no more poles in the second term; therefore, the other zeros in the third term lead to nulls in the spectrum. The general expression for zeros in the third term or nulls in frequency is

$$\pi X = \pi - \pi D + n\pi \qquad n = 1, 3, 5, \ldots \qquad (6.52)$$

$$\pi X = \pi D + n2\pi \qquad n = 0, 2, 4, \ldots \qquad (6.53)$$

For $n = 0$, the pole and zero cancel as indicated in (6.49) and (6.50), and there is no null.

The first null occurs at $n = 1$ and is of interest since this is indicator of spectral occupancy. For this case, and if $\Delta f = 0.35R_b$, in order to have a minimum BER, then (6.52) becomes

$$\frac{2\pi(f - f_c)}{R_b} = 2\pi - \frac{\pi 2\Delta f}{R_b} \qquad (6.54)$$

Solving for the frequency, f, where the null occurs gives

$$f = f_c - \Delta f + R_b \qquad (6.55)$$

Since the zeros of the numerator will also result from negative angles, the null below the carrier will occur at a frequency of

$$f = f_c + \Delta f - R_b \qquad (6.56)$$

The difference between these two frequencies is the null-to-null bandwidth, BW_{null}, and for $\Delta f = 0.35R_b$, this bandwidth is

$$BW_{null} = 1.2R_b \qquad (6.57)$$

Therefore, an estimate of transmission bandwidth, B_t, defined by the interval from 3-dB point to 3-dB point on the spectrum is

$$B_t = R_b \qquad (6.58)$$

Recent work [10] has determined the 99% bandwidth in the laboratory as a function of the peak frequency deviation. From a continuous plot in [10], a piecewise linear plot of the 99% bandwidth, B_{99}, as a function of the peak deviation, was made and is illustrated in Figure 6.10. From Figure 6.10, for a fourth-order premodulation filter and a 3-dB point equal to $0.5\,R_b$, B_{99} is given by

$$B_{99} = 1.1R_b \qquad (6.59)$$

Clearly, B_t, BW_{null}, and B_{99} are constrained to a narrow range around the spectral null. Notice from Figure 6.10 that B_t, when defined to be B_{99},

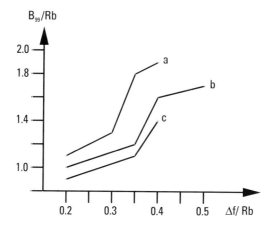

Figure 6.10 B_{99}/R_b plotted as a function of $\Delta f/R_b$ for (a) no premod filter, (b) premod filter, 3 dB = $0.7R_b$, and (c) premod filter, 3 dB = $0.5R_b$. Filter is four-pole and linear phase.

will increase rapidly as Δf is increased above $0.35R_b$. Further, (6.59) is in close agreement with IRIG 106-00 [9], which lists the 99% bandwidth as 1.16 R_b under the same circumstances.

Problems

Problem 6.1

(a) Draw a block diagram of a TDM system with four sensors. (b) If the highest frequency component in each sensor is 1,000 Hz, what is the practical sampling rate? (c) If each sample is converted into an 8-bit word, what is the system bit rate?

Problem 6.2

The bit sequence from a TDM system, such as the one in Problem 6.1, but with a bit rate of 5 Mbps, is fed into a PCM-FM system. (a) What should be the 3-dB point of the premodulation filter? (b) What should be the peak frequency deviation of the transmitter for a minimum BER? (c) What should the receiver IF bandwidth be set to? (d) What should the video filter 3-dB point be set to?

Problem 6.3

If $[S/N]_i$ = 15 dB, what is $[S/N]_o$ for the parameters of the PCM/FM system in Problem 6.2?

Problem 6.4

Discuss (6.49) and how the ratio $\dfrac{\overline{m(t)^2}}{m_p^2}$ affects the plot of Figure 6.9.

Problem 6.5

Plot (6.49) for $m(t) = \cos \omega t$ and n = 16. At what point in the carrier-to-noise ratio is it no longer productive to increase carrier power?

References

[1]   Bennet, R. B., and S. O. Rice, "Spectral Density and Autocorrelation Functions Associated with Binary Frequency-Shift Keying," *Bell Systems Tech. J.*, Vol. 42, Sept. 1963, pp. 2355–2385.

[2]   Stremler, F. G., *Introduction to Communication Systems*, 3rd Ed., Reading, MA: Addison-Wesley, 1990, p. 610, Figure 10.9.

[3]   Carden, F., and S. Ara, "PCM/FM + FM/FM Bit Error Rate Determination by Modeling and Simulation," *ITC Proceedings* Vol. XXIX, 1993, pp. 605–613.

[4]   Osborne, W., and D. Whiteman, *Optimizing PCM/FM + FM/FM Systems Using IRIG Constant Bandwidth Channels*, ECE Tech Report Series No. 92-014, Las Cruces, NM: New Mexico State University, Nov. 1992.

[5]   Carden, F., and S. Ara, *BER Determination of PCM/FM + FM/FM Systems and Coded PCM/FM Systems*, ECE Tech Report Series, No. 93-005, Las Cruces, NM: New Mexico State University, Nov. 1992.

[6]   Tjhung, T. T., and P. H. Wittke, "Carrier Transmission of Binary Data in a Restricted Band," *IEEE Transactions on Communication Technology*, Vol. Com-18, No. 4, Aug. 1970, pp. 295–304.

[7]   Law, E. L., and D. R. Hust, *Telemetry Applications Handbook*, Tech. Pub. TP000044, Weapons Instrumentation Division, Pacific Missile Test Center, Point Mugu, CA.

[8]   Cartier, D. E., "Limiter-Discriminator Detection Performance of Manchester and NRZ Coded FSK," *IEEE Trans. Aerosp. Electronics Syst.*, Vol. AES-13, Jan. 1977, pp. 62–70.

[9]   Secretariat, Range Commanders Council, *Telemetry Standards*, White Sands Missile Range, NM: RCC, IRIG 106-93. Jan. 1993, pp. A12–A14.

[10] Law, E. L., "Binary PCM/FM, Tradeoffs Between Spectral Occupancy and Bit Error Probability," *ITC Proc.*, Vol. XXX, 1994, pp. 347–355.

[11]   Roden, M. S., *Digital Communication System Design,* Englewood Cliffs, NJ: Prentice-Hall, 1988, pp. 262–270.

[12]   Mazo, J. E., and J. Salz, "Theory of Error Rates for Digital FM," *Bell Sys. Tech J.,* Vol. 45, Nov. 1966, pp. 1511–1535.

[13]   Schilling, D. L et al., "Error Rates for Digital Signals Demodulated by an FM Discriminator," *IEEE Trans. Comm. Tech.* Vol. COM-15, Aug. 1967, pp. 507–517.

[14]   Schwartz, M., W. Bennet, and S. Stein, *Communication Systems and Technology,* New York: McGraw-Hill, 1966, p. 339.

[15]   Haykin, S., *Digital Communications,* New York: Wiley, 1988, pp. 300–306.

[16]   Osborne, W. P., and M. B. Luntz, "Coherent and Noncoherent Detection of CPFSK," *IEEE Trans Comm,* Vol. Com-22, Aug. 1970, pp. 1023–1036.

[17]   Rosen, C., "System Transmission Parameters Design for Threshold Performance," *Proc. Int. Telemetering Conf,* Vol. XIX, 1988, pp. 145–182.

[18]   Tan, C. H., T. T. Tjhung, and H. Singh, "Performance of Narrow-Band Manchester Coded FSK with Discriminator Detection," *IEEE Transactions on Communications,* Vol. COM-31, May 1983, pp. 659–667.

[19]   Lathi, B. P., *Modern Digital and Analog Communication Systems,* 2nd Ed, New York: Holt, Rinehart, and Winston, 1989, p. 403.

[20]   Pelchat, M. G., "The Autocorrelation Function and Power Spectrum of PCM/FM with Random Binary Modulating Waveforms," *IEEE Trans,* Vol. SET-10, No. 1, March 1964, pp. 39–44.

7

Binary Phase-Shift Keying

Digital modulation is usually the process of modulating baseband digital signals onto a carrier wave, usually a sinusoid. This process converts a baseband signal into a bandpass signal suitable for RF or microwave transmission. At the receiver, the reverse process—demodulation—must be accomplished. Detection is the process of deciding which symbol was transmitted. While the two terms are used interchangeably, demodulation refers specifically to the removal of the carrier.

When signals are available in the receiver, which are replicas of the received symbols, particularly in phase, and are used for detection, the receiver is a coherent receiver. Typically, the receiver is phase locked to the carrier. Pulse code modulation/phase modulation or binary phase-shift keying requires a coherent receiver since the information is transmitted as two phases of the carrier. Noncoherent receivers perform demodulation without knowledge of the phase of the received signal and could not extract the information of binary phase-shift keying from the carrier. However, if the bit sequence is differentially encoded prior to transmission, then the bit sequence may be detected by a noncoherent receiver by multiplying the received carrier by a 1-bit delayed replica of the received carrier waveform.

7.1 Learning Objectives

Upon completion of this chapter, the student should know the following:

- Binary phase-shift keying waveform for signals 180° out of phase;
- Relationship between bit energy, bit period, and carrier power;
- Binary phase-shift keying modulator;
- Correlation receiver;
- How each functional block is implemented in hardware for the correlation receiver and the output waveform for each block;
- How maximum likelihood detection is achieved with hardware;
- Bit error prediction;
- General binary phase-shift keying and how the bit error rate is changed;
- Receiving hardware such as the bit synchronizer;
- Comparison of PCM/FM and binary phase-shift keying, and phase modulation and FM;
- Binary phase-shift keying spectrum.

7.2 Binary Phase-Shift Keying Model

When the ones and zeros of the bit sequence are modulated onto a carrier such that a 1 is represented by a carrier at 0° phase and a 0 by a 180° phase shift, the system modulation process is referred to as *binary phase-shift keying* (BPSK). In a pulse code modulation/phase modulation (PCM/PM) or BPSK system the probability of making wrong decisions, referred to as the BER, plays a central role in system evaluation. Prior to discussing the BER, the carrier representation of the ones and zeros for a particular modulation format will be discussed next.

The symbols, $s_1(t)$ and $s_2(t)$ representing the ones and zeros respectively, are then,

$$s_1(t) = A \cos 2\pi f_c t \qquad (7.1)$$

$$s_2(t) = -A \cos 2\pi f_c t \qquad (7.2)$$

The carrier waveforms or symbols representing $s_1(t)$ and $s_2(t)$ are shown in Figure 7.1 and are exactly 180° out of phase. Three complete carrier cycles occur during one bit period, T_b. The carrier power, C, during a bit period is

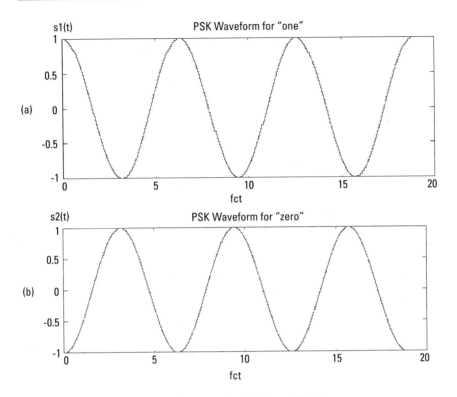

Figure 7.1 BPSK carrier symbols representing (a) 1s, and (b) 0s.

$$C = A^2/2 \text{ W}$$

The energy of one bit during this period is

$$E_b = CT_b = (A^2/2)T_b J \qquad (7.3)$$

Solving (7.3) for A gives

$$A = \sqrt{\frac{2E_b}{T_b}} \qquad (7.4)$$

Substituting this value of A into (7.1) and (7.2) gives

$$s_1(t) = \sqrt{\frac{2E_b}{T_b}} \cos 2\pi f_c t \qquad (7.5)$$

$$s_2(t) = -\sqrt{\frac{2E_b}{T_b}} \cos 2\pi f_c t \qquad (7.6)$$

Separating $\sqrt{E_b}$, gives

$$s_1(t) = \sqrt{E_b} \sqrt{\frac{2}{T_b}} \cos 2\pi f_c t \qquad (7.7)$$

$$s_2(t) = -\sqrt{E_b} \sqrt{\frac{2}{T_b}} \cos 2\pi f_c t \qquad (7.8)$$

Equations (7.7) and (7.8) correspond to transmitting a one and zero, respectively.

7.3 BPSK Generation

This form of BPSK is referred to as phase-reversal keying since the two carrier signals representing the logic ones and zeros are exactly 180° out of phase. A more general form of BPSK occurs when the phase difference between the two signals is other than 180°. This creates a residual carrier term that allows carrier tracking by a phase-lock loop (which will be discussed in a later section). Unless stated otherwise, BPSK will refer to the 180° mode. A method of generating BPSK is shown in Figure 7.2. A bit sequence represented by $\pm A$ is applied to a balanced modulator, resulting in an output of $\pm A \cos \omega_c t$, which is BPSK.

7.4 BPSK Detection by a Correlation Receiver

When binary PSK modulation is used at the transmission end, a receiver employing coherent demodulation or detection must be employed since the

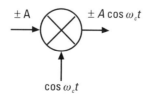

Figure 7.2 BPSK modulator.

information is contained in the carrier phase. A correlation receiver performs coherent demodulation and detection. Correlation, $C(t)$, of two signals, $r(t)$ and $s(t)$, over a period, T, is defined mathematically as

$$C(t) = \int_0^t r(t)s(t)dt \qquad 0 < t < T$$

Correlation is implemented in hardware by a multiplier and an integrator, as shown in Figure 7.3.

A BPSK correlation receiver is shown in Figure 7.4, with each block of hardware labeled with its functional purpose. The correlation receiver is so called because it correlates the received signal composed of signal plus noise with a replica of the signal. For the correlation to be achieved, it is necessary for the receiver to be phase locked to the carrier as discussed earlier.

The purpose of the correlation receiver is to reduce the received symbol to a single point or statistic that will be used by the decision device to determine which symbol was transmitted. In practice, this single point is a fixed voltage obtained by an S/H device. The decision device is a voltage

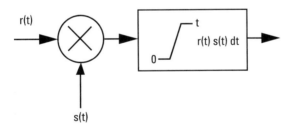

Figure 7.3 Hardware implementation of correlation.

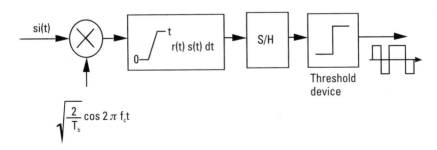

Figure 7.4 BPSK correlation receiver.

comparator that is set such that if the voltage point is above a certain level, the comparator indicates a one is received; if the voltage is below this level, a received zero is indicated. The case for no noise will be treated first.

7.4.1   No Noise

An exact replica of the carrier multiplies the received symbol, and the output of the multiplier is applied to an integrator. The output of the correlation detector (multiplier and integrator) is sampled and applied to a decision device, as shown in Figure 7.4. The decision device has a threshold and decides which symbol was received based upon whether the voltage point is above or below the threshold.

The mathematical function of each hardware block will now be discussed. The output of the multiplier is given by

$$
\begin{aligned}
e_m(t) &= \pm\sqrt{E_b}\,\sqrt{\frac{2}{T_b}}\,\cos 2\pi f_c t \times \sqrt{\frac{2}{T_b}}\,\cos 2\pi f_c t \\
&= \pm\sqrt{E_b}\,\sqrt{\frac{2}{T_b}}\,\cos^2 2\pi f_c t \qquad (7.9)\\
&= \pm\sqrt{E_b}\,\frac{2}{T_b}\,[1/2 + 1/2\cos 4\pi f_c t]
\end{aligned}
$$

The output of the multiplier, $e_m(t)$, is applied to the integrator. The integrator will integrate the double frequency term over an integer number of cycles and eliminate this term. In practice, a lowpass filter follows the integrator to ensure that this term is eliminated from the output. The output, $a(t)$, is

$$
\begin{aligned}
a(t) &= \frac{1}{T_b}\int_0^t (\pm\sqrt{E_b})\,\sqrt{\frac{2}{T_b}}\,\cos 2\pi f_c t \times \sqrt{\frac{2}{T_b}}\,\cos 2\pi f_c t\,dt \\
&= \pm\sqrt{E_b}\,\frac{2}{T_b}\int_0^t [1/2 + 1/2\cos 4\pi f_c t]\,dt \qquad (7.10)
\end{aligned}
$$

$$
a(t) = \pm\sqrt{E_b}\,\frac{1}{T_b}\,t
$$

The output of the integrator is either a positive- or negative-going ramp. The S/H device is usually set to sample the ramp whenever it reaches a maximum value, which ideally occurs whenever $t = T_b$. The upper limit of the integrator is also set to T_b. For the no noise case, the output of the integrator at $t = T_b$ is

$$a(T_b) = \pm\sqrt{E_b}$$

The S/H device is clocked to sample the output of the integrator whenever the maximum voltage is expected. For this case, the S/H samples at $t = T_b$ and the output voltage $\pm\sqrt{E_b}$ is applied to the threshold device, which nominally triggers out a crisp waveform representing a one if the voltage is greater than zero or a zero if the voltage is less than zero.

7.4.2 With Noise

The case when the received signal is contaminated with additive white Gaussian noise will be considered next. Let the white noise have a power spectral density given by $N_o/2$. The received signal and input to the correlation receiver is now given by

$$r(t) = \pm\sqrt{E_b}\,\sqrt{\frac{2}{T_b}}\,\cos 2\pi f_c t + n(t)$$

where $n(t)$ is white Gaussian noise.

The output of the integrator at $t = T_b$ due to the signal part of $r(t)$ will be the same as before, $\pm a$. For the low-noise case, the output of the integrator might look similar to the ramp shown in Figure 7.5(a); and for the high-noise case, the ramp of Figure 7.5(b) is indicative what the output might look like.

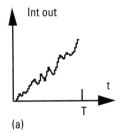

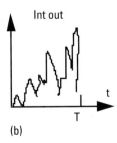

(a) (b)

Figure 7.5 Integrator output of a correlation receiver: (a) low noise, and (b) high noise.

Let the sampled voltage of the integrator at $t = T$, be represented by Z, then

$$Z = \pm\sqrt{E_b} + N$$

It can be shown that N is a Gaussian random variable with zero mean and variance, σ^2 given by [1],

$$\sigma^2 = N_o/2 \qquad (7.11)$$

Therefore, the voltage output random variable, z, that defines the sampled voltage point, Z, of the integrator is also a Gaussian random variable. Then z will have a variance of $N_o/2$ and a mean of $\pm\sqrt{E_b}$, depending upon which symbol was received.

Letting

$$a = \sqrt{E_b}$$

The conditional probability density function of z given that the symbol corresponding to $\pm a$ indicating a one or a zero was transmitted is

$$p(z|a) = (1/\sqrt{2\pi\sigma}) \exp-(1/2)((z - \sqrt{E_b})/\sigma)^2 \qquad (7.12)$$

and

$$p(z|-a) = (1/\sqrt{2\pi\sigma}) \exp-(1/2)((z + \sqrt{E_b})/\sigma)^2 \qquad (7.13)$$

The plot of these two functions is shown in Figure 7.6. This figure shows that the sampled output voltage of the integrator will fall somewhere

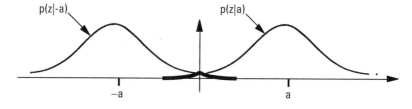

Figure 7.6 Conditional probability density functions, given a "one" or "zero" was transmitted.

along the x axis, generally in the vicinity of $\pm a$. The signal constellation for BPSK is shown in Figure 7.7.

7.5 Maximum Likelihood Detection

At the end of each symbol period when the integrator voltage is sampled, the receiver must decide which symbol was sent based upon the sampled voltage, Z. For maximum likelihood detection, conceptually, the statistic Z is substituted into (7.12) and (7.13) and the function with the largest value indicates which symbol has the maximum likelihood of having been transmitted. The test is implemented by forming a ratio of the two densities, such as

$$\lambda = p(z|a)/p(z|-a) = p(z|\sqrt{E_b})/p(z|-\sqrt{E_b}) \qquad (7.14)$$

Assuming each symbol is equally likely and the cost of all errors is the same, the received point Z is substituted for z, and if

$\lambda > 1$, choose the symbol corresponding to a;

$\lambda < 1$, choose the symbol corresponding to $-a$.

Fundamentally, this test computes the value of each conditional density at $z = Z$ and selects the density with the largest value at that point. The raw test of substituting the statistic into the functions and taking the ratio would be difficult to implement in hardware for real-time operation. Instead, the procedure is to form the likelihood test and divide the possible voltage space into two regions such that whenever the sampled value falls in a logic one interval, $\lambda > 1$, and the other region corresponding to a logic zero will be such that $\lambda < 1$. For this case, inspection of Figure 7.6 shows that whenever the sampled voltage, Z, is greater than 0V, then $\lambda > 1$, and the

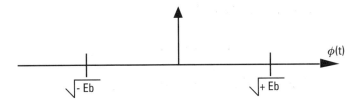

Figure 7.7 BPSK signal constellation.

voltage interval corresponding to $p(z|a) > p(z|-z)$ is established as any voltage greater than zero. A one would be declared as the most likely transmitted symbol. On the other hand, if the sampled voltage falls in the interval below zero volts, $p(z|-z) > p(z|a)$, and a zero is declared as the most likely symbol transmitted. This decision process may be implemented in hardware as a voltage comparator set to zero. For more information on maximum likelihood detection see [1–3].

7.6 Bit Errors

A bit error is made if a zero is transmitted and the sampled voltage, Z, falls above zero. The probability of this occurring is the area beneath the Gaussian curve, with mean $-a$, from zero to infinity and is given by

$$P(a|-a) = \frac{1}{\sqrt{2\pi}\sigma} \int_0^\infty \exp{-(z + a)(1/2)(1/\sigma)^2}dz \qquad (7.15)$$

To convert this expression to the Q-function let

$$x = \frac{z + a}{\sigma}$$

then

$$P(a|-a) = \frac{1}{\sqrt{2\pi}} \int_{\lambda/\sigma}^\infty \exp{-(1/2)x^2}dx \qquad (7.16)$$

$$= Q(a/\sigma)$$

Then by symmetry, the probability of mistaking $-a$ for $+a$ is

$$P(-a|a) = Q(a/\sigma) \qquad (7.17)$$

The total probability of an error is

$$P_e = P(a)P(-a|a) + P(-a)P(a|-a) \qquad (7.18)$$

$$P_e = 1/2Q(a/\sigma) + (1/2)Q(a/\sigma) \tag{7.19}$$

$$= Q(a/\sigma) \tag{7.20}$$

Since $a = \sqrt{E_b}$, (7.20) becomes

$$P_e = Q\left(\frac{\sqrt{E_b}}{\sigma}\right) \tag{7.21}$$

7.6.1 Bit Error Probability in Terms of N_0 and E_b

Because white noise is presented, it can be shown that the variance of the noise is given by [1]

$$\sigma^2 = N_0/2$$

Equation (7.21) may be written as

$$P_e = Q\left(\sqrt{\frac{2E_b}{N_o}}\right) \tag{7.22}$$

From (7.3), it can be seen that the bit energy, E_b, is related to carrier power by

$$E_b = CT_b \tag{7.23}$$

For an ideal IF bandpass filter with bandwidth B_{IF}, the noise power, N_p, in the IF is given by

$$N_p = N_o B_{IF} \tag{7.24}$$

Under ideal conditions and NRZ data format, $B_{IF} = R_b$ (this is usually not the case unless raised cosine filtering is used); then

$$N_p = N_o R_b \tag{7.25}$$

Using the above three relationships, (7.22) may be written as

$$P_e = Q\left(\sqrt{\frac{2C}{N_p}}\right) \tag{7.26}$$

This is an important equation in that it relates the BER to the carrier-to-noise ratio in the IF, C/N_p, which is determined by link analysis.

7.7 BPSK Modulation

A balanced modulator or a diode quad type multiplier may be used to implement the phase-reversal keyed modulation or BPSK as given by (7.1) and (7.2). A multiplier is shown in Figure 7.8.

7.8 BPSK in General

Both the 180° mode of BPSK or PRK and BPSK with a phase separation other than 180° require coherent demodulation; that is, a receiver with a carrier recovery system is necessary. However, since no carrier is present in the PRK signal, a PLL will not track the carrier, and a more sophisticated loop must be employed. A general BPSK signal may be generated with a carrier component, which a PLL will track and produce a carrier for synchronous or coherent demodulation. While this method does create a carrier component, it also extracts power from the modulation component degrading the BER. A general BPSK model will be discussed next.

7.8.1 Model for General BPSK

The general expression for BPSK is given by

$$s_i(t) = \sqrt{2E/T}\cos(\omega_c t + b\delta) \tag{7.27}$$

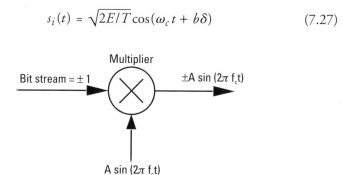

Figure 7.8 Phase reversal BPSK modulator.

where

$b = \pm 1;$

$0 < \delta < \pi/2$ usually within $15°$ of $\pi/2$.

A carrier index for BPSK is defined as

$$m = \cos \delta \tag{7.28}$$

or

$$\delta = \cos^{-1} m \quad \text{where} \quad 0 < m < 1 \tag{7.29}$$

where $0 < m < 1$. Using a trigonometric identity and this carrier index, the expression for $s_i(t)$ becomes

$$s_i(t) = \sqrt{2E/T} \cos \omega_c t \cos(b \cos^{-1} m) - \sqrt{2E/T} \sin \omega_c t \sin(b \cos^{-1} m) \tag{7.30}$$

But,

$$\cos(\pm \cos^{-1} m) = m$$

and

$$\sin(\pm \cos^{-1} m) = \pm\sqrt{1 - m^2} = b\sqrt{1 - m^2}$$

Then (7.30) becomes

$$s_i(t) = \sqrt{2E/T}\, m \cos \omega_c t - b\sqrt{2E(1 - m^2)/T} \sin \omega_c t \tag{7.31}$$

The first term is the carrier term, and the second is the modulation term, which shows the modulation energy reduced by $(1 - m^2)$. The larger the carrier index, the larger the component and if $m = 0$, then the expression reverts to PRK. The probability for a bit error becomes

$$P_b = Q\left(\sqrt{\frac{2E_b(1 - m^2)}{N_o}}\right) \tag{7.32}$$

Table 7.1 shows the additional power required as a function of δ.

In actual systems employing premodulation filters, a carrier deviation of $\pm\pi/2$ will not produce a carrier null. A deviation of 100° to 110° is required depending upon the premodulation filter, the order and the 3-dB point of the filter.

7.9 Actual Receiving Hardware

There are a number of excellent off-the-shelf receivers available capable of receiving BPSK and several that will handle quadrature phase-shift keying. With the appropriate RF tuner, the receivers will handle both the S-band and the L-band frequencies.

7.9.1 The Receiver

Functional blocks of the actual hardware are shown in Figure 7.9. For a coherent receiver, such as one used to receive BPSK, the cos ωt term multiplying the received carrier signal must be locked in phase to the carrier. This is usually achieved by a loop. If the loop is a suppressed carrier tracking loop, such as a Costas loop, the received signal can be a suppressed carrier signal such as the BPSK signal with a 180° phase difference defined in (7.1) and (7.2). Modern receivers such as the Microdyne 750 and 1400 and the Scientific Atlantic 930B and 930C will demodulate suppressed carrier BPSK. If the loop in the receiver is an ordinary phase-lock loop, then it is necessary to design the RF signal such that there is a residual carrier term.

The Costas loop or a squaring circuit produces a 180° ambiguity since the minus sign is eliminated by multiplying or squaring. The receiver must resolve this problem. The mark and space codes will prevent the ambiguity but will have about twice the BER.

Table 7.1
Additional Power Required for Same BER Due to Carrier Component

δ (Degrees)	Additional PWR in dB
88	0.005
85	0.033
80	0.113
75	0.3
70	1.8

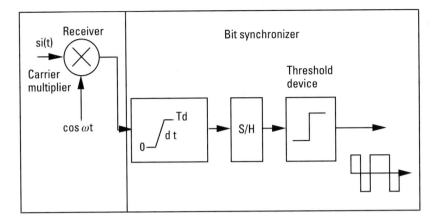

Figure 7.9 The correlation receiver composed of the receiver and bit synchronizer.

In most receivers, the loop bandwidth may be selected from a menu. The loop bandwidth is set from 0.01 to 0.001% of the bit rate. The smaller the loop bandwidth, the higher the carrier-to-noise ratio in the loop but the longer the acquisition time. For example, loop bandwidths available in the Microdyne 1400 are 100; 300; 1,000; and 3,000 Hz.

7.9.2 The Bit Synchronizer and Matched Filter Implementation

Functionally and conceptually, the correlation receiver is composed of the receiver and bit synchronizer. Usually in actual hardware and as shown in Figure 7.9, the correlation receiver is composed of the left block in the figure, which is the receiver, and the right block, containing the other functional operations, is the bit synchronizer. Each of these two blocks is a separate piece of hardware.

For NRZ, the output of the receiver carrier multiplier is shown in Figure 7.10. This output is a distorted and noisy signal representing the received bit sequence. This is the input to the bit synchronizer, which must make a decision each bit period as to whether a zero or one was received. Once this decision has been made, the bit synchronizer outputs a clean, crisp bit sequence, as shown in Figure 7.10.

The correlation receiver and the matched filter receiver are equivalent. Specifically, the integrator and S/H, at $t = T_b$, are equivalent to a matched filter, sampled at the output. Most bit synchronizers, in fact, use matched filters, which are matched to the anticipated waveform coming from the receiver.

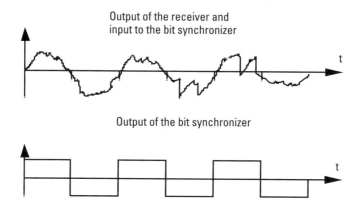

Output of the receiver and
input to the bit synchronizer

t

Output of the bit synchronizer

t

Figure 7.10 Input-output waveforms of a bit synchronizer.

7.10 Comparison of BERs for BPSK and PCM/FM

For BPSK, the probability of a bit error, P_b, is the same as a symbol error. If the bit errors are averaged over a large number of bits, the BER and P_b are used interchangeably; therefore, to designate that the error is explicitly a bit error, the expression is

$$P_b = Q\left(\sqrt{\frac{2E_b}{N_o}}\right) \qquad (7.33)$$

For noncontinuous PCM/FM and noncoherent reception, the BER is given by

$$P_b = 0.5 \exp-(0.5E_b/N_o) \qquad (7.34)$$

For PCM/FM and noncoherent reception, which is almost always used, the probability for a bit error is

$$P_b = 0.5 \exp-(0.7E_b/N_o) \qquad (7.35)$$

Equations (7.33), (7.34), and (7.35) are plotted in Figure 7.11.

Noncoherent but continuous PCM/FM was chosen over coherent BPSK because the receiver is much simpler in that a carrier-tracking loop is not necessary. The simpler receiver, which did not require a locked loop for operation, was an important consideration in the early days since all the

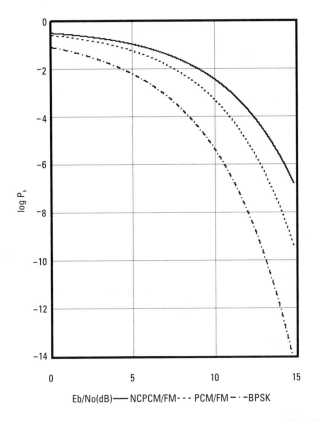

Figure 7.11 Probability of a bit error, plotted as a function of E_b/N_0 BPSK, PCM/FM, and NCPCM/FM.

data would be lost during the period when the loop was out of lock. Also, for a short flight, the time required for the loop to lock up was of concern.

As developments in loops made the receivers employing them more reliable and less expensive than the earlier models, the new receivers now have the capability to operate in the coherent mode and support BPSK. This is of importance, since for the same BER less power must be transmitted. Alternately, for the same transmitted power and same BER, the bit rate may be increased. As the demand for video signals increases, requiring greater bit rates, this is an important consideration.

In comparing continuous PCM/FM, referred to as simply PCM/FM, and BPSK, approximately 1.5 dB more transmitted power is required for PCM/FM for the same BER. In deciding between PCM/FM and BPSK, the relative error rates must be considered; however, other considerations such as actual bandwidth requirements, equipment availability, and design constraints must also be considered. These topics will be discussed in later sections.

7.11 Q-Function

The Q-function is defined to be

$$Q(a) = \text{Prob}(y > a) = \frac{1}{\sqrt{2\pi}} \int_a^\infty \exp-(y^2/2)dy \qquad (7.36)$$

That is, for a Gaussian distribution with mean equal to zero and variance equal to one, $Q(a)$ is the area beneath the tail of the curve from a to infinity. For more information on the Q-function, tables, and relationships between the Q-function and the error functions see [4].

7.12 BPSK PSD

Since PRK has a double sideband suppressed carrier modulation format, the transmission PSD is the modulating signal's two-sided spectrum translated to the carrier frequency. For the modulating signal of $a \pm A$ bit stream, the baseband modulating signal, has a PSD of sinc^2, as given by (8.34) in Chapter 8, which is translated to the carrier frequency to obtain the modulation PSD. By using the null-to-null points of the sinc^2 function as a definition of bandwidth, the transmission bandwidth becomes $2R_b$, which gives a spectral efficiency of 0.5 bps/Hz. Using the frequency interval from the 3-dB point to the 3-dB point on the envelope of the sinc function, the defined transmission bandwidth is then approximately R_b, which gives a spectral efficiency of 1.

The PSD of general BPSK, when there is not a 180° phase shift between the mark and space, is similar to PRK except there is a carrier component.

7.12.1 Spectrum in a Practical System

The spectrum in an actual system is much more complex than the two sinc curves discussed above, due to spectral spreading [5–7]. Spectral spreading is the result of hard-limiting and operating with a nonlinear amplifier. In order to constrain the transmission bandwidth to R_b, the premodulation filter must be a well-designed realizable raised cosine filter [5] with a α of 0.3. Further, if the signal is used to drive a nonlinear power amplifier prior to transmission, spectral spreading will result.

7.13 Overall Comparison Between PCM/FM and BPSK

In deciding between PCM/FM and BPSK, a number of important factors must be considered. BERs, transmission bandwidth, and practical hardware implementation are relevant factors.

7.13.1 BER

For the same BER, PCM/FM requires about 1.5 dB more transmitted power than BPSK. This is an important consideration, since for the same $[C/N_p]$, the same system will support approximately a 50% higher bit rate using BPSK at the same BER. It is important to note that the BPSK receiver employing coherent demodulation will require time for the loop to lock; if the loop loses lock, data will be lost during these intervals. This does not occur for a PCM/FM receiver.

7.13.2 Transmission Bandwidth

The transmission bandwidth required for PCM/FM is a well-established bandwidth, approximately equal to the bit rate R_b. This is achieved with a premodulation linear phase filter with a 3-dB corner frequency set equal to $0.7R_b$. A large number of PCM/FM systems have been designed, implemented in hardware, and deployed. Their performance was as expected with respect to bandwidth requirements. That is, a spectral efficiency of 1 bit/Hz is normally achieved.

The bandwidth requirements for BPSK and QPSK in satellite communications systems are well known, but in telemetering systems they are not as well known or documented. To achieve the 1-bit/Hz spectral efficiency, a realizable, well-designed raised cosine filter must be employed as a premodulation filter. Spectral spreading caused by the use of a nonlinear amplifier prior to transmission may well negate the bandwidth-limiting work of the premodulation filter. Filtering after modulation at S-band is difficult to achieve.

For BPSK, according to IRIG 106-93 *Telemetry Standards* document [8], the 99% spectral containment bandwidth, B_{99}, obtained with a lowpass premodulation filter with a 3-dB frequency of $0.7R_b$, is given by

$$B_{99}(\text{BPSK}) = 1.5R_b$$

For PCM/FM under the same conditions,

$$B_{99}(\text{PCM/FM}) = 1.16R_b$$

Certainly, BPSK is going to require more transmission bandwidth than PCM/FM when both use the same linear phase premodulation filter. See Chapter 9 for more details.

7.13.3 Hardware Considerations

The vast majority of commercially available off-the-shelf transmitters are designed and built for use in PCM/FM telemetering systems. The parameters given on the specification sheets are also given for PCM/FM applications. Further, if the transmitter has a built in premodulation filter, it is normally a linear phase filter.

7.14 General PM Modulation

The general equation for phase modulation is

$$\varphi_{pm}(t) = A\cos(2\pi f_c t + k_p m(t)) \tag{7.37}$$

where

phase modulator sensitivity constant = k_p rad/V;

$$m(t) = \text{modulating signal.}$$

If single tone modulation is employed, that is, if

$$m(t) = a\cos 2\pi f_m t \tag{7.38}$$

then the expression becomes

$$\varphi_{pm}(t) = A\cos(2\pi f_c t + k_p a\cos 2\pi f_m t) \tag{7.39}$$

$$= A\cos(2\pi f_c t + M_{pm}\cos 2\pi f_m t) \tag{7.40}$$

where $M_{pm} = k_p a$ = peak phase deviation = modulation index for phase modulation.

Since k_p has units of radians per volt and a is voltage, M_{pm} has units of radians. As mentioned in Section 7.7, a balanced modulator or a diode quad type multiplier may be used to implement the phase-reversal keyed modulation as given by (7.1) and (7.2). If it is desired to generate BPSK with a residual carrier, it is necessary to use a PM modulator, as shown in Figure 7.12, with ak_p equal to the desired carrier phase offset. Say we want to have a carrier phase offset of 80°. Then ak_p = 80, and the expression becomes

$$\varphi_{pm}(t) = A \cos(2\pi f_c t \pm 80°) \tag{7.41}$$

7.15 Comparison of PM and FM Modulation

Consider a modulating signal, $m(t)$, with a peak value of m_p, then for FM, the peak frequency deviation is given by

$$\Delta f_{fm} = k_f m_p \tag{7.42}$$

where k_f is the sensitivity constant of the FM modulator, (hertz per volt).

For FM, peak frequency deviation is independent of the spectrum of $m(t)$. For PM, the peak frequency deviation is

$$\Delta f_{pm} = k_p m_p' \tag{7.43}$$

where m_p' is the peak value of the derivative of $m(t)$.

For PM, the peak frequency deviation is highly dependent upon the spectrum of $m(t)$. For example consider single-tone modulation with

$$m(t) = a \cos 2\pi f_m t$$

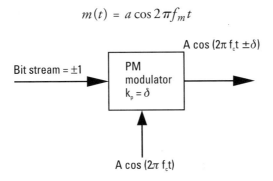

Figure 7.12 PM modulator with carrier phase offset.

For FM the peak deviation is

$$\Delta f_{fm} = k_f m_p = k_f a \qquad (7.44)$$

Whereas for PM, it is

$$\Delta f_{pm} = k_p m_p' = k_p a f_m \qquad (7.45)$$

7.16 FM Modulation Employing a PM Modulator

It is sometimes desirable to employ a PM modulator to perform FM modulation. For example, the PM modulator may be more shock resistant. The general expression for FM is

$$\phi_{fm}(t) = A \cos(2\pi f_c t + \Delta f \int m(t) dt) \qquad (7.46)$$

The general expression for PM is

$$\phi_{pm}(t) = A \cos(2\pi f_c t + k_p m(t)) \qquad (7.47)$$

7.16.1 Hardware to Achieve FM with a PM Modulator

For a PM modulator to perform as an FM modulator, the modulating signal must be passed through an integrator as shown in Figure 7.13. When $m(t)$ of (7.47) is replaced by the integral of $m(t)$, (7.46) results.

The integrator in Figure 7.13 may be approximated by a simple single-pole RC lowpass network, as shown in Figure 7.14.

The transfer function for this network is given by

$$H(j2\pi f) = \frac{2\pi f_1}{j2\pi f + 2\pi f_1} \qquad (7.48)$$

where

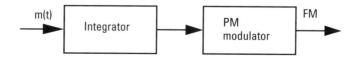

Figure 7.13 PM modulator using an integrator to generate FM.

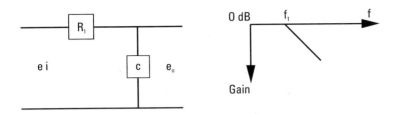

Figure 7.14 Single-pole lowpass network.

$$f_1 = 1/2\pi RC$$

For $f \gg f_1$, then

$$H(j2\pi f) \approx 1/j2\pi f$$

The network performs as an integrator in the time domain on frequencies well above the corner frequency.

Example 1

The Microcom TM transmitter, T4XO, is an S-band PM transmitter built to operate under extreme stress. Currently, one of its uses is as an FM/PCM transmitter. The Biφ bit format is used and passed through the single-pole lowpass RC network shown in Figure 7.14 prior to modulating the transmitter. The corner frequency, f_1, is set to

$$f_1 = 0.1R_b$$

From Figure 5.7(b), it can be seen that the spectral distribution of the Biφ format is well above f_1; hence, the RC network performs as an integrator. Further, since the spectrum has no DC, there is no problem with passing the bit sequence through the integrator.

Example 2

The transmitter of Example 1 is being used for FM/FM. Modulating the subcarriers onto the carrier without an integrator does not cause a problem since the phase of the subcarriers, whether sine or cosine, is not of importance since noncoherent detection is used. It must be remembered that since this is a PM modulator, the peak deviation of a subcarrier is given by

$$\Delta f = ak_p f_{si}$$

That is, the deviation is proportional to (1) the amplitude of the subcarrier, a, (2) the sensitivity constant, k_p, and (3) the frequency of the modulating subcarrier, f_{si}. In setting up the preemphasis schedule implementation, f_{si} must be taken into account.

7.17 Differential Phase-Shift Keying

When loops are used to extract the carrier in order to perform coherent demodulation, the phase of the carrier due to modulation is lost in the squaring circuit or in the Costas loop. This creates a phase ambiguity of 180° for this type of carrier tracking circuit. The phase-lock loop tracks the carrier and does not create the phase ambiguity. If the Costas loop or the squaring circuit is used, the ambiguity must be resolved. One way of resolving this problem is to use differential encoding.

In the implementation of differential encoding, the encoded bit d_k, is generated by comparing the present bit, b_k, with $d_k - 1$. If there is no difference, d_k is set equal to one, otherwise to zero. In terms of Boolean algebra, this can be expressed as

$$d_k = b_k d_{k-1} + \overline{b_k}\ \overline{d_{k-1}}$$

The differential encoder is shown in Figure 7.15. The waveforms of the original bit stream, b_k, and the resulting differential encoded sequence, d_k, are shown in Figure 7.16.

To obtain the original bit sequence, b_k, the receiver must apply reverse logic to the differentially encoded sequence, d_k. This may be achieved at baseband or at the differential phase-shift keying (DPSK) level. Figure 7.17

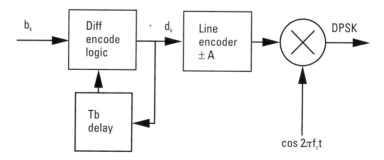

Figure 7.15 DPSK encoder.

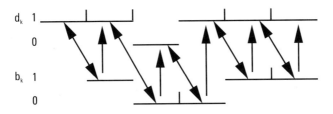

Figure 7.16 Logic waveforms for differential encoding.

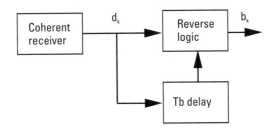

Figure 7.17 DPSK receiver with baseband recovery of b_k.

depicts the blocks necessary to recover b_k and Figure 7.18 illustrates the waveforms.

For the recovery of the information sequence at baseband, a carrier recovery loop must be used to obtain a replica of the carrier in order to multiply the received signal in the coherent receiver.

The recovery of b_k from the DPSK waveform at the RF level can be achieved by delaying the modulated carrier waveform one bit period and multiplying the received waveform by the delayed signal, as shown in Figure 7.19.

Since the received signal is

$$r(t) = \pm\cos 2\pi f_c t$$

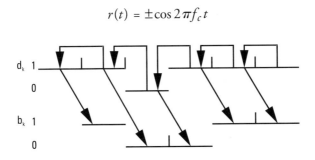

Figure 7.18 Waveforms for DPSK baseband recovery of b_k.

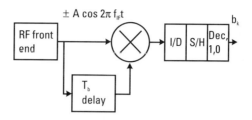

Figure 7.19 DPSK receiver with carrier recovery of b_k.

it can be seen that if the current signal and the delayed signal are of the same sign, $a +$ one results. If they are of different signs, $a -$ one results. This recovers the original b_k sequence. Since the signal multiplying the received signal is obtained from the received signal by a 1-bit delay, a carrier recovery loop does not have to be used, and certainly this is an advantage. While this type of receiver being noncoherent does not employ a loop with the attendant complexity, the delay network, which must accommodate the variations in the bit rate at the carrier frequency or at the IF frequency, is more difficult to build than the delay at baseband.

Problems

Problem 7.1

Draw a block diagram of a BPSK transmitter system.

Problem 7.2

Sketch BPSK waveforms 180° out of phase, represented by sine waves.

Problem 7.3

Draw a block diagram of a BPSK receiver and describe the function of each block.

Problem 7.4

Discuss how each block of Problem 7.3 can be implemented in hardware.

Problem 7.5

By using the necessary trig identity, show that the output of the multiplier of Figure 7.4 is $[1/2 + 1/2 \cos 4\pi f_c t]$.

Problem 7.6

Show that the output of the integrator in Figure 7.4 does not contain the $\cos 4\pi f_c t$ term.

Problem 7.7

Determine the output of a multiplier if one input is $\cos 2\pi f_c t$ and the other term is (a) $\cos 2\pi f_c t$, and (b) $\cos(2\pi f_c t + \delta)$.

Problem 7.8

Determine the output of an integrator after one second if the input is (a) $s_1(t) = 2$, (b) $s_1(t) = 4$, and (c) $s_1(t) = t$.

Problem 7.9

If $a = 1$ and $-a = -1$ and $\sigma = 0.1$, compute the likelihood ratio λ, if the received point Z is (a) 0.6, (b) 1, (c) −0.6, and (d) 0.

Problem 7.10

For BPSK, compute P_b if E_b/N_o is (a) 9.2, (b) 9.6, (c) 10.2, and (d) 10.6.

Problem 7.11

For general BPSK, compute P_b if E_b/N_o is 9.6 and δ takes on the values given in Table 7.1.

Problem 7.12

From Figure 7.10, find P_b for the three types of systems if $E_b/N_o = 10$ dB.

Problem 7.13

In Problem 7.12, if the IF bandwidth is set equal to R_b, in the three systems, what is the carrier-to-noise ratio in the three systems?

Problem 7.14

For the same condition in Problem 7.13, what is the required carrier-to-noise ratios in the three systems for a $P_b = 1 \times 10^{-5}$?

Problem 7.15

Discuss the function of the bit synchronizer.

References

[1]   Haykin, S., *Digital Communications,* New York: Wiley, 1988.

[2]   Zimmer, R. E., and R. L. Peterson, *Digital Communications and Spread Spectrum Systems,* New York: Macmillan, 1985, pp. 186–207.

[3]   Lathi, B. P., *Modern Digital and Analog Communication Systems,* 2nd Ed, New York: Holt, Rinehart and Winston, 1989.

[4]   Carlson, A. B., *Communication Systems,* 3rd Ed, New York: McGraw-Hill, 1986.

[5]   Feher, K., *Digital Communications,* Englewood Cliffs, NJ: Prentice-Hall, 1983.

[6]   Devieux, C., "Spectral Spreading," COMSAT J. *Tech. Rev.,* Sept. 1974.

[7]   Harris, R. A., "Transmission Analysis and Design for the ECS Systems," *Proc. 4th Int. Conf., Digital Satellite Communications,* INTELSAT, Montreal, Canada: Oct. 23, 1978.

[8]   Secretariat, Range Commanders Council, *Telemetry Standards,* White Sands Missile Range, NM: RCC, IRIG 106-93, and IRIG 106-00.

8

Binary Digital Communication Systems

The basic digital communication concepts presented in Chapter 4 are expanded upon in this chapter. Details are presented for digital communication systems in which the encoder input and output signals are binary, that is, $E = 2$, and $M = 2$. These systems are referred to as binary digital communications systems, and are classified by the type of modulation/demodulation method used. Three categories characterize these methods: ASK, FSK, and PSK. Only FSK and PSK and their derivatives are covered herein. ASK has relatively poor performance, and is omitted. PCM/FM, which was covered in Chapter 5, is a variant of FSK and is referred to in this chapter.

8.1 Learning Objectives

Upon completion of this chapter, the reader should understand the following:

- Binary frequency-shift keying and its derivatives: continuous-phase frequency-shift keying, minimum-shift keying, and Gaussian minimum-shift keying;
- BPSK and its derivative differential phase-shift keying;
- Mathematical modeling of binary frequency-shift keying, BPSK, and their derivatives;

- Generation and detection of binary frequency-shift keying, BPSK, and their derivatives;
- Composite and component waveforms of binary frequency-shift keying, BPSK, and their derivatives;
- The spectrum and bandwidth efficiency of binary frequency-shift keying, BPSK, and their derivatives;
- BERs and power efficiency of binary frequency-shift keying, BPSK, and their derivatives;
- Performance comparison of binary frequency-shift keying, BPSK, and their derivatives.

8.2 FSK

In FSK modulation, the frequency f of the carrier is shifted between two different frequencies corresponding to the digital source symbols. If the encoder output $m(t)$ is a binary signal ($M = 2$), then this method is often called binary frequency-shift keying (BFSK), or simply FSK. If $M > 2$, this process is known as M-ary FSK.

8.2.1 BFSK

8.2.1.1 Dual-Carrier Generation

In BFSK, the binary source symbols 0 and 1 cause two sinusoidal signals that differ in frequency to be transmitted. These two carrier signals $s_1(t)$, and $s_2(t)$, are specified by

$$s_{FSK}(t) = s_i(t) = \sqrt{\frac{2E_b}{T_b}} \cos(2\pi f_i t) = A \cos(2\pi f_i t) \qquad (8.1)$$

$$0 < t < T_b \qquad i = 1, 2$$

where E_b is the transmitted energy per symbol, T_b is the symbol duration, and A is the peak carrier voltage. Note that $A^2 = 2E_b / T_b$.

The waveform specified by (8.1) may be generated by using the transmitter shown in Figure 8.1(a). The basis functions $\phi_1(t)$ and $\phi_2(t)$ correspond to f_1 and f_2 respectively. When the digital source output $e_j = 0$, $m_k = 0$, the lower channel is switched off, and the upper channel is switched on. The result then, is that source binary 0 causes $s_1(t) = \phi_1(t)$ to be transmitted with frequency f_1. Conversely, a source binary 1 causes $s_2(t) = \phi_2(t)$ to be transmitted with frequency f_2. A typical composite waveform generated by this process is shown in Figure 5.2(a). The upper and lower

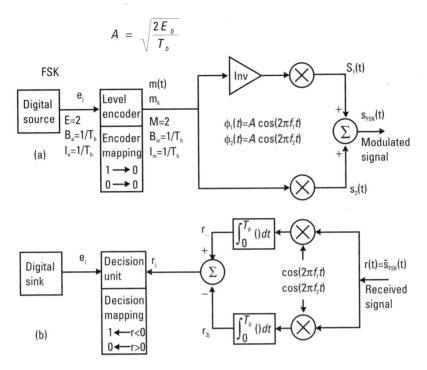

Figure 8.1 BFSK system block diagrams: (a) transmitter, and (b) coherent receiver.

channel switched waveforms $s_1(t)$, and $s_2(t)$ forming the composite waveform are shown in (b) and (c) of this same figure. The digital source generates a new binary symbol every T_b seconds, for a baud rate of $B_e = 1/T_b$. Since $E = 2$, the source bit rate I_e equals the source baud rate B_e. The level encoder generates one symbol for every bit generated by the source (1 bit/symbol), and $B_m = B_e$. It also follows from (4.11) and (4.12) that $I_m = I_e$.

8.2.1.2 Coherent Detection

The BFSK receiver shown in Figure 8.1(b) correlates the received signal with two locally generated coherent reference signals $\cos(2\pi f_1 t)$ and $\cos(2\pi f_2 t)$. The output r_{ij}, of correlator i when $s_j(t)$ is transmitted is

$$r_{ij} = A \int_0^{T_b} \cos(2\pi f_i t) \cos(2\pi f_j t)dt \tag{8.2}$$

$$= \frac{A}{2} \int_0^{T_b} \cos[2\pi(f_i + f_j)t]dt + \frac{A}{2} \int_0^{T_b} \cos[2\pi(f_i - f_j)t]dt$$

where an identity for $\cos(a)\cos(b)$ is used to obtain the sum and difference frequency terms. Integrating over the period 0 to T_b yields

$$r_{ij} = \frac{AT}{2}\left[\frac{\sin[2\pi(f_i-f_j)T_b]}{2\pi(f_i-f_j)T_b} + \frac{\sin[2\pi(f_i+f_j)T_b]}{2\pi(f_i+f_j)T_b}\right] \quad i,j = 1, 2 \tag{8.3}$$

Define the frequency deviation of the transmitted signal as

$$\Delta f \equiv \frac{|f_2-f_1|}{2} = \frac{f_d}{2} \tag{8.4}$$

and the carrier frequency as

$$f_c \equiv \frac{f_2+f_1}{2} \tag{8.5}$$

then (8.3) becomes

$$r_{11} = r_{22} = \frac{AT}{2}\left[1 + \frac{\sin(4\pi f_c T_b)}{4\pi f_c T_b}\right] \tag{8.6}$$

$$r_{12} = r_{21} = \frac{AT}{2}\left[\frac{\sin(4\pi\Delta f T_b)}{4\pi\Delta f T_b} + \frac{\sin(4\pi f_c T_b)}{4\pi f_c T_b}\right]$$

Making the assumption that the carrier frequency is large compared to the bit rate ($f_c \gg 1/T_b$), the second terms in (8.6) become small, and we have the signal autocorrelated outputs as

$$r_{11} = r_{22} = \frac{AT}{2} \tag{8.7}$$

and the signal cross-correlated outputs as

$$r_{12} = r_{21} = \frac{AT}{2}\left[\frac{\sin(4\pi\Delta f T_b)}{4\pi\Delta f T_b}\right] \tag{8.8}$$

The output r_j, of the correlation detector in Figure 8.1(b), when $s_j(t)$ is transmitted is

$$r_j = r_{1j} - r_{2j} \qquad j = 1, 2 \tag{8.9}$$

Substituting (8.7) and (8.8) into (8.9) gives the detected signal as

$$r_j = [-(-1)^j]\frac{AT}{2}\left[1 - \frac{\sin(4\pi\Delta f T_b)}{4\pi\Delta f T_b}\right] \qquad j = 1, 2 \tag{8.10}$$

When the received signal $r(t)$ is $s_1(t) = \phi_1(t)$, $j = 1$ and $r_1 > 0$, and the decision unit selects a binary 0 as the received symbol. Since this corresponds to a binary 0 transmitted symbol, the receiver correctly detects the transmitted symbol. In a similar way, transmission of a binary 1 symbol ($j = 2$) causes $r_2 < 0$, resulting in a binary 1 symbol being detected.

The minimum probability of error occurs when the detected signal given by (8.10) is maximized. This occurs when $4\pi\Delta f T_b = 3\pi/2$, or when $\Delta f = 3/8T_b$. Thus, for unfiltered coherently detected FSK, the minimum BER occurs when $\Delta f = 0.375/T_b$. For noncoherent detection, the minimum BER as determined experimentally in (5.6), occurs when $\Delta f \cong 0.35/T_b$.

8.2.1.3 Orthogonal Signaling

For orthogonal signaling, the cross-correlation parameter (8.8) must be zero, yielding the solution

$$2\Delta f_o = f_{d,o} = \frac{n}{2T_b} \tag{8.11}$$

where n is some nonzero integer. This requires that the transmitted carrier frequency difference be a multiple of one-half the source baud rate.

During the symbol transmission time T_b, $n_i = f_i T_b$ cycles of the carrier occur, and

$$f_i = \frac{n_i}{T_b} \qquad i = 1, 2 \tag{8.12}$$

Substituting (8.12) into (8.11) yields the requirement for orthogonal signaling as

$$|n_1 - n_2|_o = \frac{n}{2} \tag{8.13}$$

For $n_i \gg 1$, (8.13) requires that the difference of two large numbers be precisely controlled to ensure orthogonal signaling. The detection process performance degrades rapidly as n departs from the required integer value. While the signal with frequency f_i is transmitted, the carrier phase is

$$\theta_i(t) = 2\pi f_i t = 2\pi \left(\frac{n_i}{T_b}\right) t \qquad (8.14)$$

The change in phase from the beginning of symbol i transmission to the end of transmission is

$$\Delta \theta_i = \theta_i(T_b) - \theta_i(0) = 2\pi \left(\frac{n_i}{T_b}\right) T_b = 2\pi n_i \qquad i = 1, 2 \quad (8.15)$$

The difference between the change in phase caused by the transmission of a 1 and a 0 symbol is

$$\theta_d \equiv |\Delta \theta_1 - \Delta \theta_2| = 2\pi |n_1 - n_2| \qquad (8.16)$$

From (8.13) the requirement for orthogonal signaling, substituted into (8.16) gives

$$\theta_{d,O} = 2\pi |n_1 - n_2| = 2\pi \frac{n}{2} = \pi n \qquad (8.17)$$

where n is some nonzero integer.

An example, with $n_1 = 4$ and $n_2 = 2$, is shown in Figure 4.9(b). This satisfies the condition for orthogonal signaling in (8.10) with $n = 4$. Note that an integer number of cycles of each frequency occurs, and the signal is continuous. Also, (8.16) gives the difference between the change in phase caused by the transmission of a 1 and a 0 symbol as $\theta_d = 4\pi$, implying that the phase is also continuous during symbol transitions.

8.2.1.4 Single-Carrier, Dual-Baseband Signal Representation

From (8.4) and (8.5) the frequencies f_1 and f_2 are given by

$$f_i = f_c + (-1)^i \Delta f \qquad i = 1, 2 \qquad (8.18)$$

Let

$$A = \sqrt{\frac{2E_b}{T_b}} \qquad (8.19)$$

Substituting (8.18) and (8.19) into (8.1) gives

$$s_{FSK}(t) = A \cos[2\pi(f_c + (-1)^i \Delta f)t] \qquad (8.20)$$

Using a trigonometric identity for $\cos(a + b)$, $\cos(-a)$, and $\sin(-a)$, (8.20) becomes

$$s_{FSK}(t) = s_i(t) = i_{FSK}(t) \cos(2\pi f_c t) + q_{FSK}(t) \sin(2\pi f_c t) \quad (8.21)$$

where

$$i_{FSK}(t) = A \cos[2\pi \Delta f t] \qquad (8.22)$$

is the in-phase baseband signal, and

$$q_{FSK}(t) = (-1)^i A \sin[2\pi \Delta f t] \qquad (8.23)$$

is the quadrature baseband signal.

Figure 8.2 gives a plot of $s_{FSK}(t)$, $i_{FSK}(t)$, and $q_{FSK}(t)$ as defined by (8.21), (8.22), and (8.23). The modulating signal $m(t)$ corresponds to the digital source sequence $\mathbf{e} = \{1,0,1,1,0,1,0,0\}$, with $T_b = 0.25$ seconds between symbols, or 4 bps. The carrier frequency $f_c = 8$ Hz, the frequency deviation $\Delta f = 2$ Hz, and the carrier amplitude is $A = 1$. Also $f_d = 2\Delta f = 4$, which satisfies (8.11) for orthogonal signaling with $n = 2f_d T_b = 2$.

A plot of the baseband signal $q(t)$ as a function of $i(t)$, called the *signal space diagram,* sometimes called a *vector diagram,* is shown. The locus of the plot forms a circle of radius $A = 1$, centered at the origin. This corresponds to a constant-amplitude sinusoid of varying frequency and phase.

8.2.2   Continuous-Phase Frequency-Shift Keying

8.2.2.1   Sunde's FSK

The orthogonal signaling requirement specified by (8.12) may be established by

$$f_i = \frac{n_c + i}{T_b} \qquad i = 1, 2 \qquad (8.24)$$

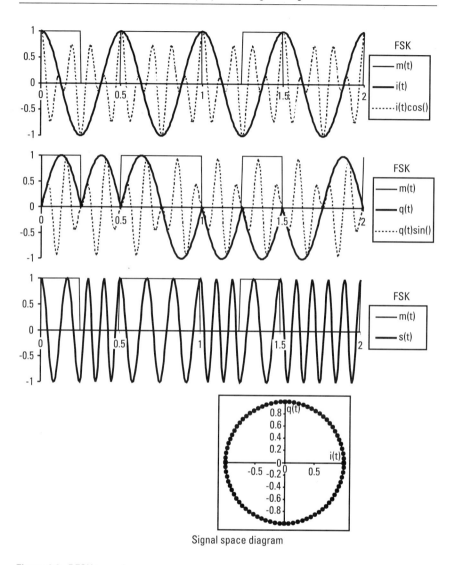

Signal space diagram

Figure 8.2 BFSK waveforms.

This method is known as Sunde's FSK [1, p. 512], and corresponds to $n_i = n_c + i$ in (8.12). The difference in frequency is

$$f_d = |f_1 - f_2| = \frac{1}{T_b} \tag{8.25}$$

This frequency difference is equal to the digital signal baud rate. Comparing (8.25) with (8.11), we see that the orthogonal signaling requirement is satisfied with integer $n = 2$. With $n_1 = n_c + 1$, and $n_2 = n_c + 2$, $|n_1 - n_2| = 1$, and from (8.17), $\theta_d = 2\pi$. Since symbol transitions take place at the same phase angle (multiples of 2π), the signal phase is continuous during symbol transmission. This is an example of a continuous-phase frequency-shift keying (CPFSK) modulation method.

8.2.2.2 Minimum-Shift Keying

The minimum frequency difference for which orthogonal signaling occurs is given from (8.11) with $n = 1$, as

$$|f_1 - f_2| \equiv f_{d\min} = \frac{1}{2T_b} \qquad (8.26)$$

In this case the frequency difference is half the digital signal baud rate. Substituting (8.26) into (8.4) gives the frequency deviation as

$$\Delta f_{\min} \equiv \frac{1}{4T_b} \qquad (8.27)$$

The transmitted signal is obtained by substituting (8.27) into (8.21), (8.22), and (8.23).

$$s_{\min}(t) = A \cos\left[\frac{\pi t}{2T_b}\right] \cos(2\pi f_c t) + (-1)^i A \sin\left[\frac{\pi t}{2T_b}\right] \sin(2\pi f_c t) \qquad (8.28)$$

From (8.17) with $n = 1$, we obtain the phase difference as $\theta_d = \pi$. Thus, the transmission of a 1 symbol increases the carrier phase by $\pi/2$ radians during time T_b, while a 0 symbol decreases the phase by $\pi/2$ radians during the same time. Since the carrier phase changes by $\pi/2$ radians during symbol transmission, there are four possible phase states at the beginning of the next symbol transmission: 0, $\pi/2$, π, and $-\pi/2$. The initial phase $\theta(0)$, as a function of initial in-phase signal $i(0)$, and the quadrature signal $q(0)$ is given in Table 8.1.

Introducing I and Q as the polarity of the in-phase and quadrature baseband signals respectively, (8.28) becomes

$$s_{MSK}(t) = i_{MSK}(t) \cos(2\pi f_c t) + q_{MSK}(t) \sin(2\pi f_c t) \qquad (8.29)$$

where

Table 8.1
Initial Phase θ (0), As a Function of i(0) and q(0)

θ (0)	0	$\pi/2$	π	$-\pi/2$
i(0)	1	0	−1	0
q(0)	0	1	0	−1

$$i_{MSK}(t) = IA \cos\left[\frac{\pi t}{2T_b}\right] \qquad q_{MSK}(t) = QA \sin\left[\frac{\pi t}{2T_b}\right] \qquad (8.30)$$

and A is defined in (8.19). This modulation method is called *minimum-shift keying* (MSK) since the frequency shift is the minimum that will maintain orthogonal signaling, and the signal phase is continuous from symbol to symbol. A block diagram of an MSK transmitter based on (8.29) and (8.30) is shown in Figure 8.3(a). The corresponding coherent receiver is given in Figure 8.3(b).

When $e_k = 1$ the phase is increased by $\pi/2$ radians, and the phase is decreased by $\pi/2$ radians when $e_k = 0$. The values of I and Q depend on the phase at the beginning of transmission, and the symbol to be transmitted. The required values to maintain continuous phase are listed in Figure 8.3(c).

An example of MSK waveforms for $A = 1$, $f_1 = 2$, and $f_2 = 4$ is shown in Figure 8.4. The signals $s_{MSK}(t)$, $i_{MSK}(t)$, and $q_{MSK}(t)$ as defined by (8.29), and (8.30) are shown. The modulating signal $m(t)$ corresponds to the digital source sequence $\mathbf{e} = \{1,0,1,1,0,1,0,0\}$, with $T_b = 0.25$ seconds between symbols, or 4 bps. As can be seen, the transmitted signal $m(t)$ is continuous in phase, making smooth transitions when a new symbol is presented.

The signal space diagram forms a semicircle of radius $A = 1$, centered at the origin. This corresponds to a constant-amplitude sinusoid of varying frequency and continuous phase. In this case, the phase varies from 0° to 180°, because of the specific sequence $\mathbf{e}$. In general, a longer sequence will cause the phase to vary the full 360°, forming a circle for the signal space diagram.

8.2.2.3 Gaussian Minimum-Shift Keying

The spectral bandwidth of CPFSK may be modified with a pulse-shaping filter applied to the baseband signals $i(t)$ and $q(t)$. The filter commonly used is a lowpass filter called a Gaussian filter, which produces low spectral components outside the FSK main lobe. The *Gaussian minimum-shift keying*

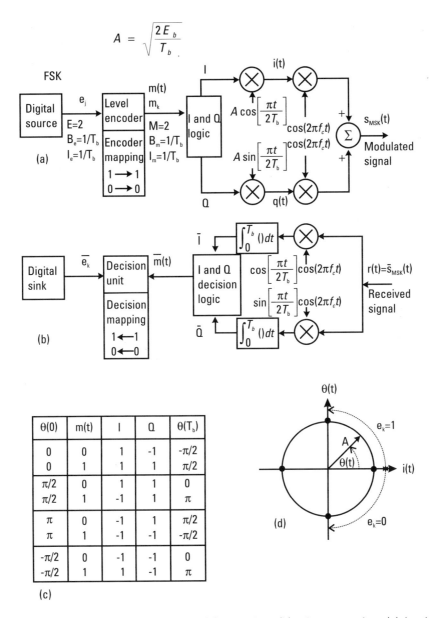

Figure 8.3 MSK system block diagrams: (a) transmitter, (b) coherent receiver, (c) *I* and *Q* logic, and (d) signal space diagram.

(GMSK) method, used in the European and North American (GSM) digital cellular systems is such an example.

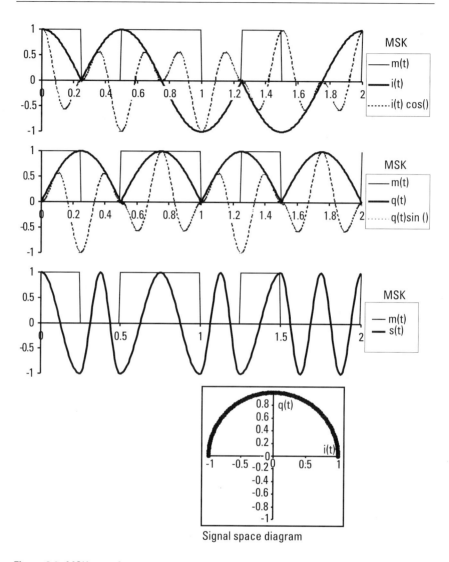

Signal space diagram

Figure 8.4 MSK waveforms.

8.2.2.4 PCM/FM

FSK may be generated by applying the binary signal $m(t)$ to a VCO, as described in Chapter 5. The limited frequency response of the VCO to the input signal $m(t)$ is a form of filtering that results in a gradual (rather than instantaneous) change in the signal frequency. This results in a continuous phase transmitted signal. Thus PCM/FM may be viewed as a form of CPFSK.

8.3 BPSK, a Generalized Model

8.3.1 A Single-Carrier Dual-Baseband Representation for BPSK

Figure 8.5 is a plot of $s_{PSK}(t)$ and the in-phase signal $i_{PSK}(t)$. The modulating signal $m(t)$ corresponds to the digital source sequence $\mathbf{e} = \{1,0,1,1,0,1,0,0\}$, with $T_b = 0.25$ seconds between symbols, or 4 bps. The carrier frequency $f_c = 8$ Hz, and the carrier amplitude is $A = 1$.

The plot of the baseband signals $q(t)$ as a function of $i(t)$, called the *signal space diagram,* is shown in the figure. Note that the quadrature baseband

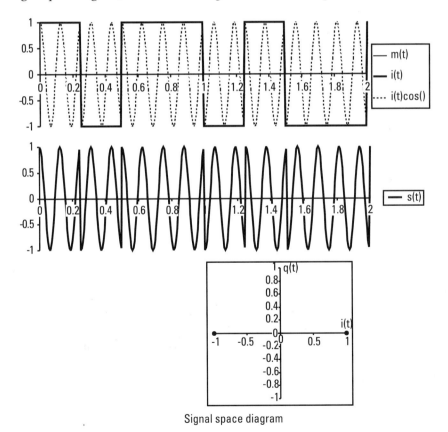

Signal space diagram

Figure 8.5 BPSK waveforms.

signal component $q(t)$ is zero, and the in-phase component is either +1 or −1 during transmission. When a different symbol is presented by the source, the phase of the carrier changes by 180°. This creates a phase discontinuity in the transmitted signal and induces an amplitude variation if the transmitted signal is filtered.

In BPSK, the binary source symbols 0 and 1 cause two sinusoidal signals that differ in phase to be transmitted. These two carrier signals $s_1(t)$, and $s_2(t)$ may be specified by

$$s_{PSK}(t) = s_i(t) = m_i\sqrt{\frac{2E_b}{T_b}}\cos(2\pi f_c t) = i_{PSK}(t)\cos(2\pi f_c t)$$

(8.31)

$$0 < t < T_b \qquad i = 1, 2$$

where E_b is the transmitted energy per symbol, and T_b is the symbol duration. Note that $A^2 = 2E_b/T_b$. When the digital source generates a binary 1 ($e_j = 1$), $m_I = m_1 = 1$, and an in-phase sinusoidal signal is transmitted. Conversely, a binary 0 source output ($e_j = 0$), results in $m_k = m_2 = -1$, and causes an out-of-phase sinusoidal signal to be transmitted.

The waveform specified by (8.31) may be generated by the transmitter shown in Figure 8.6(a). The digital source generates a new binary symbol every T_b seconds, for a baud rate of $1/T_b$. Since E = 2, the source bit rate equals the source baud rate. The level encoder generates one symbol for every bit generated by the source (1 bit/symbol), and $B_e = B_m$.

The BPSK receiver shown in Figure 8.6(b) correlates the received signal with the locally generated coherent reference signal $\cos(2\pi f_c t)$. This carrier reference signal is often recovered from the received signal by a Costas loop. If the received signal $r(t)$ results from a binary 1 transmission ($e_i = 1$), then the correlator output is greater than zero, and the decision unit selects a binary 1 as the received symbol. Transmission of a binary 0, causes the correlator output to be less than zero, resulting in a binary 0 being detected.

8.3.2 Differential Phase-Shift Keying

Phase-locked loops and the Costas loop previously mentioned, used to recover the carrier reference signal, introduce a phase ambiguity. The recovered carrier is either in phase or 180° out of phase with the transmitted carrier. Thus, the detected signal may (or may not) be inverted compared with the source signal. A training sequence, known to both the transmitter and receiver, may be sent by the transmitter to allow the receiver to resolve the ambiguity.

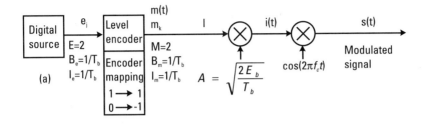

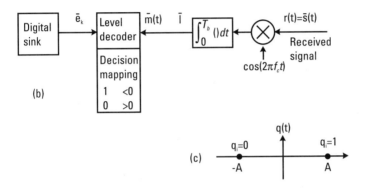

Figure 8.6 BPSK system block diagrams: (a) transmitter, (b) coherent receiver, and (c) signal space diagram.

Another solution is to design an encoding/decoding scheme that will give the same decoded output regardless of whether the received data is inverted. One common method is termed *differentially encoded phase-shift keying* (DEPSK). The encoder compares two consecutive source data bits and sends a logic 1 when there is a change in state, and a logic 0 when there is a no change in state. This process is easily implemented with an exclusive-or gate and 1-bit delay buffer. The decoding process is equally simple to implement using a second exclusive-or gate and a 1-bit delay. A drawback to this paradigm is that single bit errors often turn into double bit errors after decoding. Differential phase-shift keying (DPSK) uses the DEPSK paradigm for transmission. The receiver integrates differential decoding with data demodulation.

8.4 Performance

8.4.1 Filtering and Nonlinear Amplification

As can be seen from Figure 8.5, the unfiltered PSK signal has a constant amplitude envelope. Filtering is used to constrain the modulation bandwidth and reduce noise. Such filtering prevents an "instantaneous" change in phase required by unfiltered PSK. Because of this filtering, phase discontinuities of the modulated signal are manifested as amplitude variations in the filtered PSK waveform [2, p. 128]. The larger the phase discontinuities, the larger the amplitude variation will be.

RF power amplifiers in radio systems and higher power lasers in optical systems exhibit a nonlinear relationship between input power and output power. These nonlinear amplifiers convert DC power to AC power efficiently, and are important in space applications where electrical power is limited. When filtered PSK signals pass through these nonlinear devices, the amplitude variations produce intermodulation products. This can result in what is commonly termed *spectral regrowth*. These impairments can be mitigated by minimizing the amplitude variation (ensuring that phase transitions are continuous), as in the CPFSK methods previously discussed.

8.4.2 Bandwidth Efficiency: Spectrum, Filtering, and Bandwidth

8.4.2.1 Spectrum of FSK

The bandwidth required for FSK signal transmission increases with an increase in $f_d = 2\Delta f$, the separation between the frequencies representing the symbol states. Additionally, FSK using continuous-phase transitions between symbols will have much lower sidelobe energy and bandwidth than for the discontinuous case. Details can be found in Lucky et al. [3]. A metric of the spectral efficiency of a modulation method is I_e / B_{tnn}, the number of bps the source can send per hertz of channel bandwidth. The larger this metric is, the more efficient the modulation process.

8.4.2.2 PSD for Sunde's FSK

Sunde's FSK is binary FSK with a frequency separation of $f_d = 1/T_b$ as given by (8.25), and no pulse shaping. The baseband PSD of Sunde's FSK for $A = 1$ is given by [2, p. 122]. Replacing f by $f - f_c$ in that equation, multiplying by A^2, and dividing by 2 yields the carrierband one-sided PSD as

$$G(f)_{Sunde's\text{-}FSK} = \frac{A^2}{8} \left[\delta\left(f - f_c - \frac{1}{2T_b} \right) + \delta\left(f - f_c + \frac{1}{2T_b} \right) \right] \quad (8.32)$$

$$+ \frac{2A^2 T_b}{\pi^2} \left[\frac{\cos(\pi(f - f_c)T_b)}{4(f - f_c)^2 T_b^2 - 1} \right]^2$$

A plot of the power spectral density given in (8.32) normalized to unity amplitude at the carrier frequency, is given in Figure 8.7. The normalized frequency $(f - f_c)T_b$ is plotted as the x-axis variable.

There are two discrete components in the spectrum at the two symbol frequencies of $f \pm 0.5/T_b$. These are useful in recovering the carrier component for coherent detection. The first zero of the cosine term at $f - f_c = 0.5/T_b$ is offset by a zero in the denominator occurring at this same frequency. Thus, the first zero of the continuous spectrum occurs when the cosine term in the numerator undergoes its second zero (when $\pi(f - f_c)T_b = \pm 3\pi/2$), at $f = f_c \pm 1.5/T_b$ as can be seen in Figure 8.7. The null-to-null bandpass transmission bandwidth then, is $B_{tnn} = 3/T_b$ Hz. Since $I_e = 1/T_b$ bps, the spectral efficiency is $I_e/B_{tnn} = 0.334$ bps/Hz. By comparison, the null-to-null bandwidth of PCM/FM as given in (6.57) is $1.2R_b = 1.2/T_b$, yielding a spectral efficiency of $I_e/B_{tnn} = 0.833$ bps/Hz.

8.4.2.3 PSD for MSK

MSK is CPFSK with no pulse shaping and a frequency separation given by (8.26) as $f_d = 1/2T_b$. The baseband PSD of MSK for $A = 1$ is given by [2, p. 123]. Replacing f by $f - f_c$ in that equation, multiplying by A^2, and dividing by 2 yields the carrierband one-sided PSD as

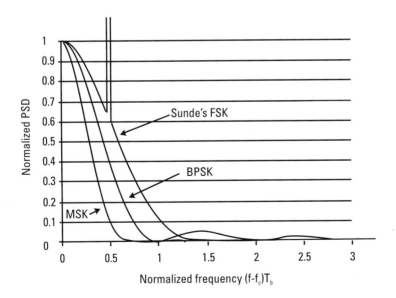

Figure 8.7 Normalized PSD as a function of normalized frequency $(f - f_c)T_b$.

$$G(f)_{MSK} = \frac{8A^2 T_b}{\pi^2} \left[\frac{\cos(2\pi(f - f_c)T_b)}{16(f\ f_c)^2 T_b^2 - 1} \right]^2 \qquad (8.33)$$

A plot of this PSD is shown in Figure 8.7. Unlike Sunde's FSK, this spectrum has no discrete components and a much narrower main lobe, as would be expected with the narrower frequency separation between symbols. The first zero of the spectrum occurs when $2\pi(f - f_c)T_b = \pm 3\pi/2$, or at $f = f_c \pm 3/(4T_b)$. The null-to-null bandpass bandwidth then, is $B_{tnn} = 3/2T_b$ Hz, yielding a spectral efficiency of 1.5 bps/Hz.

8.4.2.4   PSD for BPSK

BPSK can be viewed as an ASK signal with the carrier amplitudes as $+A$ and $-A$ (rather than $+A$ and 0 for ASK). The carrierband power spectral density is given by [4]. Replacing f by $f - f_c$ in that equation, and dividing by 2 yields the carrierband one-sided PSD as

$$G(f)_{BPSK} = \frac{A^2 T_b}{2} \left[\frac{\sin \pi(f - f_c)T_b}{\pi(f - f_c)T_b} \right]^2 \qquad (8.34)$$

A plot of this PSD is also shown in Figure 8.7. The first zero of the spectrum occurs when $\pi(f - f_c)T_b = \pi$, or at $f = f_c \pm 1/T_b$. Therefore the null-to-null bandpass bandwidth is $B_{tnn} = 2/T_b$ Hz, yielding a spectral efficiency of 1/2 bps/Hz. If the phase changes are abrupt at the symbol boundaries, then, just like FSK, the occupied bandwidth will be much larger than for smooth transitions between phase states, implying the need for shaping of the modulation waveform.

A summary of the bandwidth requirements of the modulation methods discussed in this chapter and PCM/FM discussed in Chapter 5, is given in Table 8.2. B_{nn} is the null-to-null bandwidth, and B_{99} is the bandwidth containing 99% of the unfiltered transmitted power. Additionally, the percent of power contained in the null-to-null bandwidth is also given. The power parameters were computed by numerical integration of the PSD functions including frequencies up to $80/T_b$ above and below the carrier frequency.

As can be seen PCM/FM has the best null-to-null bandwidth efficiency, followed by MSK, BPSK, and Sunde's FSK. However, MSK and PCM/FM exhibit the best 99% bandwidth efficiency, followed by Sunde's FSK, and BPSK.

Table 8.2
Bandwidth Comparison of Binary Modulation Methods

Modulation Method	B_{nn} (Hertz)	% power in B_{nn}	B_{99} (Hertz)
PCM/FM	$1.2/T_b$	99+	$1.16/T_b$
MSK	$1.5/T_b$	99.5	$1.16/T_b$
BPSK	$2.0/T_b$	90.6	$16.5/T_b$
Sunde FSK	$3.0/T_b$	99.8	$2.0/T_b$

8.4.3   Power Efficiency: BER Performance

The probability of error results that follow assume that the received signal is corrupted by additive white Gaussian noise (AWGN). The white noise has a two-sided power spectral density of $N_o/2$ as discussed in Section 6.4.2, and variance of $\sigma^2 = N_o/2$ as presented in (6.11). Additionally, no signal distortion resulting from bandwidth limitations is assumed. The probability that a received symbol is detected in error (P_s) is the same as the probability that a bit is detected in error (P_b), since one symbol corresponds to 1 bit in binary digital modulation systems.

Detailed derivations for each result are not included, but are similar to that given in Chapter 6 for BPSK. Anticipating the need, we now define the complementary error function and its relation to the Q-function as defined in (6.36)

$$erfc(x) \equiv \frac{2}{\sqrt{\pi}} \int_{x}^{\infty} e^{-z^2} dz = 2Q(\sqrt{2}x) \qquad (8.35)$$

8.4.3.1   BER Performance for BFSK

The *theoretical* performance for coherent FSK is given by [1, pp. 515, 540] as

$$P_{b,FSK,C} = \frac{1}{2} erfc\left(\sqrt{\frac{E_b}{2N_o}} \right) \qquad (8.36)$$

and as

$$P_{b,FSK,NC} = \frac{1}{2} \exp\left(-\frac{E_b}{2N_o} \right) \qquad (8.37)$$

for noncoherent detection.

Equation (8.37) is the same result given by (5.15) for noncontinuous and noncoherently detected PCM/FM. Section 5.4.1 describes PCM/FM as being continuous due to the limited dynamic response of the VCO. Equation (5.16) shows better experimental performance for continuous PCM/FM than that indicated by (5.15). The E_b/N_o penalty of the simpler noncoherent detection method is only about 1 dB at acceptable error rates. As a result, the simpler, noncoherent FSK modems are commonly used for many low-speed telemetry, telephone and radio applications in the marketplace.

8.4.3.2 BER Performance for MSK

The theoretical performance for coherent MSK is given by [1, p. 531] as

$$P_{b,MSK,C} = \frac{1}{2} erfc\left(\sqrt{\frac{E_b}{N_o}} \right) \qquad (8.38)$$

8.4.3.3 BER Performance for BPSK

Using (8.35), the result given in (6.22) for coherent BPSK bit error probability may be rewritten as

$$P_{b,BPSK,C} = \frac{1}{2} erfc\left(\sqrt{\frac{E_b}{N_o}} \right) \qquad (8.39)$$

8.4.3.4 BER Performance for DPSK

The theoretical performance for coherent DPSK is given by [1, p. 544] as

$$P_{b,DPSK} = \frac{1}{2} \exp\left(-\frac{E_b}{N_o} \right) \qquad (8.40)$$

The probability of a bit being detected in error (P_b), is equivalent to the average BER. The unfiltered, theoretical BER performance for the major binary modulation schemes specified by (8.36) to (8.40) is given in Figure 8.8. The BER decreases monotonically as the received energy per bit to noise spectral density ratio, E_b/N_o, increases.

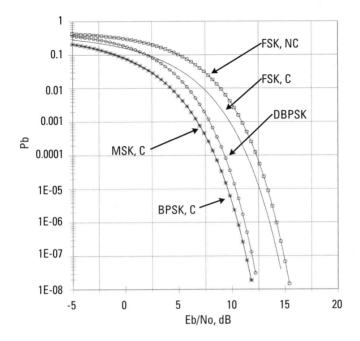

Figure 8.8 BER performance for unfiltered binary modulation methods.

Coherently detected BPSK and MSK exhibit the best performance, resulting in the lowest BER for a given E_b/N_o. DBPSK gives the next best results, followed by coherently detected FSK, and noncoherently detected FSK. It should be noted that at high E_b/N_o, noncoherently detected FSK performs almost as well as coherently detected FSK. The MSK receiver processes the received signal over two consecutive offset bit periods, having a detection process with memory. The other methods process on a bit-by-bit basis and have no memory. The extra information provided by the memory feature allows MSK to perform better than FSK.

8.4.4 Power and Bandwidth Efficiency Comparison

The ratio E_b/N_o is a measure of the power efficiency of the system. The smaller the ratio, the less energy required per bit in the presence of noise for a given BER. I_e/B_{tnn} represents the bandwidth efficiency (bps/Hz) of the system. The larger this ratio, the greater the bandwidth efficiency. Table 8.3 summarizes the power and bandwidth efficiency for the binary modulation methods covered in this chapter.

Table 8.3
Power and Bandwidth Efficiency Comparison of Binary Modulation Methods

Modulation Method	$P_e = 10^{-6}$ E_b/N_o	$P_e = 10^{-8}$ E_b/N_o	I_e/B_{tnn} (bps/Hz)
MSK, C	10.5	15.8	0.67
BPSK, C	10.5	15.8	0.50
DBPSK	11.13	17.5	0.34
FSK, C	22.5	31.0	0.34
FSK, NC	26.0	35.0	0.34

Problems

Problem 8.1

Describe the FSK modulation process. What is M?

Problem 8.2

Consider a BFSK modulator with $f_1 = 90$ KHz, $f_2 = 110$ KHz, and a digital source with a baud rate of 2,000 symbols/second. The transmitted energy per bit $E_b = 0.002$J. Find (a) the bit duration T_b, (b) the peak carrier voltage A, and (c) the two carrier signals $s_1(t)$, and $s_2(t)$.

Problem 8.3

What are f_d, f_c, and Δf for the FSK signal specified in Problem 8.2?

Problem 8.4

Find the receiver correlator outputs r_1 and r_2 shown in Figure 8.1(b) for the FSK signal specified in Problem 8.2.

Problem 8.5

Consider the FSK signal specified in Problem 8.2. Find the in-phase baseband signal using (8.22), and find the quadrature baseband signal using (8.23).

Problem 8.6

Find new values of f_1 and f_2 so that the FSK signal specified in Problem 8.2 becomes a MSK signal. Keep f_c at 100KHz and find Δf.

Problem 8.7

Consider a BPSK modulator with f_c = 100 kHz, and a digital source with a baud rate of 2,000 symbols/second. The transmitted energy per bit E_b = 0.002J. Find (a) the bit duration T_b, (b) the peak carrier voltage A, and (c) the two carrier signals $s_1(t)$, and $s_2(t)$.

Problem 8.8

Consider the two carrier signals $s_1(t)$, and $s_2(t)$ for the PSK signal specified in Problem 8.7. Assume a bit sequence of m_k = {1, −1, 1}. How many carrier cycles occur during the transmission of each bit? Make a simple sketch of the transmitted signal.

Problem 8.9

Find the receiver correlator output I, for the PSK receiver shown in Figure 8.6, for the PSK signal specified in Problem 8.7.

Problem 8.10

A binary source in a DPSK system generates the binary sequence 0 0 0 1 1 0 1 1. Determine the differential encoder output as described in Section 8.3.2.

Problem 8.11

Compute the null-to-null bandwidth B_{nn} and the 99% power bandwidth B_{99} for (a) the FSK system specified in Problem 8.2, and (b) the PSK system specified in Problem 8.7. Which of these is more bandwidth efficient?

Problem 8.12

Which of the modulation methods given in Table 8.3 has the smallest (a) null-to-null bandwidth B_{nn} and (b) the smallest 99% power bandwidth B_{99}?

Problem 8.13

Which of the modulation methods given in Table 8.3 has (a) the best power efficiency, (b) the worst bandwidth efficiency?

Problem 8.14

Which of the modulation methods given in Table 8.3 has (a) the best bandwidth efficiency, (b) the worst power efficiency?

References

[1]   Haykin, S., *Communication Systems* 3rd Ed., New York: Wiley, 1994.

[2]   Bateman, A., *Digital Communication* New York: Addison-Wesley, 1988.

[3]   Lucky, R. W., J. Saly, and S. J. Welden, *Principles of Data Communication* New York: McGraw-Hill, 1968.

[4]   Couch, L. W., *Digital and Analog Communication Systems,* 6th Ed., Upper Saddle River, NJ: Prentice-Hall, 2001, pp. 341–342.

9

M-ary Digital Communication Systems

The binary digital communication systems presented in Chapter 8, in which $M = 2$, are extended in this chapter. Details are presented for digital communication systems in which the encoder output signal may assume more than two levels (i.e., $M > 2$). These systems are referred to as M-ary digital communications systems, and are classified by the type of modulation/demodulation method used. Four categories characterize these methods: *M-ary amplitude-shift keying* (M-ary ASK), *M-ary frequency-shift keying* (M-ary FSK), *M-ary phase-shift keying* (M-ary PSK), and *M-ary amplitude and phase-shift keying*. M-ary ASK exhibits relatively poor BER performance, and is sensitive to channel amplitude nonlinearities, resulting in few practical applications. Additionally it may be considered as a special case of the last category, and is thus not considered in this chapter.

M-ary communication methods are preferable to binary methods to send data over bandwidth-constrained bandpass channels. This occurs in practice when binary modulation and demodulation does not provide sufficient throughput. The designer can increase throughput by increasing M, at the cost of increased power to maintain acceptable BER.

9.1 Learning Objectives

Upon completion of this chapter, the reader should understand the following:

- M-ary FSK and its orthogonal and nonorthogonal implementations;

- M-ary PSK and its derivatives;
  - Quadrature phase-shift keying;
  - Offset quadrature phase-shift keying;
  - Interference and jitter-free quadrature phase-shift keying;
  - Feher-patented quadrature phase-shift keying;
  - Enhanced Feher-patented quadrature phase-shift keying;
- M-ary amplitude and phase-shift keying and its derivatives;
  - Quadrature amplitude modulation;
  - Amplitude and phase keying;
- Mathematical modeling of M-ary FSK, M-ary PSK, and quadrature amplitude modulation and derivatives;
- Generation and detection of M-ary FSK, M-ary PSK, and quadrature amplitude modulation and derivatives;
- Composite and component waveforms of M-ary FSK, M-ary PSK, and quadrature amplitude modulation and derivatives;
- Spectrum and bandwidth efficiency of M-ary FSK, M-ary PSK, and quadrature amplitude modulation and derivatives;
- BER and power efficiency of M-ary FSK, M-ary PSK, and quadrature amplitude modulation and derivatives;
- Performance comparison of M-ary FSK, M-ary PSK, and quadrature amplitude modulation and derivatives.

9.2 M-ary FSK

In M-ary FSK, the frequency f of the carrier is shifted between M different frequencies corresponding to the digital source symbols, where $M > 2$. The case in which $M = 2$ is known as BFSK, and is covered in Chapter 8. The M carrier signals $s_1(t)$, $s_2(t) \ldots s_M(t)$ are specified by

$$s_{M\text{-}ary\ FSK}(t) = s_i(t) = \sqrt{\frac{2E_s}{T_s}} \cos(2\pi f_i t) \qquad (9.1)$$

$$0 < t < T_s \qquad i = 1, 2, \ldots, M$$

where E_s is the transmitted energy per symbol, and T_s is the symbol duration. Note that $A^2 = 2E_b/T_b$.

The waveform specified in (9.1) may be generated by using the transmitter shown in Figure 9.1(a). The encoder selects n consecutive bits from the digital source and encodes them into one of M symbols, where $M = 2^n$.

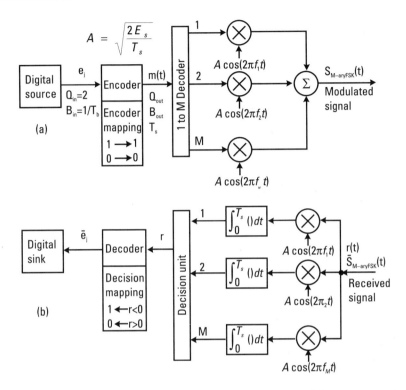

Figure 9.1 M-ary FSK system block diagrams: (a) transmitter, and (b) coherent receiver.

This encoder output symbol $m_k = i$, causes output i of the 1 of M decoder to be 1, with all other outputs 0. This, in turn, causes the signal $s_i(t) = A\cos(2\pi f_i t)$ to be selected and transmitted. The digital source generates a new binary symbol every T_b seconds, for a baud rate of $B_e = 1/T_b$. Since $E = 2$, the source bit rate I_e equals the source baud rate B_e. The level encoder generates one symbol for every n bits generated by the source (n bit/symbol), and therefore the encoder output baud rate is $B_m = B_e/n$. It also follows from (4.11) and (4.12) that $I_m = I_e$.

The M-ary FSK coherent receiver shown in Figure 9.1(b) correlates the received signal with M locally generated coherent reference signals $\cos(2\pi f_i t)$, $i = 1, 2, \ldots M$. The decision unit selects the correlator i with the largest output r_i, corresponding to the symbol transmitted. The decoder then generates the n bits corresponding to symbol i.

9.2.1 M-ary Nonorthogonal FSK

By spacing the frequencies close together, it is possible to squeeze more symbols into a given bandwidth, and hence improve bandwidth efficiency.

However this causes a decrease in noise immunity as a result of the symbol waveforms $s_i(t)$ no longer being orthogonal. In the limit, as the frequencies become very close to each other, the correlator outputs are nearly identical, and the error rate is unacceptably high.

9.2.2 M-ary Orthogonal FSK

For orthogonal signaling, the cross-correlation between signals must be zero. This occurs when

$$f_i = f_c + \frac{i}{2T_s} \tag{9.2}$$

where $i = 1, 2, \ldots M$, and $f_c = n_c/2T_s$, for n_c a fixed integer. The signals are orthogonal over the symbol period T_s. Note that the unfiltered channel bandwidth is proportional to the encoder output baud rate $B_m = 1/T_s$. Substituting (9.2) into (9.1), yields

$$s_{M\text{-ary FSK}}(t) = s_i(t) = \sqrt{\frac{2E_s}{T_s}} \cos\left(2\pi f_c t + \frac{2\pi i t}{2T_s}\right) \tag{9.3}$$

for orthogonal signaling.

9.2.2.1 Channel Bandwidth for Orthogonal M-ary FSK

For M-ary signaling, the symbol duration is

$$T_s = nT_b = (\log_2 M)T_b \tag{9.4}$$

For orthogonal M-ary FSK, adjacent signals need only be separated by a frequency difference of $1/(2T_s)$ to maintain orthogonality, as specified in (9.2). Since there are M signals, the frequency span is $(M-1)/2T_s$ plus another $1/T_s$ Hz on each end of the spectrum to include the null points. Using (9.4), we have the approximate bandwidth as

$$B_{t, M\text{-ary FSK}} = \frac{2}{T_s} + \frac{M-1}{2T_s} = \frac{M+3}{2T_b \log_2 M} \tag{9.5}$$

Equivalently, the bandwidth efficiency is

$$\frac{I_e}{B_{t,M\text{-}ary\ FSK}} = \frac{2\log_2 M}{M+3}\frac{\text{bps}}{\text{Hz}}$$

(9.6)

The efficiency decreases nonlinearly with M, having a maximum value of 0.4 bps/Hz when $M = 2$. This is reasonably close to the value of 0.34 bps/Hz given in Section 8.4.2.2, for BFSK.

9.2.2.2 Symbol Error Probability for Orthogonal M-ary FSK

As a result of the nonlinear modulation process, the error probability for M-ary FSK is difficult to derive, and must be calculated numerically for the general case. An upper bound for the probability of symbol error is given by [1] as

$$P_s \leq \frac{1}{2}(M-1)erfc\left(\sqrt{\frac{E_s}{2N_o}}\right)$$

(9.7)

For fixed M, this bound becomes increasingly tight as E_s/N_o is increased. Additionally, for $M = 2$, the bound becomes equality, as given in (8.36) for BFSK.

9.3 M-ary PSK

9.3.1 M-ary PSK Introduction

In M-ary PSK modulation, the phase of the carrier is shifted between M different phases corresponding to the digital source symbols, where $M > 2$. The case where $M = 2$ is known as binary PSK and is covered in Chapter 8. The M carrier signals $s_1(t)$, $s_2(t)$, ... $s_M(t)$ for orthogonal M-ary PSK are specified by

$$s_{M\text{-}ary\ FSK}(t) = s_i(t) = \sqrt{\frac{2E_s}{T_s}}\cos\left(2\pi f_c t + \frac{2\pi}{M}(i-1)\right)$$

(9.8)

$$0 < t < T_s \quad i = 1, 2, \ldots, M$$

where E_s is the transmitted energy per symbol, T_s is the symbol duration, and $f_c = n_c/T_s$ for some fixed integer n_c.

The waveform specified by (9.8) may be generated by using the transmitter similar to that shown in Figure 9.1(a), with $s_i(t)$ replaced by (9.8).

The encoder selects n consecutive bits from the digital source and encodes them into one of M symbols, where $M = 2^n$. This encoder output symbol $m_k = i$, causes output i of the 1 of M selector to be 1, with all other outputs 0. This, in turn, causes the signal $s_i(t)$ to be selected and transmitted. The digital source generates a new binary symbol every T_b seconds, for a baud rate of $B_e = 1/T_b$. Since $E = 2$, the source bit rate I_e equals the source baud rate B_e. The level encoder generates one symbol for every n bits generated by the source (n bit/symbol), and therefore the encoder output baud rate is $B_m = B_e/n = 1/T_s$. It also follows from (4.11) and (4.12) that $I_m = I_e$.

The M-ary PSK coherent receiver is similar to that shown in Figure 9.1(b), with $s_i(t)$ replaced by (9.8). The received signal is correlated with M locally generated coherent reference signals $s_i(t)$, $i = 1, 2, \ldots M$. The decision unit then selects the correlator i with the largest output r_i, corresponding to the symbol transmitted. The decoder then generates the n bits corresponding to symbol i.

9.3.1.1 Channel Bandwidth for Orthogonal M-ary PSK

From (8.34), the null-to-null bandpass transmission bandwidth for binary PSK is $B_{tnn} = 2/T$ Hz, where T is the symbol duration. Using (9.4), we have the channel approximate bandwidth as

$$B_{t,\text{M-ary FSK}} = \frac{2}{T_b \log_2 M} \tag{9.9}$$

Equivalently, the bandwidth efficiency is

$$\frac{I_e}{B_{t,\text{M-ary FSK}}} = \frac{\log_2 M}{2} \frac{\text{bps}}{\text{Hz}} \tag{9.10}$$

The efficiency increases nonlinearly with M, having a minimum value of 1/2 bps/Hz for the binary case when $M = 2$. This is the same as given in section 8.4.2.4 for BPSK.

9.3.1.2 Symbol Error Probability for Orthogonal M-ary PSK

An upper and lower bound for symbol error probability is given in [2] as

$$\frac{1}{2} erfc\left(\sqrt{\frac{E_s}{N_o}} \sin\left(\frac{\pi}{M}\right)\right) \le P_s < erfc\left(\sqrt{\frac{E_s}{N_o}} \sin\left(\frac{\pi}{M}\right)\right) \tag{9.11}$$

The upper bound becomes very tight for fixed M as E_s/N_o becomes large, and for M large. The lower bound is exact for $M = 2$, giving the same result as (8.39) for BPSK.

9.3.2 Quaternary Phase-Shift Keying ($M = 4$)

9.3.2.1 Quadrature Phase-Shift Keying

Setting $M = 4$ in (9.8) for M-ary PSK, yields four symbols corresponding to carrier phases 0, $\pi/2$, π, and $3\pi/2$. This modulation method in which the symbols are separated by $\pi/2$ radians, is known as *quadrature phase-shift-keying* (QPSK) [also known as *quadriphase-shift keying*]. In $\pi/4$ QPSK, the constellation is rotated $\pi/4$ radians, and transmitted signal is

$$s_{QPSK}(t) = s_k(t) = \sqrt{\frac{2E_s}{T_s}} \cos\left(2\pi f_c t + \frac{\pi}{4}(2k - 1) \right) \qquad (9.12)$$

$$0 < t < T_s \quad k = 1, 2, 3, 4$$

The carrier phase takes on one of four equally spaced values corresponding to carrier phases $\pi/4$, $3\pi/2$, $5\pi/4$, and $7\pi/4$. Introducing I_k and Q_k as the polarity of the in-phase and quadrature baseband signals respectively, and using a well-known trigonometric identity, (9.12) becomes

$$s_{QPSK}(t) = i_{QPSK}(t)\cos(2\pi f_c t) + q_{QPSK}(t)\sin(2\pi f_c t) \qquad (9.13)$$

where

$$i_{QPSK}(t) = I_k = \sqrt{\frac{2E_s}{T_s}} \cos\left[(2k - 1)\frac{\pi}{4} \right] \qquad (9.14)$$

$$q_{QPSK}(t) = Q_k = -\sqrt{\frac{2E_s}{T_s}} \sin\left[(2k - 1)\frac{\pi}{4} \right]$$

A block diagram of a QPSK transmitter based on (9.13) and (9.14) is shown in Figure 9.2(a). The corresponding coherent receiver is given in Figure 9.2(b). The values of I_k and Q_k depend on the encoder output state k and are given in (9.14). The encoder output state in turn, is a function of two consecutive encoder input bits and is specified in Figure 9.2(c). The

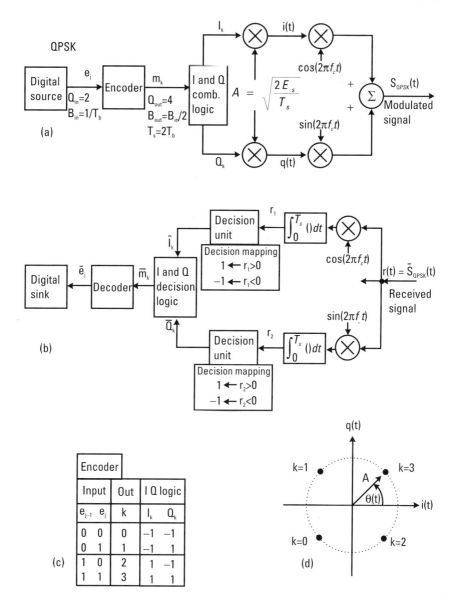

Figure 9.2 QPSK system block diagrams: (a) transmitter, (b) coherent receiver, (c) I_k and Q_k logic, and (d) signal space diagram.

signal space diagram showing the four symbols spaced $\pi/2$ radians apart is given in Figure 9.2(d).

An example of QPSK waveforms for $A = 1$, a symbol period of

$T_s = 0.5$ seconds, and $f_c = 2/T_s = 4$ Hz is shown in Figure 9.3. The signals $s_{QPSK}(t)$, $i_{QPSK}(t)$, and $q_{QPSK}(t)$ as defined by (9.13) and (9.14) are shown. The modulating signal $m(t)$ corresponds to the digital source sequence

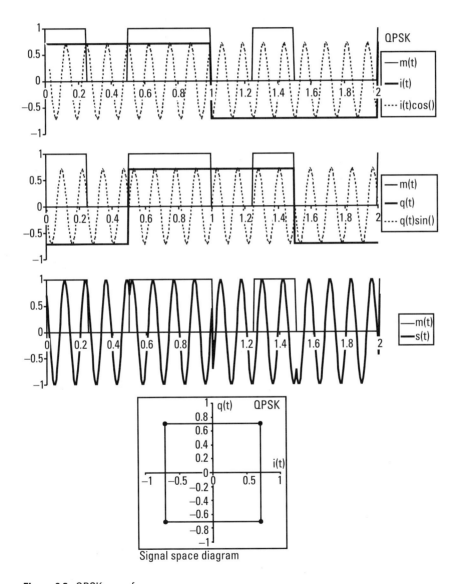

Signal space diagram

Figure 9.3 QPSK waveforms.

e = {1,0,1,1,0,1,0,0}, with T_b = 0.25 seconds between symbols, or 4 bps. As can be seen, the transmitted signal $m(t)$ is not continuous in phase, making abrupt transitions when a new symbol is presented.

The signal space diagram forms a square, centered at the origin. Each symbol has a distance of A = 1 from the origin. Transitions abruptly change the phase by $\pi/2$ radians, as represented by the straight lines connecting the four corners of the diagram. In general, a longer bit sequence will cause transitions diagonally as well, corresponding to a phase discontinuity of π radians. For example, when the symbol for k = 1 is followed by the symbol for k = 2, the phase instantaneously decreases by π radians. This phase discontinuity in the transmitted signal induces an amplitude variation if the transmitted signal is filtered. When this time-varying envelope is amplified by a nonlinear power amplifier, in regard to amplitude modulation (AM) and phase modulation (PM), AM-to-AM and AM-to-PM conversion occurs, causing signal degradation [3].

9.3.2.1.1   Channel Bandwidth for QPSK

The null-to-null transmission bandwidth may be determined from (9.10), for M-ary PSK, with M = 4. This gives a bandwidth efficiency of

$$\frac{I_e}{B_{t,QPSK}} = \frac{\log_2 4}{2} = 1 \frac{\text{bps}}{\text{Hz}} \tag{9.15}$$

9.3.2.1.2   Symbol Error Probability for QPSK

The probability of QPSK symbol error is given by [1] as

$$P_s = erfc\left(\sqrt{\frac{E_s}{2N_o}}\right) - \frac{1}{4} erfc^2\left(\sqrt{\frac{E_s}{2N_o}}\right) \tag{9.16}$$

For $E_s/2N_o \gg 1$, the second term is small and may be ignored.

9.3.2.2   Offset Quadrature Phase-Shift Keying

The π radian phase discontinuity in QPSK signal induces an amplitude variation if the transmitted signal is filtered. Offset quadrature phase-shift keying is QPSK in which the allowed transition times for the I_k and Q_k components are offset by a 1/2 symbol interval ($T_s/2$). This offset greatly reduces the induced amplitude modulation on the offset quadrature phase-shift keying (OQPSK) signal compared with that on the corresponding QPSK signal. The amplitude modulation is reduced because the maximum phase discontinuity is reduced from π radians to $\pi/2$ radians, since only one of the I_k and Q_k components can change at a time. Modifying (9.13) to incorporate this change, we have

$$s_{OQPSK}(t) = i_{QPSK}(t)\cos(2\pi f_c t) + q_{QPSK}\left(t - \frac{T_s}{2}\right)\sin(2\pi f_c t)$$

$$(9.17)$$

where

$$i_{OQPSK}(t) = I_k = \sqrt{\frac{2E_s}{T_s}}\cos\left[(2k-1)\frac{\pi}{4}\right] \qquad (9.18)$$

$$q_{OQPSK}(t) = Q_k = -\sqrt{\frac{2E_s}{T_s}}\sin\left[(2k-1)\frac{\pi}{4}\right]$$

A block diagram of an OQPSK transmitter based on (9.17) and (9.18) is shown in Figure 9.4(a). The corresponding coherent receiver is given in Figure 9.4(b). The values of I_k and Q_k depend on the encoder output state k as given in (9.18). The encoder output state in turn, is a function of two consecutive encoder input bits and is specified in Figure 9.4(c). The signal space diagram showing the four symbols spaced $\pi/2$ radians apart is given in Figure 9.4(d).

An example of OQPSK waveforms for $A = 1$, a symbol period of $T_s = 0.5$ seconds, and $f_c = 2/T_s = 4$ Hz is shown in Figure 9.5. The signals $s_{OQPSK}(t)$, $i_{OQPSK}(t)$, and $q_{OQPSK}(t)$, as defined by (9.17) and (9.18), are shown. The modulating signal $m(t)$ corresponds to the digital source sequence $\mathbf{e} = \{1,0,1,1,0,1,0,0\}$, with $T_b = 0.25$ seconds between symbols, or 4 bps. As can be seen, the transmitted signal $m(t)$ is not continuous in phase, making abrupt transitions when a new symbol is presented.

The signal space diagram forms a square, centered at the origin. Each symbol has a distance of $A = 1$ from the origin. Transitions abruptly change the phase by $\pi/2$ radians, as represented by the straight lines connecting the four corners of the diagram. In general, a longer bit sequence will cause transitions with phase discontinuity limited to $\pi/2$ radians. This phase discontinuity is half of that of QPSK resulting in significantly reduced amplitude variation if the transmitted signal is filtered. The channel bandwidth and symbol error probability are the same as for QPSK.

9.3.2.3   Interference and Jitter-Free OQPSK

The $\pi/2$ phase discontinuity in OQPSK is a direct result of the instantaneous change in the baseband signals $i(t)$ and $q(t)$. This is readily apparent from

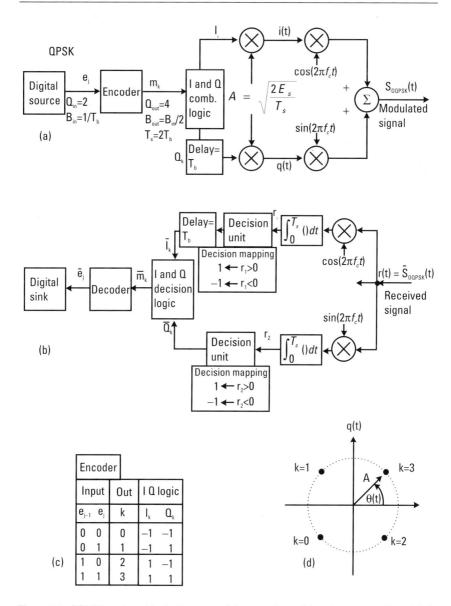

Figure 9.4 OQPSK system block diagrams: (a) transmitter, (b) coherent receiver, (c) I_k and Q_k logic, and (d) signal space diagram.

the waveforms in Figure 9.5. By filtering these baseband waveforms prior to multiplication with the carrier component, the change in carrier phase

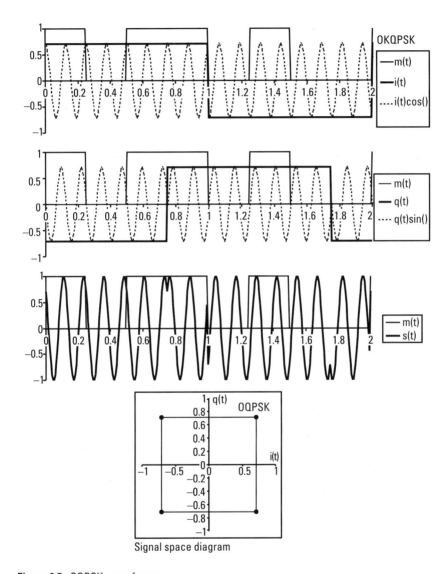

Figure 9.5 OQPSK waveforms.

will be more gradual. The result is a smaller induced amplitude variation in the filtered QPSK waveform.

Such a technique, introduced by Feher et al. [3], is the intersymbol-interference and jitter-free offset quadrature phase-shift keying (IJF-OQPSK) modulation method. This method employs nonlinear switched filters known as Feher's nonlinear processor (or filter) [4] to assure that the baseband signals $i_{IJF\text{-}OQPSK}(t)$ and $q_{IJF\text{-}OQPSK}(t)$ make smooth transitions in phase. The transitions are made by switching in the appropriate sinusoidal signal when a change in amplitude occurs and maintaining constant amplitude when no change occurs. Modifying (9.17) and (9.18) to incorporate this change, we have

$$s_{IJF\text{-}OQPSK}(t) = i_{IJF\text{-}OQPSK}(t)\cos(2\pi f_c t) + q_{IJF\text{-}OQPSK}\left(t - \frac{T_s}{2}\right)\sin(2\pi f_c t)$$

(9.19)

and

$$I_k = \sqrt{\frac{2E_s}{T_s}}\cos\left[(2k-1)\frac{\pi}{4}\right]$$

(9.20)

$$Q_k = -\sqrt{\frac{2E_s}{T_s}}\sin\left[(2k-1)\frac{\pi}{4}\right]$$

A block diagram of an IJF-OQPSK transmitter based on Feher's nonlinear filter is shown in Figure 9.6(a). The corresponding coherent receiver is given in Figure 9.6(b). The values of I_k and Q_k depend on the encoder output state k as given in (9.20). The I_k and Q_k combinational logic output state, in turn, is a function of two consecutive encoder input bits and is specified in Figure 9.6(c). Consecutive values of I_k and Q_k assume ± 1 values, and are then nonlinearly filtered by the I sequential logic and switch, and the Q sequential logic and switch respectively. The switch outputs are defined in Figure 9.6(d). Note that the switch output is determined by two consecutive inputs, and thus has memory. This is in contrast to OQPSK, which is memoryless. The signal space diagram showing all possible transitions is given in Figure 9.6(e).

An example of IJF-OQPSK waveforms for $A = 1$, a symbol period of $T_s = 0.5$ seconds, and $f_c = 2/T_s = 4$ Hz is shown in Figure 9.7. The signals $s_{IJF\text{-}QPSK}(t)$, $i_{IJF\text{-}QPSK}(t)$, and $q_{IJF\text{-}QPSK}(t)$ as defined by (9.19) and (9.20) are shown. The modulating signal $m(t)$ corresponds to the digital source sequence $\mathbf{e} = \{1,0,1,1,0,1,0,0\}$, with $T_b = 0.25$ seconds between symbols, or 4 bps. As can be seen, the baseband signals $i_{IJF\text{-}QPSK}(t)$, and $q_{IJF\text{-}QPSK}(t)$ make smooth transitions between extremes, in contrast to the abrupt transitions of

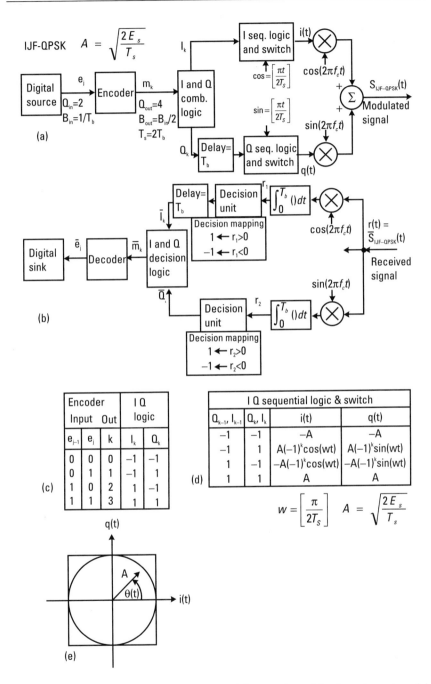

Figure 9.6 IJF-OQPSK system block diagrams: (a) transmitter, (b) coherent receiver, (c) I_k and Q_k logic, (d) IQ sequential logic and switch, and (e) signal space diagram.

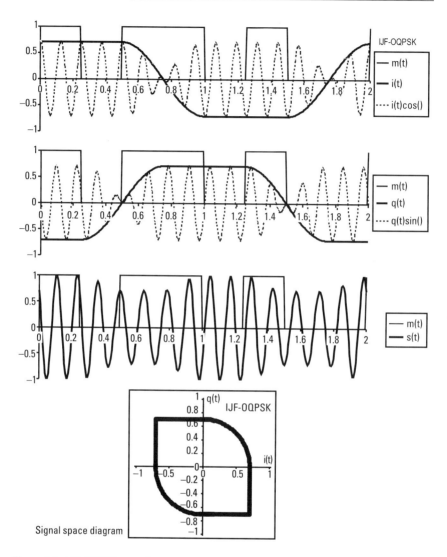

Figure 9.7 IJF-OQPSK waveforms.

OQPSK. This is manifested in the continuous nature of the transmitted waveform $s_{IJF-QPSK}(t)$.

The signal space diagram is continuous, exhibiting smooth transitions as new input symbols are presented. The maximum value of the carrier amplitude is 1, while the minimum value is 0.707. This translates to a 3-dB envelope fluctuation of the transmitted signal $s_{IJF-QPSK}(t)$.

Studies by Feher et al. [3], indicate that the IJF-OQPSK modulation scheme has superior bandwidth efficiency and error probability to QPSK, OQPSK, and MSK in narrowband hard-limited channels

9.4 Feher-Patented Quadrature Phase-Shift Keying

As described above, IJF-QPSK has a 3-dB envelope fluctuation. With the expressed purpose of reducing this fluctuation to 0 dB, Feher-patented quadrature phase-shift keying (FQPSK) has been patented [5] and reported in the recent literature [6, 7]. Conceptually this technique is the same as the cross-correlated phase-shift keying (XPSK) modulation technique introduced by Kato and Feher [8]. This technique was in turn a modification of the IJF-QPSK method described previously.

Kato and Feher achieved the 3-dB reduction by introducing an intentional but controlled amount of cross correlation between the I_k and Q_k channels. This cross correlation was applied to the IJF-QPSK baseband signals prior to modulation onto the I_k and Q_k carriers. This transformation was implemented by mapping in each half symbol the 16 possible combinations of the I_k and Q_k baseband waveforms present in the IJF-QPSK signal, into a new set of 16 waveform combinations. These new waveforms were chosen in such a way that the baseband signals are time continuous and the envelope is constant. However, the first derivative, or the slope, of the cross-correlator output of FQPSK is discontinuous at the half-symbol transition point for a random input sequence [9]. Since the smoother the modulating signal—that is, for those with continuous derivatives—the faster the side lobes of the spectral density decay, research is being pursued to develop modulation signals with continuous or smooth slopes, resulting in faster spectral roll off.

Feher developed a group of additional modulation methods, referred to generally as the FQPSK family, that have both good BER versus E_b/N_o performance, and well-controlled spectral characteristics [10]. One member of this family is FQPSK-B [11], which includes proprietary-designed filtering for better spectral control [1, 2, 12]. The filter implementation for better spectral roll off involves additional post low-pass filtering of the cross-correlator output [13].

Simon and Yan [14], in order to achieve a continuous slope of the modulating signal, have recast the original characterization of FQPSK into a mapping performed directly on the input I_k and Q_k data sequences every full symbol interval T_s. To do so they defined 16 waveforms $s_I(t)$; $I = 0$, 1, 2, . . . 15 over the interval $-T_s/2 \leq t \leq T_s/2$. These waveforms are specified by

$$s_0(t) = -s_0(t) = A, \quad -T_s/2 \leq t \leq T_s/2$$

$$s_1(t) = -s_9(t) = \begin{cases} A, & -T_s/2 \leq t \leq 0 \\ 1 - (1-A)\cos^2\dfrac{\pi t}{T_s}, & 0 \leq t \leq T_s/2 \end{cases}$$

$$s_2(t) = -s_{10}(t) = \begin{cases} 1 - (1-A)\cos^2\dfrac{\pi t}{T_s}, & -T_s/2 \leq t \leq 0 \\ A, & 0 \leq t \leq T_s/2 \end{cases}$$

$$s_3(t) = -s_{11}(t) = 1 - (1-A)\cos^2\dfrac{\pi t}{T_s}, \quad -T_s/2 \leq t \leq T_s/2 \quad (9.21a)$$

$$s_4(t) = -s_{12}(t) = A\sin\dfrac{\pi t}{T_s}, \quad -T_s/2 \leq t \leq T_s/2 \quad\quad\quad (9.21b)$$

$$s_5(t) = -s_{13}(t) = \begin{cases} A\sin\dfrac{\pi t}{T_s}, & -T_s/2 \leq t \leq 0 \\ \sin\dfrac{\pi t}{T_s}, & 0 \leq t \leq T_s/2 \end{cases}$$

$$s_6(t) = -s_{14}(t) = \begin{cases} \sin\dfrac{\pi t}{T_s}, & -T_s/2 \leq t \leq 0 \\ A\sin\dfrac{\pi t}{T_s}, & 0 \leq t \leq T_s/2 \end{cases}$$

$$s_7(t) = -s_{15}(t) = \sin\dfrac{\pi t}{T_s}, \quad -T_s/2 \leq t \leq T_s/2$$

The particular $i(t)$ and $q(t)$ waveforms chosen for any particular T_s signaling interval on each channel depend on the most recent data transition on that channel, as well as the two most recent successive transitions on the other channel. Tables 9.1 and 9.2 specify the details where d_{Ik} is the data sequence on the I channel, and d_{Qk} the data sequence for the Q channel.

Modifying (9.19) and (9.20) to incorporate the modifications, we have

$$s_{FQPSK}(t) = i_{FQPSK}(t)\cos(2\pi f_c t) + q_{FQPSK}\left(t - \frac{T_s}{2}\right)\sin(2\pi f_c t)$$

$$(9.22)$$

Table 9.1
Mapping for $i(t)$ in the Interval $(n - 1/2)T_s < t < (n - 1)T_s$

$\|d_{In} - d_{In-1}\|/2$	$\|d_{Qn-1} - d_{Qn-2}\|/2$	$\|d_{Qn} - d_{Qn-1}\|/2$	$i(t)$
0	0	0	$I_n s_0(t - nT_s)$
0	0	1	$I_n s_1(t - nT_s)$
0	1	0	$I_n s_2(t - nT_s)$
0	1	1	$I_n s_3(t - nT_s)$
1	0	0	$I_n s_4(t - nT_s)$
1	0	1	$I_n s_5(t - nT_s)$
1	1	0	$I_n s_6(t - nT_s)$
1	1	1	$I_n s_7(t - nT_s)$

Table 9.2
Mapping for $q(t)$ in the Interval $nT_s < t < (n + 1)T_s$

$\|d_{Qn} - d_{Qn-1}\|/2$	$\|d_{In} - d_{In-1}\|/2$	$\|d_{In+1} - d_{In}\|/2$	$q(t)$
0	0	0	$Q_n s_0(t - nT_s)$
0	0	1	$Q_n s_1(t - nT_s)$
0	1	0	$Q_n s_2(t - nT_s)$
0	1	1	$Q_n s_3(t - nT_s)$
1	0	0	$Q_n s_4(t - nT_s)$
1	0	1	$Q_n s_5(t - nT_s)$
1	1	0	$Q_n s_6(t - nT_s)$
1	1	1	$Q_n s_7(t - nT_s)$

and

$$I_k = \sqrt{\frac{2E_s}{T_s}} \cos\left[(2k - 1)\frac{\pi}{4}\right] \qquad (9.23)$$

$$Q_k = -\sqrt{\frac{2E_s}{T_s}} \sin\left[(2k - 1)\frac{\pi}{4}\right]$$

A block diagram of a FQPSK transmitter based on the reformulation by Simon and Yan [14] is shown in Figure 9.8(a). The corresponding

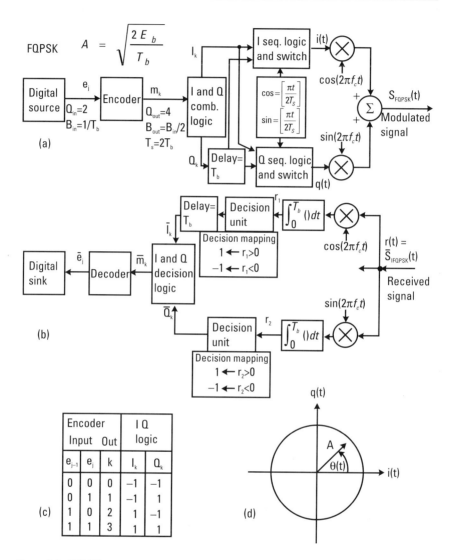

Figure 9.8 FQPSK system block diagrams: (a) transmitter, (b) coherent receiver, (c) I_k and Q_k logic, and (d) signal space diagram.

coherent receiver is given in Figure 9.8(b). The values of I_k and Q_k depend on the encoder output state k as given in (9.23), and is a function of two consecutive encoder input bits specified in Figure 9.8(c). Consecutive values of I_k and Q_k assume ±1 values, and are then nonlinearly filtered by the I sequential logic and switch, and the Q sequential logic and switch respectively. The switch outputs are defined in Table 9.1. Note that the switch output is determined by three consecutive inputs, and thus has memory. This is in contrast to OQPSK, which is memoryless.

An example of FQPSK waveforms for $A = 1$, a symbol period of $T_s = 0.5$ seconds, and $f_c = 2/T_s = 4$ Hz is shown in Figure 9.9. The signals $s_{FQPSK}(t)$, $i_{FQPSK}(t)$, and $q_{FQPSK}(t)$ as defined by (9.22), and (9.23) are shown. The modulating signal $m(t)$ corresponds to the digital source sequence $\mathbf{e} = \{1,0,1,1,0,1,0,0\}$, with $T_b = 0.25$ seconds between symbols, or

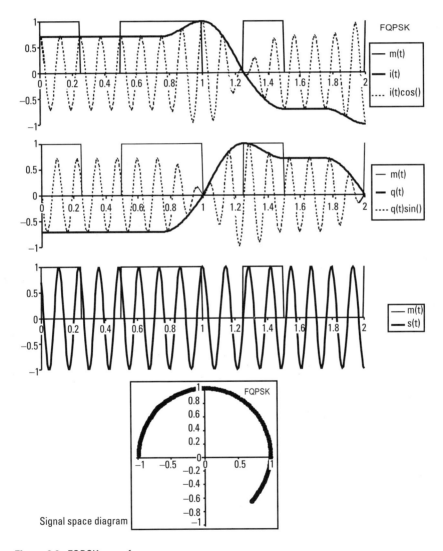

Signal space diagram

Figure 9.9 FQPSK waveforms.

4 bps. As can be seen, the baseband signals $i_{FQPSK}(t)$, and $q_{FQPSK}(t)$ make smooth transitions between extremes, in contrast to the abrupt transitions of OQPSK. This is manifested in the continuous nature of the transmitted waveform $s_{FQPSK}(t)$.

The signal space diagram is continuous, exhibiting smooth transitions as new input symbols are presented. The carrier amplitude is constant at $A = 1$, translating to zero envelope fluctuation of the transmitted signal $s_{IJF-QPSK}(t)$.

9.4.1   Laboratory, Hardware, and Flight Testing of Newly Developed Spectrally Efficient Modulation Techniques

Required bit rates in range telemetry are increasing dramatically, resulting in a surge of research to develop modulation techniques that have greater spectral efficiency than the 35-year-old workhorse of the telemetry industry, NRZ PCM/FM. The Advanced Range Telemetry (ARTM) program, the DoD Director of Test, System Engineering and Evaluation/Test facilities and Resources, and the Telemetry Group of the Range Commanders Council are funding such research programs, as well as the CCSDS-SFCG, which sponsored studies at JPL.

One such research effort was conducted at Naval Air Warfare Center Weapons Division, (NAWCWD), Point Mugu and the Air Force Flight Test Center. PCM/FM, FQPSK-B, and FQPSK-S were evaluated with respect to the BER and spectral occupancy in a laboratory and compared. The modulated signals were generated and, in order to show compatibility with existing range equipment, processed with a Microdyne model 1200MRA receiver, with the modulated carrier being removed at the IF and demodulated with a prototype FQPSK demodulator. The results were presented at the 1997 International Telemetry Conference [10]. Figure 9.10 (Figure 3 in [10]) shows the actual spectral plot of NRZ PCM/FM, FQPSK-B, and GMSK with BTb = 0.3. Figure 9.11 (Figure 5 in [10]) shows the BER versus E_b/N_o for NRZ PCM/FM and FQPSK-B. The two significant conclusions for this work are:

1. The 99.99% bandwidths of filtered FQPSK-B are approximately one-half of the corresponding bandwidths of optimized PCM/FM, *even when the signal is nonlinearly amplified.*

2. The E_b/N_o required for a BER of 1×10^{-5} for nonoptimized FQPSK-S was approximately 12 dB, which is approximately the same as the limiter discriminator detected PCM/FM.

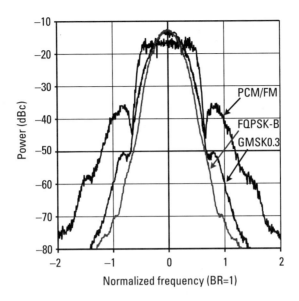

Figure 9.10 Spectral plot for NRZ PCM/FM, FQPSK-B, and GMSK. (Developed by, and courtesy of Gene Law, Naval Air Warfare Center Weapons Division.)

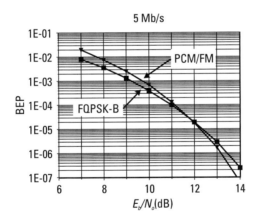

Figure 9.11 BER versus E_b/N_0 for NRZ PCM/FM and FQPSK-B. (Developed by, and courtesy of Gene Law, Naval Air Warfare Center Weapons Division.)

Actual flight testing and evaluation of FQPSK-B and PCM/FM [15] was conducted at NAWCWD with the following results:

1. The 99% bandwidth of FQPSK-B was about 66% that of optimum filtered PCM/FM, slightly higher than the laboratory results.

2. The effects of NLA were not significant.

3. FQPSK-B requires a slightly higher E_b/N_o for a BER of 1×10^{-6}.

4. Unlike PCM/FM, which does not have a polarity ambiguity problem, FQPSK-B does. To solve this, differential encoding was employed.

5. FQPSK-B was compatible with virtually all existing equipment with which it was tried. The one exception was receivers with excessive phase noise.

The goal of the testing and evaluation program at NAWCWD was to develop IRIG standards. This was accomplished, and specifications are to be included in the IRIG 106-00 document [15]. Chapter 2 of IRIG-106-00 has a description of FQPSK-B, as well as a discussion of the specification for and requirement of both differential encoding and data randomization when FQPSK-B is employed in tests involving range telemetry.

Some of the modulation formats that were considered for use as being spectrally efficient were MSK, GMSK, OQPSK, and FQPSK-B. Results clearly favored FQPSK-B. There are receivers in the market that have an optional FQPSK-B demodulator, such as both the Microdyne RCB-2000 and the DR 200. However, for the interim, test ranges have added an FQPSK-B demodulator to existing receivers that have no FQPSK-B option, with the modulated signal being obtained at the receiver IF. Most generic OQPSK receivers have the ability to demodulate FQPSK-B, although certainly not optimally.

FQPSK-B transmitters and demodulators are in various stages of development under ARTM contracts [16], and some are now on the market. Herly-Vega has a transmitter contract, while RF Networks, L3-Microdyne, and L3-Ayden have demodulator contracts. Nova, Inc. is developing a transmitter and receiver based on constant envelope, multi-h, continuous phase modulation [16]. Computer simulations indicate that this type modulation will be 50% more bandwidth-efficient than FQPSK.

9.5 M-ary Quadrature Amplitude Modulation, Amplitude and Phase Keying

Thus far, we have considered digital modulation methods in which only one of the three carrier parameters—amplitude, frequency, or phase—is modulated. A logical extension is to vary two parameters for each new symbol. This section covers two digital modulation methods that modulate both the amplitude and phase of the carrier. The in-phase and quadrature components of the modulated signal are permitted to be independent.

9.5.1 M-ary QAM

In M-ary QAM, the in-phase and quadrature components of the modulated signal are permitted to independently assume values from the same set of L discrete amplitudes. This forms an $L \times L$ constellation in the state signal space corresponding to the $M = L^2$ symbols. With $M = 16$ for example, each component may assume one of four levels, resulting in 4×4 square lattice for the signal space diagram.

9.5.1.1 Channel Bandwidth for QAM

The symbol duration $T_s = LT_b$, and the bandwidth is the same as for an M-ary ASK system, which has a bandwidth of $B_{tnn} = 2/T$ Hz, where T is the symbol duration. Thus, we have the channel bandwidth as

$$B_{t,QAM} = \frac{2}{T_s} = \frac{2}{LT_b} = \frac{2}{\sqrt{M}\,T_b} \tag{9.24}$$

Equivalently, the bandwidth efficiency is

$$\frac{I_e}{B_{t,QAM}} = \frac{\sqrt{M}}{2} \frac{\text{bps}}{\text{Hz}} \tag{9.25}$$

9.5.1.2 Symbol Error Probability for QAM

The probability of M-ary QAM symbol error is given by [1] as

$$P_s = 2\left(1 - \frac{1}{\sqrt{M}}\right) erfc\left(\sqrt{\frac{3E_{av}}{2(M-1)N_o}}\right) \tag{9.26}$$

where E_{av} is the average energy per symbol.

9.5.2 Amplitude and Phase Keying

In some applications, the square lattice constraint imposed by QAM modulation is removed, allowing the designer to place symbol constellation points anywhere on the signal space diagram. This method is used to maximize the phase difference between symbols where phase distortion is significant, such as on telephone lines.

9.6 Performance

9.6.1 BER Versus Symbol Error Rate

To determine the BER from the symbol error rate, two approaches may be taken, depending on the encoder mapping from the bit sequence to the symbol signal space points.

One approach is to assume that adjacent symbol signal space points correspond to a change of 1 bit in encoder input. Such is the case, for example, if a Gray code is used to map binary words to the signal space. Assuming that the probability of mistaking an adjacent point for the actual point is more probable than for a nonadjacent point, it follows that the most probable number of bit errors for each symbol error is one. Since there are $n = \log_2 M$ bits corresponding to each symbol signal space point, we have the bit error probability as

$$P_b = \frac{P_s}{\log_2 M} \tag{9.27}$$

A second approach is to assume that all symbols are equally likely and occur with probability $P_s/(M-1)$. It is straightforward to show that

$$P_b = \frac{M}{2(M-1)} P_s \tag{9.28}$$

9.6.2 Power Efficiency: BER Performance

The probability of error results that follow assume that the received signal is corrupted by AWGN. The white noise has a two-sided power spectral density of $N_0/2$ as discussed in Section 6.4.2, and variance of $\sigma^2 = N_0/2$ as presented in (6.11). Additionally, no signal distortion resulting from bandwidth limitations is assumed. The probability of a bit being detected in error (P_b) is equivalent to the average BER. Figure 9.12 shows the BER performance for unfiltered M-ary FSK, as determined from (9.7), as an upper bound, and (9.28) to determine P_b from P_s. In a similar way, (9.11) is used as an upper bound for the BER of unfiltered M-ary PSK plotted in Figure 9.13. P_b is found from P_s by using (9.27).

In comparing these figures it can be seen that M-ary PSK is more power efficient that M-ary FSK for $M = 2$, and $M = 4$. However for $M > 4$, M-ary FSK outperforms M-ary PSK, with the difference becoming

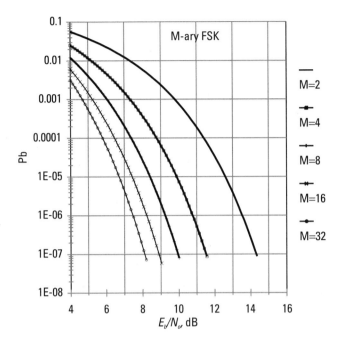

Figure 9.12 BER performance for unfiltered M-ary FSK.

larger as M increases. The reason for this disparity can be explained in terms of the signal space for each. The M signal points for M-ary PSK lie on a circle of radius A, while each signal point for M-ary FSK lies on a single axis at a distance A from the origin. Therefore, as M is increased, the signal points for M-ary PSK become closer to adjacent points, while M-ary FSK signal points maintain the same energy distance. As a result, M-ary PSK becomes more susceptible to noise than does M-ary FSK as M increases, and hence has a lower BER for a given E_s/N_0.

9.6.3 Bandwidth Efficiency: Spectrum, Filtering, and Bandwidth

Comparing M-ary FSK to M-ary PSK on the basis of power efficiency (BER performance) does not tell the complete story. The other part of the story is bandwidth efficiency. Figure 9.14 is a plot of the bandwidth efficiency of M-ary FSK and M-ary PSK modulation techniques as defined by (9.6) and (9.10).

The comparison illustrated by Figure 9.14 shows that the better power efficiency performance of M-ary FSK is obtained at the expense of bandwidth efficiency. M-ary PSK has greater bandwidth efficiency than does M-ary

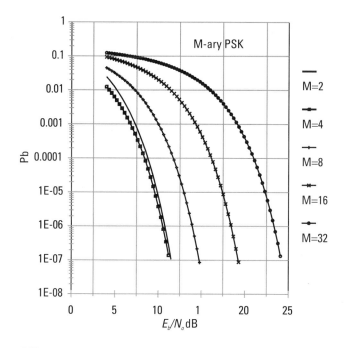

Figure 9.13 BER performance for unfiltered M-ary PSK.

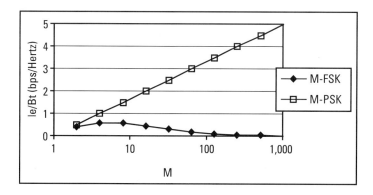

Figure 9.14 Bandwidth efficiency for M-ary FSK and M-ary PSK.

FSK. As M increases M-ary PSK increases in bandwidth efficiency, while M-ary FSK decreases in bandwidth efficiency, with the difference becoming more pronounced. The additional bandwidth requirement for M-ary FSK should not be surprising, since each frequency signal added as M increases requires additional bandwidth.

9.6.4   Power and Bandwidth Efficiency Comparison

The ratio E_b/N_o is a measure of the power efficiency of the system. For a smaller ratio, less energy is required to transmit a bit of information for a given BER. I_e/B_{tnn} represents the bandwidth efficiency (bps/Hz) of the system. The larger this ratio, the greater the bandwidth efficiency. Table 9.3 summarizes the power and bandwidth efficiency for M-ary FSK and M-ary PSK modulation methods covered in this chapter. The bandwidth efficiencies were determined from equations (9.6) and (9.10) with M ranging from 2 to 256. The power efficiency data was obtained from Figure 9.12 and 9.13 with $P_e = 10^{-6}$.

Claud Shannon [17] determined the theoretical capacity of a communication channel corrupted with AWGN. In the late 1940s, he proved that signaling schemes exist such that error-free transmission can be achieved at any rate lower than capacity. The normalized error-free capacity is given by the Shannon-Hartley equation as

$$\frac{C}{B} = \log_2\left(1 + \frac{E_b}{N_o}\frac{R}{B}\right) \tag{9.29}$$

where C is the channel capacity in bps, B is the transmission bandwidth in Hertz, E_b the energy per bit of the received signal in joules, N_o the single-sided noise spectral density in watts/hertz, and R the data rate in bps. By setting $R = C$, in (9.29) and solving for E_b/N_o we find that the upper bound for error-free transmission is

Table 9.3
Power and Bandwidth Efficiency Comparison of M-ary FSK and M-ary PSK

M	Bandwidth Efficiency I_e/B_{tnn} M-FSK	I_e/B_{tnn} M-PSK	Power Efficiency ($P_e = 10^{-6}$) E_b/N_o M-FSK	E_b/N_o M-PSK
2	0.400	0.500	13.5	10.5
4	0.571	1.000	10.8	10.5
8	0.545	1.500	9.3	14.0
16	0.421	2.000	8.2	18.5
32	0.285	2.500	7.5	23.4
64	0.179	3.000	6.9	28.5
128	0.106	3.500	6.4	33.8
256	0.061	4.000	6.0	39.2

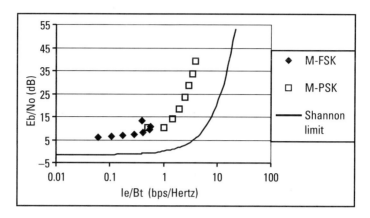

Figure 9.15 Comparison of M-ary FSK and M-ary PSK with Shannon's capacity bound.

$$\frac{E_b}{N_o} = \frac{2^{I_e/B_t} - 1}{I_e/B_t} \tag{9.30}$$

where the ratio C/B has been replaced by I_e/B_{tnn}.

A plot of the Shannon capacity boundary is shown in Figure 9.15, along with the M-ary FSK and M-ary PSK modulation methods given in Table 9.3. The power efficiency/bandwidth efficiency tradeoff is obvious for each modulation method. As M increases the two methods follow opposite directions parallel to (but far from) the channel capacity bound. A general conclusion is that M-ary PSK has better bandwidth efficiency and is useful in applications in which bandwidth is limited. M-FSK has better power efficiency, important in applications with limited power.

Problems

Problem 9.1

Consider an M-ary FSK transmitter with a carrier frequency f_c = 100 KHz, a symbol rate $1/T_s$ = 1,000 symbols per second, and with M = 3 levels. The transmitted energy per symbol is E_s = 0.01J. (a) Find an equation for the three transmission signals $s_1(t), s_2(t), s_3(t)$. (b) How many bits of information (n) are sent with each symbol transmission? (c) What is the bit duration (T_b)? (d) What is the approximate transmission bandwidth? (e) If N_o = 0.005J, what is an upper bound for the probability of symbol error?

Problem 9.2

Consider an M-ary PSK transmitter with a carrier frequency f_c = 100 KHz, a symbol rate $1/T_s$ = 1,000 symbols per second, and with $M = 4$ levels. The transmitted energy per symbol is E_s = 0.01J. (a) Find an equation for the four transmission signals $s_1(t)$, $s_2(t)$, $s_3(t)$, $s_4(t)$. (b) How many bits of information (n) are sent with each symbol transmission? (c) What is the bit duration (T_b)? (d) What is the approximate transmission bandwidth? (e) If N_o = 0.005J, what is an upper bound for the probability of symbol error?

Problem 9.3

Consider a QPSK transmitter with a carrier frequency f_c = 100 KHz, a symbol rate $1/T_s$ = 1,000 symbols per second. The digital source outputs the bit sequence {0,1,1,1,0,0} and the transmitted energy per symbol is E_s = 0.01J. (a) How many bits of information (n) are sent with each symbol transmission? (b) What is the bit duration (T_b)? (c) Find the sequence of encoder output symbols m_1, m_2, m_3, and m_4 as defined in encoder table in Figure 9.2. (d) Sketch the in-phase baseband signal $i(t)$, and the quadrature baseband signal $q(t)$. (e) What is the approximate transmission bandwidth?

Problem 9.4

What is the advantage of OQPSK when compared with QPSK?

Problem 9.5

What is the advantage of IJF-OQPSK when compared with OQPSK and QPSK?

Problem 9.6

What is the advantage of FQPSK when compared with OQPSK, IJF-OQPSK, and QPSK?

References

[1]   Haykin, S., *Communication Systems,* 3rd Ed. New York: Wiley, 1994, p. 553.

[2]   Ziemer, R., and R. Peterson, *Introduction to Digital Communication,* 2nd Ed. Upper Saddle River, NJ: Prentice-Hall, 2001, p. 226.

[3]   Le-Ngoc, T., K. Feher, and H. Pham Van, "New Modulation Techniques for Low-Cost Power and Bandwidth Efficient Satellite Earth Stations," *IEEE Transactions on Communications,* Vol. 30, No. 1, Jan. 1982.

[4]   Feher, K., *Digital Communications Satellite/Earth Station Engineering,* Englewood Cliffs, NJ: Prentice-Hall, 1983, p. 118.

[5]   Feher, K., and S. Kato, U.S. patent 4,567,602; Feher, K, U.S. patent 5,491,457; Feher, K, U.S. patent 5,784,402.

[6]   Feher, K., *Wireless Digital Communications: Modulation and Spread Sprectrum Applications,* Englewood Cliffs, NJ: Prentice-Hall, 1995.

[7]   Feher, K., "FQPSK Doubles Spectral Efficiency of Operational Telemetry Systems," *European Telemetry Conference, ETC 98,* Garmish-Patternk, Germany, May 5–8, 1998.

[8]   Kato, S., and K. Feher, "A New Cross-Correlated Phase-Shift Keying Modulation Technique," *IEEE Transactions on Communications,* Vol. 31, No. 5, May 1983.

[9]   Simon, M. K., and T. Y. Yan, "Performance Evaluation and Interpretation of Unfiltered Feher-Patented Quadrature-Phase-Shift Keying (FQPSK)," *CCSDS TMO Progress Report 42-137,* May 15, 1999.

[10] Law, E., and K. Feher, "FQPSK Versus PCM/FM for Aeronautical Telemetry Applications; Spectral Occupancy and Bit Error Probability Comparisons," *ITC Proc,* 1997.

[11] Digicom, Inc. document, "FQPSK-B, Revision A1 Digcom-Feher Patented Technology Transfer Document, Jan. 15, 1999. This document can be obtained under a license from: Digicom, Inc., 44685 Country Club Drive, El Macero, CA 95618.

[12] Tsou, H., S. Darden, and T. Y. Yan, "An Off-line Coherent FQPSK-B Software Reference Receiver," Consultative Committee for Space Data Systems (CCSDS). Authors are with JPL/NASA. Pasadena, CA. For a reprint of this paper, contact the authors for CCSDS P1E number 00/07, submission May 15, 2000, Annapolis, MD. This paper is a modified version of: Tsou, H., S. Darden, and T. Y. Yan, "An Off-line Coherent FQPSK-B Software Reference Receiver," CalTech-JPL/NASA, Pasadena, CA. European Telemetry Conference ETC 2000, Garmisch-Partenkirchen, Germany, May 22–25, 2000.

[13] Gao, W., and K. Feher, "FQPSK: A Bandwidth and RF Power Efficient Technology for Telemetry Applications," *ITC Proc,* 1997.

[14] Simon, M., and T. Tan, "Unfiltered FQPSK: Another Interpretation and Further Enhancements," *Applied Microwave and Wireless,* February 2000, pp. 76–96.

[15] Secretariat, Range Commanders Council, *Telemetry Standards,* White Sands Missile Range, NM: RCC, IRIG 106-00.

[16] Law, E., IRIG FQPSK-B Standardization Progress Report, NAWCWD, Pt. Mugu, CA.

[17] Shannon, C., "A Mathematical Theory of Communications," *Bell Syst. Tech Journal,* Vol. 27, July 1948, pp. 379–423, and Oct. 1948, pp. 623–656.

10

Spread Spectrum Communication Systems

In the modulation/demodulation methods presented thus far, the goal has been to communicate information from source to sink efficiently in a band-limited channel corrupted by AWGN. The performance metrics have been in terms of bandwidth efficiency (bps/Hz) and energy efficiency (received signal-to-noise ratio for a given BER). To the contrary, spread spectrum methods have a goal of increasing the transmission bandwidth, making this method bandwidth-inefficient.

Although many bandlimited communication systems may be accurately modeled as AWGN channels, in some applications, we also need to consider other systems that do not fit this model. Antijam capability, interference rejection, and low probability of intercept (LPI) capability are important examples. For example, a military communication system may be jammed by a continuous wave (CW) or by a modulated carrier near the transmitter's center frequency. This interfering noise signal cannot be modeled as an AWGN process. A spread spectrum system is effective in mitigating this type of man-made interference.

Multiple users share a band of frequencies in cellular telephone and personal communication applications, where there is not enough available bandwidth to assign a permanent frequency channel to each user. In this case, spread spectrum techniques allow simultaneous use of a wide frequency band via code-division multiple-access (CDMA) techniques. In addition to this multiple-access capability, spread spectrum is able to minimize multipath reception in which the multiple delayed copies of the transmitted signal are received by the demodulator.

This chapter presents spread spectrum modulation and demodulation techniques that may be used to provide multiple access and to mitigate the detrimental effects of the interference described above.

10.1 Learning Objectives

Upon completion of this chapter, the reader should understand the following:

- The concept of spread spectrum;
- Direct sequence spread spectrum;
- Modeling and implementation of direct sequence BPSK spread spectrum;
- Power spectral density of direct sequence BPSK spread spectrum;
- Direct sequence QPSK spread spectrum;
- Modeling and implementation of frequency hop spread spectrum;
- Slow-frequency-hopping spread spectrum;
- Fast-frequency-hopping spread spectrum.

10.2 Introduction to Spread Spectrum

To be considered spread spectrum, a communication system must satisfy two criteria:

1. The bandwidth of the transmitted signal $s(t)$ needs to be much greater than that of the message signal $m(t)$.
2. The relatively wide bandwidth of $s(t)$ is caused by an independent modulating waveform $c(t)$ called the *spreading signal*. This same signal is used by the receiver to despread the received signal in order to recover the message signal $m(t)$.

A wideband spread spectrum signal is generated from a data modulated carrier by modulating the carrier a second time using a wideband spreading signal. The transmitted radio frequency spread spectrum signal may be represented by

$$s(t) = \text{Re}(b(t)e^{j\omega_c t}) \qquad (10.1)$$

where $b(t)$ is the complex baseband envelope of signal $s(t)$ prior to translation to the carrier frequency ω_c. In many cases, the baseband signal is formed by a product that is a function of $m(t)$ and $c(t)$ as

$$b(t) = b_m(t)b_c(t) \qquad (10.2)$$

The following are some of the most common types of SS signals:

- *Direct sequence* (DS). Spreading is accomplished by phase modulation with $b_c(t) = c(t)$, where $c(t)$ is a polar NRZ waveform having values ± 1.
- *Frequency-hop* (FH). Spectrum spreading is accomplished by rapid changing of the carrier frequency. Here, $b_c(t)$ is of the FM type, where there are $M = 2^k$ hop frequencies determined by the k-bit words obtained from the spreading code waveform $c(t)$.
- *Hybrid.* Techniques that include both DS and FH.

DS and FH systems are explained and illustrated in the following sections.

10.3 Direct Sequence Spread Spectrum

10.3.1 DS/BPSK Spread Spectrum

BPSK is often employed as the spreading modulation in a *direct sequence spread spectrum* (DSSS) system. Such modulation can be represented mathematically as a multiplication of the carrier by $m(t)$ and $c(t)$ which take on the values ± 1. The result is a $\pm 180°$ phase shift of the carrier, as in BPSK. In this case, the baseband envelope (10.2) becomes

$$b_{DS/BPSK}(t) = A_c m(t)c(t) \qquad (10.3)$$

where A_c is the peak carrier amplitude. The corresponding transmitted signal is obtained by substituting (10.3) into (10.1), yielding

$$s_{DS/BPSK}(t) = A_c m(t)c(t)\cos(\omega_c t) = c(t)s_{BPSK}(t) \qquad (10.4)$$

where $\omega_c = 2\pi f_c$, and $s_{BPSK}(t)$ is a BPSK signal given by

$$s_{BPSK}(t) = A_c m(t) \cos(\omega_c t) \qquad (10.5)$$

A block diagram of a direct sequence BPSK (DS/BPSK) spread spectrum transmitter is shown in Figure 10.1(a). The digital data source information is level encoded into a sequence of symbols m_k, each assuming ±1 values. A new symbol is generated every T_b seconds, forming the modulating signal $m(t)$, which assumes this sequence of ±1 symbols. This signal, in turn, is multiplied by the carrier oscillator signal to produce the BPSK signal. The spreading code generator signal $c(t)$ then multiplies the BPSK signal to generate the DS/BPSK spread spectrum signal $s_{DS/BPSK}(t)$. The pulse width of $c(t)$, denoted by T_c, is called a *chip interval* (as contrasted with the bit interval T_b).

An example of transmitter signals for $A_c = 1$, and the data sequence $m_k = \{1,-1,1\}$ is shown in Figure 10.2. Here $T_b = 1$ second, $T_c = 0.25$ second, and the carrier frequency $f_c = 8$Hz. The data signal $m(t)$ is shown in part (a) of Figure 10.2, and the corresponding BPSK signal as defined by (10.5) is shown in part (b). The spreading code generator signal sequence $c(t) = \{1,-1,1,-1,-1,1,-1,1,-1,1,1,-1\}$ shown in part (c) then multiplies the BPSK signal to generate the spread spectrum signal shown in (d) of the

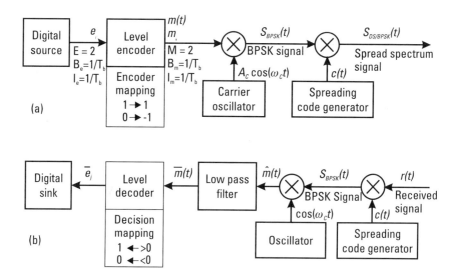

Figure 10.1 DS/BPSK spread spectrum system block diagram: (a) transmitter, and (b) receiver.

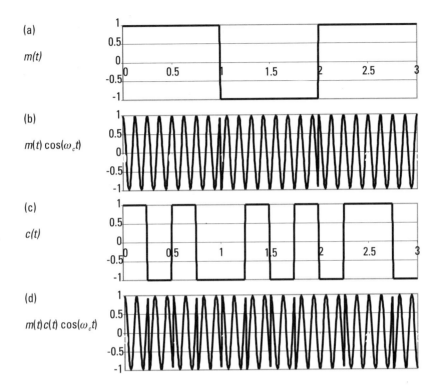

Figure 10.2 DS/BPSK spread spectrum transmitter signals (a) data, (b) BPSK signal, (c) spreading signal, and (d) spread spectrum signal.

figure. Note that the transitions in the spreading signal $c(t)$ cause additional transitions in the transmitted signal, and thus "spreads" the spectrum. In practical systems $T_c \ll T_b$, and $f_c \gg 1/T_b$, causing the spreading effect to be more pronounced than that illustrated in Figure 10.2.

A receiver that recovers the data from the DS/BPSK spread spectrum signal is shown in Figure 10.1(b). The received signal is assumed to be corrupted by additive noise, and is given by

$$r(t) = s_{DS/BPSK}(t) + n(t) = A_c m(t)c(t) \cos(\omega_c t) + n(t) \quad (10.6)$$

The recovered BPSK signal obtained by multiplying the received signal $r(t)$, as defined in (10.6), by $c(t)$ is

$$\bar{s}_{BPSK}(t) = A_c m(t)c^2(t) \cos(\omega_c t) + c(t)n(t) \quad (10.7)$$

Assuming the dispreading front end is driven by a spreading code generator in synchronism with the transmitter spreading code $c(t)$, we have $c^2(t) = 1$, since $c(t) = \pm 1$. Thus (10.7) reduces to

$$\tilde{s}_{BPSK}(t) = A_c m(t) \cos(\omega_c t) + c(t) n(t) \qquad (10.8)$$

where the first term is the recovered BPSK signal, and the second term is the noise component. The signal given by (10.8) is then multiplied by a replica of the transmitted carrier, yielding

$$\hat{m}(t) = A_c m(t) \cos^2(\omega_c t) + c(t) n(t) \cos(\omega_c t) \qquad (10.9)$$

Using the trigonometric identity $\cos^2(\omega_c t) = 1/2[1 + \cos(2\omega_c t)]$, (10.9) becomes

$$\hat{m}(t) = \frac{A_c}{2} m(t) + \frac{A_c}{2} \cos(2\omega_c t) + c(t) n(t) \cos(\omega_c t) \qquad (10.10)$$

The lowpass filter then allows the baseband component $m(t)$ in (10.10) to be level decoded into the transmitted data sequence of symbols m_k, while rejecting the two terms centered at twice the carrier frequency.

It is significant to note that the received noise component $n(t)$ in (10.10) is multiplied by spreading signal $c(t)$, thus reducing its power spectral density, and is simultaneously translated to twice the carrier frequency. This greatly reduces the influence of interfering noise on the recovered baseband signal $m(t)$. This noise rejection property is a significant advantage of spread spectrum systems.

10.3.1.1   PSD of DS/BPSK

By comparing (10.4) for the DS/BPSK transmitted signal and (10.5) for the BPSK transmitted signal, it can readily be seen that the DS/BPSK signal is equivalent to the BPSK signal with $m(t)$ replaced with $m(t) c(t)$. Assuming that the spreading signal *chip duration* T_c is smaller than the *bit symbol duration* T_b, the product signal $m(t) c(t)$ will have transitions every T_c seconds, and it is this parameter that defines the high frequency components of the signal. Thus, the PSD for the transmitted DS/BPSK signal may be obtained from the BPSK PSD equation by replacing T_b with T_c. Performing this operation on the BPSK PSD baseband signal given in (8.34) and simultaneously translating to the carrier frequency (replacing f with $f - f_c$), yields

$$G(f)_{DS/BPSK} = \frac{A_c^2 T_c}{8} \left[\frac{\sin \pi (f - f_c) T_c}{\pi (f - f_c) T_c} \right]^2 \qquad (10.11)$$

where A_c is the carrier amplitude, f_c the carrier frequency, and T_c is the chip duration.

A plot of $G(f)_{DS/BPSK}$ for positive frequencies with $A_c = 9$, $T_b = 0.1$ second, and $f_c = 100$ Hz is shown in Figure 10.3. Three different cases are shown, $T_c = T_b$, $T_c = T_b/5$, and $T_c = T_b/10$. The case in which $T_c = T_b$ corresponds to no signal spreading with a PSD the same as for BPSK. With $T_c = T_b/5$, the PSD is reduced by a factor of 5 and the bandwidth increased by a factor of 5. A similar effect is observed for $T_c = T_b/10$, but the factor is 10. Thus we observe that the spreading modulation $c(t)$ reduces the level of the PSD by a factor equal to the ratio T_b/T_c. The bandwidth of the transmitted signal is increased by this same factor. In practice $T_c \ll T_b$, and this effect is more pronounced than is illustrated in Figure 10.3. The wide bandwidth and low-PSD level characteristics of spread spectrum are useful in environments requiring antijam capability, interference rejection, and LPI.

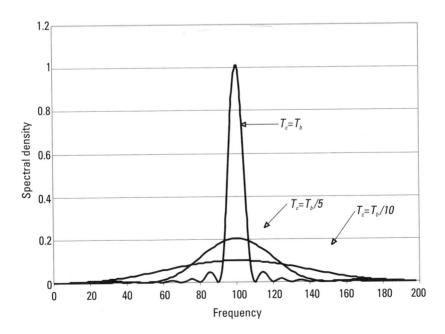

Figure 10.3 DS/BPSK spread spectrum power spectral density.

10.3.2 DS/QPSK Spread Spectrum

QPSK was introduced in Chapter 9 as a bandwidth-efficient alternative to PSK, transmitting the same bit rate using only one-half the bandwidth. The use of this alternative for spread spectrum transmission is contrary to the expressed purpose of increasing the bandwidth of the signal. However, QPSK in spread sprectrum applications has the advantage of being more difficult to detect, where LPI is important. Additionally, QPSK is less sensitive to some types of jamming or interference.

The DS/QPSK spread spectrum signal may be generated by summing two DS/PSK signals defined in (10.4). The in-phase component is obtained by multiplying the modulating signal $m(t)$ by the carrier $\cos(\omega_c t)$ and the spreading signal $c_1(t)$. This is added to the quadrature component obtained by multiplying the modulating signal $m(t)$ by the quadrature carrier $\sin(\omega_c t)$, and the spreading signal $c_2(t)$. The resulting transmitted signal is

$$s_{DS/QPSK}(t) = A_c m(t) c_1(t) \cos(\omega_c t) + A_c m(t) c_2(t) \sin(\omega_c t)$$

$$(10.12)$$

The spreading waveforms $c_1(t)$ and $c_2(t)$ are assumed to be chip synchronous, but are otherwise independent. This independence contributes to the LPI feature of this technique. The receiver is similar to that in Figure 10.1(b) where the received signal split into two paths, each being processed by the corresponding despreading signal and carrier oscillator.

10.3.3 Examples of DSSS

The Tracking and Data Relay Satellite (TDRS) is used to relay communications and tracking information between the NASA space shuttle and a ground station at White Sands, New Mexico. One link used is the S-band forward link (TDRS to shuttle) operating at one of two carrier frequencies (2,106.406 and 2,287.5 MHz). The transmitter employs a BPSK spread spectrum modulation technique. The data rate is either 32 Kbps or 72 Kbps with BPSK Manchester modulation and rate 1/3 convolutional coding. This results in a symbol rate of 96 Kbps or 216 Kbps. The spreading code has a rate of 11,232 Mchips/s, giving a spreading factor of 11,232M/216 K = 117,000 chips per symbol.

The Global Positioning System (GPS) allows users to find their position on the Earth by measuring their range to four to six satellites whose position is known. The 24 GPS satellites are positioned in circular orbits so that a GPS user anywhere on Earth should be able to receive signals from at least four satellites. Two DSSS signals, called L1 and L2 are transmitted on frequencies 1,575.42 MHz and 1,227.60 MHz. L1 is a DS/QPSK spread spectrum signal. The *I*-channel uses a Gold (C/A code) code with a chip rate of 1.023 MHz and period of 0.001 second. A different Gold code is used for each satellite, so the ground receiver can identify which satellite signal is being received. The *Q*-channel of L1 is spread by a long code (P code) with a chip rate of 10.230 MHz. Both the C/A and *P* channels are modulated by a 1,500-bit, 50 bps signal containing position and other information. The L2 signal is a DS/PSK spread spectrum signal using the P code at a chip rate of 10.23 MHz. This L2 signal is intended for military users.

10.4 Frequency-Hop Spread Spectrum

In *frequency-hop spread spectrum* (FHSS) the carrier hops from one frequency to another within the transmission bandwidth. Spectrum spreading is accomplished by rapid changing of the carrier frequency. Here, $b_c(t)$ in (10.2) is of the FM type, where there are $H = 2^k$ hop frequencies determined by the k-bit words obtained from the spreading code waveform $c(t)$.

A block diagram of the transmitter is shown in Figure 10.4(a). The digital data causes the M-ary FSK modulator to encode a sequence of n data bits into an output signal that assumes one of $M = 2^n$ frequencies. Let the duration of this symbol be T_s, and the bit duration be T_b. Then

$$T_s = nT_b \qquad (10.13)$$

The M-ary FSK signal is then translated in frequency (up-converted) by the frequency synthesizer output $h(t)$, which assumes one of the $H = 2^k$ hop frequencies every T_c seconds. The bandpass filter removes the difference frequencies of the multiplier output while allowing the sum frequencies to pass through and be transmitted.

In the receiver, the frequency hopping is removed by multiplying (down-converting) the received signal with synthesized signal that is hopping synchronously with the transmitter frequency synthesizer.

Frequency hopping does not cover the entire bandwidth instantaneously, but sequentially places the modulated carrier at predetermined locations within the bandwidth. This leads us to classify this method according to the rate at which the hops occur:

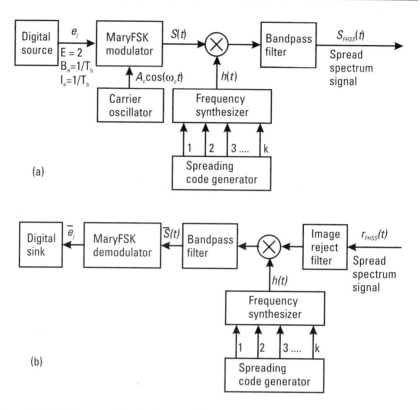

Figure 10.4 FHSS system block diagram: (a) transmitter, and (b) receiver.

1. Slow-frequency hopping, in which several symbols are transmitted on each frequency hop. Here the hop duration is an integer multiple of the symbol duration ($T_c \geq T_s$).

2. Fast-frequency hopping, in which several hops in frequency occur during the transmission of one symbol. Here the symbol duration is an integer multiple of the hop duration ($T_s \geq T_c$).

10.4.1 Slow-Frequency Hopping

In slow-frequency hopping spread spectrum, the hop duration is an integer multiple of the symbol duration, so we have

$$T_h = mT_s \qquad (10.14)$$

where m is a nonzero positive integer. The spacing of the M-ary FSK signals is assumed to be $1/T_s$ Hz, which assures that the signals are orthogonal as described in Chapter 9. The implication is that a transmitted symbol will not produce "crosstalk" in adjacent symbol detection units of the demodulator. Since the M-ary FSK signal takes on one of M distinct frequencies spaced $1/T_s$ Hz apart, the approximate bandwidth of this signal is

$$B_{MFSK} = M/T_s \qquad (10.15)$$

The spacing of the H hop frequencies is set to B_{MFSK} given in (10.15) in order to avoid overlap of the translated M-ary FSK signal and avoid gaps in the transmitted signal. Thus the approximate bandwidth of the slow-frequency-hop spread spectrum signal is

$$B_{SFHSS} = HM/T_s \qquad (10.16)$$

The output of a slow-frequency-hop spread spectrum system transmitter for $m = 3$ is illustrated in Figure 10.5. The time axis scale is normalized to T_b seconds. The instantaneous frequency of the transmitted spectrum is shown for $M = 4$ ($n = 2$) and $H = 8$ ($k = 3$). That is, the M-ary FSK signal assumes one of four different frequencies, and the frequency synthesizer output hops between eight different frequencies.

As can be seen, two consecutive data bits cause the M-ary FSK signal to assume one of the four frequencies every T_s seconds (two normalized time units). After three consecutive symbols ($3T_s$ seconds), the frequency synthesizer hops to one of the eight hop frequencies causing the transmitted signal to hop to one of the 32 frequencies. The transmission bandwidth as given by (10.16) is $8 \times 4/T_s = 32/T_s$.

An understanding of what is known as processing gain can be obtained by considering the effect of a noise jammer. With no spreading ($B_{SFHSS} = B_{MFSK}$) the jammer will likely place a noiselike signal centered on the carrier frequency, having the same bandwidth (B_{MFSK}) as the transmitted signal. For a given jamming signal power, say P_J, the receiver operates at a certain BER. With spread spectrum, the transmission bandwidth is increased by a factor of B_{SFHSS}/B_{MFSK}. This forces the jammer to increase P_J by the same factor in order to cause the same certain bit error rate. This ratio is known as the processing gain of spread spectrum.

10.4.2 Fast-Frequency Hopping

For fast-frequency hopping, the symbol duration is an integer multiple of the hop duration, so we have

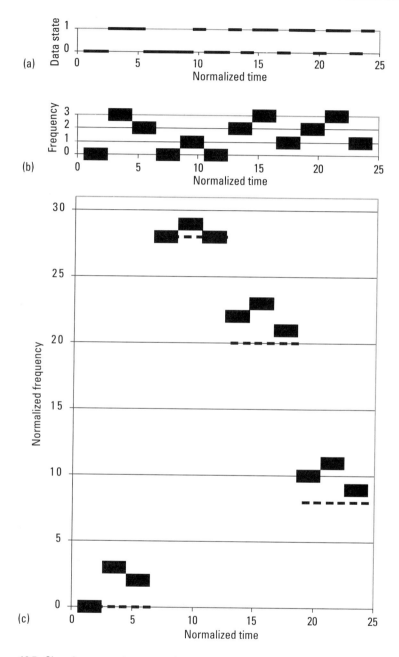

Figure 10.5 Slow-frequency-hop spread spectrum system instantaneous frequency: (a) Data bit sequence, (b) M-ary FSK signal, and (c) transmitted signal.

$$T_s = lT_h \qquad (10.17)$$

where l is a nonzero positive integer. The output of the M-ary FSK modulator is one of M frequencies, as for slow-frequency hopping, but each tone is subdivided into l chips. Since the chip duration T_h is shorter than the symbol duration T_s, it is this smaller duration that determines the transmission bandwidth. Replacing T_s by T_h in (10.16) gives the approximate fast-frequency-hopping bandwidth as

$$B_{FFHSS} = HM/T_h \qquad (10.18)$$

The output of a fast-frequency-hop spread spectrum system transmitter for $l = 2$ is illustrated in Figure 10.6. The instantaneous frequency of the transmitted spectrum is shown for $M = 4$ ($n = 2$) and $H = 8$ ($k = 3$). That is, the M-ary FSK signal assumes one of four different frequencies, and the frequency synthesizer signal hops between eight different frequencies.

As can be seen, two consecutive data bits cause the M-ary FSK signal to assume one of the four frequencies every T_s seconds. During transmission of a symbol, the frequency synthesizer hops twice (every $T_s/2$ seconds), to one of the eight hop frequencies. This causes the transmitted signal to hop to one of the 32 frequencies within the $32/T_h$ Hz bandwidth.

10.5 CDMA

Spread spectrum systems are able to provide multiple users of an assigned bandwidth access to the frequency resource. This multiple access capability is an alternative to frequency-division and time-division multiple access techniques. Users are each assigned unique spreading codes so their transmitted signal can be separated from others in the receiver. Since the channel resource is divided among users according a unique spreading code assignment, this method is known as CDMA.

A primary goal in CDMA design is to find a set of spreading codes so as many users as possible can use the assigned frequency resources. A tradeoff in this design process is the increase in crosstalk interference as the number of users increases. A detailed discussion is beyond the scope of this chapter, but the essential result is that a set of codes known as Gold codes are quite useful for CDMA [1, 2]. These codes exhibit low cross-correlation which translates to low crosstalk interference.

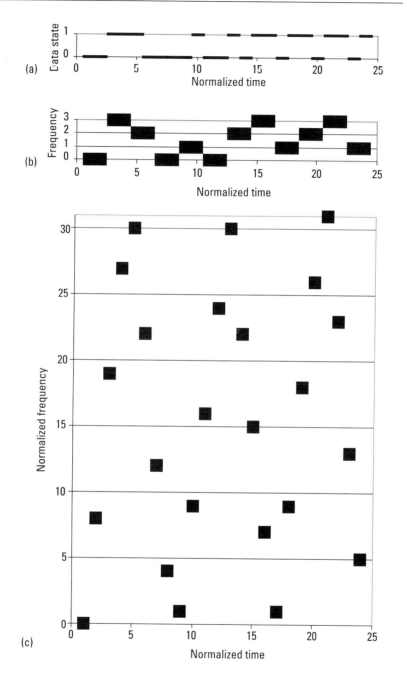

Figure 10.6 Fast-frequency-hop spread spectrum system instantaneous frequency: (a) data bit sequence, (b) M-ary FSK signal, and (c) transmitted signal.

One application of CDMA is the GPS system previously described in section 10.3.3. One of the second-generation cellular systems employs CDMA for multiple user access. The nominal voice signal data rate is 8,600 bps, and, after encoding, the binary symbol rate is 19,200 bps. Details can be found in [2].

Problems

Problem 10.1

Describe two applications for spread spectrum.

Problem 10.2

What two criteria must a spread spectrum signal satisfy?

Problem 10.3

List three types of spread spectrum.

Problem 10.4

Consider a DS/BPSK system with $A_c = 1$, data sequence $m_k = \{-1,1\}$ and spreading code generator signal sequence $c(t) = \{1,-1,1,-1,-1,1,-1,1,-1,1\}$. The bit duration $T_b = 1$ second, the chip duration $T_c = 0.2$ second, and the carrier frequency $f_c = 8$Hz. (a) What is the bit rate and the chip rate? (b) Sketch the data signal $m(t)$. (c) Sketch the spreading signal $c(t)$. (d) Sketch the product signal $m(t)c(t)$. (e) Sketch the spread spectrum signal $m(t)c(t)\cos(\omega_c t)$.

Problem 10.5

Assume that the spectrum for DS/BPSK, given by (10.11), is represented by a simplified rectangular spectrum that has a constant amplitude proportional to T_c extending from $f_c - 1/T_c$ to $f_c + 1/T_c$. Sketch this spectrum for $T_c = T_b$, $T_c = T_b/5$, and $T_c = T_b/10$. What do you conclude about the PSD amplitude and the bandwidth as T_c is decreased (the chip rate increased)?

References

[1] Haykin, S., *Communication Systems,* 3rd Ed., New York: Wiley, 1994.

[2] Ziemer, R., and R. Peterson, *Introduction to Digital Communication,* 2nd Ed., Upper Saddle River, NJ: Prentice-Hall, 2001.

11

Antennas and Link Analysis

This chapter introduces the basic operation of antennas and their use in a telemetry system. The function of transmitter and receiver antennas, which are an integral part of the overall system, is to direct the telemetry data from the object under test toward the ground station. To introduce antenna theory, a simplified form of the radiation integral is presented in a form familiar to the telecommunications engineer. Important parameters are discussed so that specification of the antennas can be done within the context of the RF portion of the telemetry system. The chapter concludes with a detailed discussion of link analysis including the impact of multipath fading.

11.1 Learning Objectives

Chapter 11 addresses the terminal characteristics of antennas and how they operate in a telemetry link. Upon completing this chapter the reader should understand the following:

- The basic RF components of the telemetry link;
- Fundamental antenna concepts such as gain and directivity;
- Antenna radiation patterns;
- Antenna impedance and mismatch;
- Antenna polarization and mismatch;
- The Friis transmission formula in its most practical form.

11.2 Telemetry Link Overview and Components

The objective of this section is to outline the RF operation of a telemetry system; that is, calculation of the received power using system parameters such as transmitter output power, antenna gain, frequency of operation, and range. The discussion is limited to components in the RF path between the transmitter and receiver; it does not address the details of transmitter operation or receiver design. Subsequent sections of this chapter will present further details of antenna gain, impedance, and polarization mismatch. The topics are tailored to telemetry engineers whose area of expertise is other than antennas and RF/microwave systems. For the interested reader there are comprehensive textbooks that cover this material in greater detail [1–6].

The basic RF components of the telemetry system shown in Figure 11.1 consist of: transmit (Tx) cable or waveguide, transmit antenna, receive antenna, and receive (Rx) cable or waveguide. For the purposes of this discussion, it will be assumed that devices such as isolators are contained within the transmitter, and that filters or preamplifiers are included as part of the receiving system.

An essential part of the telemetry system is the path; that is, the medium between the two antennas. In this discussion it is assumed that the two antennas are in free space, Earth is not present, and the transmitter/receiver separation distance is sufficiently great to ensure plane wave theory adequately represents the electromagnetic propagation. Real world considerations such as multipath fading due to reflections from the ground and/or other obstructions will be the topic of Section 11.5.

The objective is to determine the power entering the receiver, P_{rec}, in terms of the transmitter power, P_s, and the characteristics of the intervening

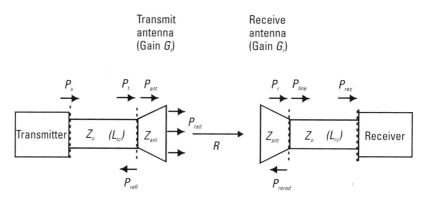

Figure 11.1 RF components of a telemetry system.

system. It is observed that the power at the terminals of the receiver can be increased when: (1) P_s is increased; (2) the cable losses represented by the loss factors L_{tc} and L_{rc} are minimized; and/or (3) the transmit and receive antenna gains, G_t and G_r, respectively, are greater. The received power must be large enough to provide a good quality signal but limited such that it does not overload the receiver.

Tools are presented in Chapters 11, 12, and 13 to incorporate real-world factors so that the signal and noise characteristics of the telemetry system can be quantified. Specification of parameters to optimize performance of telemetry systems is the subject of Chapter 14.

11.2.1 EIRP and Transmit Power Density

The transmitter is connected to the antenna terminals by a transmission line of characteristic impedance, Z_0, as shown in Figure 11.1. RF components are usually designed for $Z_0 = 50\ \Omega$ systems. In practice the transmission line, or cable, has non-negligible loss, and the terminal impedance of the antenna, Z_{ant}, is not perfectly matched to Z_0. Thus, the power radiated by the system, P_{rad}, will be reduced by three mechanisms: (1) cable attenuation, A_{tc}, (2) impedance mismatch producing a nonzero reflection coefficient, Γ_t, and (3) dissipative losses within the antenna structure itself, represented by the radiation efficiency, e_{rad}.

The loss factor, L_{tc}, is proportional to the power dissipated along the transmission line and is always greater than or equal to one. The terms P_s and P_t are the power available from the source and the output end of the transmission line, respectively.

$$L_{tc} = \frac{P_s}{P_t} \tag{11.1}$$

The attenuation, A_{tc} in dB, is defined in terms of the loss factor as

$$A_{tc} = 10 \log_{10}(L_{tc}) \tag{11.2}$$

The power available to a matched load at the end of the transmission line is P_t; however, due to the impedance mismatch at the antenna/cable interface, some power will be reflected, P_{refl}. The difference between P_t and P_{refl} is the power delivered to the antenna, P_{ant}. Impedance mismatch is quantified in terms of the voltage reflection coefficient looking into the terminals of the transmit antenna. It is defined as

$$\Gamma_t = \frac{Z_{ant} - Z_o}{Z_{ant} + Z_o} \tag{11.3}$$

In general, Γ_t is a complex quantity and the magnitude is bounded by $0 \le |\Gamma_t| \le 1$ for passive antennas. The power reflection coefficient, $|\Gamma_t|^2$, is the ratio of P_{refl} to P_t. Taking the difference between P_t and P_{refl} yields P_{ant}.

$$P_{ant} = P_t(1 - |\Gamma_t|^2) \tag{11.4}$$

The reflected power must be absorbed by the transmitter and should be minimized for proper system operation. An antenna will not radiate all power delivered to it; that is, P_{rad} will be less than P_{ant} because of dissipative losses resulting from the presence of nonideal conductors and dielectrics. Radiation efficiency accounts for this loss and is given as

$$e_{rad} = \frac{P_{rad}}{P_{ant}} \tag{11.5}$$

The range is $0 \le e_{rad} \le 1$ with unity the desired value.

Consider a transmit antenna that radiates equally in all directions (an isotropic antenna or point source) and operates in free space. It is convenient to assume that the radiation efficiency of this fictitious antenna is one so that $P_{rad} = P_{ant}$ and this power is delivered equally to all portions of space. The radial power density, or average Poynting vector, $\overline{S}_{iso}$, in the outward direction, $\hat{r}$, at a fixed range $r = R$ may then be written as

$$\overline{S}_{iso}(R) = \hat{r}\,\frac{P_{ant}}{4\pi R^2} \tag{11.6}$$

If the range, R, is specified in meters, the units of $\overline{S}_{iso}$ are W/m^2. The purpose of the antenna in Figure 11.1 is twofold: (1) it is a transition from the guided wave in the cable to the free space wave propagating toward the ground station, and (2) it is used to direct the energy toward desired portion(s) of space.

Antenna gain, which is dimensionless, will be formally defined in Section 11.3.3. Here it is simply stated that an increase in gain, G_t, will cause a concomitant increase in the average power density, $\overline{S}_{ave}$, at the peak of the beam for fixed P_{ant} and R, relative to that of an isotropic transmit antenna. Thus,

$$\overline{S}_{ave}(R) = \hat{r} \frac{P_{ant} G_t}{4\pi R^2} \tag{11.7}$$

The inverse square law dependence of the Poynting vector on range is evident in both (11.6) and (11.7). To contrast the two equations, Figure 11.2 depicts the average Poynting vector as a function of direction at a fixed range, R, for two cases: (a) $\overline{S}_{iso}$, for an isotropic source, and (b) $\overline{S}_{ave}$, for a directive antenna. The distance of the surface from the origin is proportional to the power density in that direction. Gain for the lossless isotropic antenna in Figure 11.2(a) is $G_t = 1$, whereas the gain is much greater than unity for the directive antenna pattern depicted in Figure 11.2(b).

A passive antenna increases the power density in a particular direction (resulting in a greater value of gain) only by reducing the power delivered to other regions of space.

One way to characterize the transmitter is in terms of its effective isotropic radiated power, *EIRP*. In the direction of the mainbeam peak the *EIRP* is defined as

$$EIRP = P_{ant} G_t \tag{11.8}$$

That is, in the direction of peak radiation depicted in Figure 11.2(b), the power density appears to be an amount of power equal to $P_{ant} G_{ant}$ watts radiated uniformly in all directions. However, as discussed in the sections to follow, the power density of a directive antenna cannot be uniform throughout space.

11.2.2 Received Power

The transmitted wave propagates toward the ground station where it is received by an antenna. P_r is the power available to a matched load connected to the receive antenna terminals. The plane wave impinging upon the receive antenna at a distance R from the transmitter has a power density $\overline{S}_{ave}(R)$ with units W/m^2. The received power will be proportional to the effective collecting area, A_{er}, of the receive antenna as illustrated in Figure 11.3. It is usually termed effective aperture and is the effective area normal to the radial direction, $\hat{r}$.

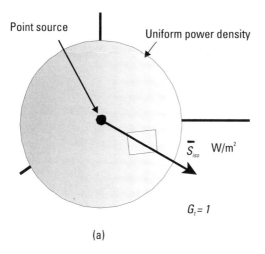

(a)

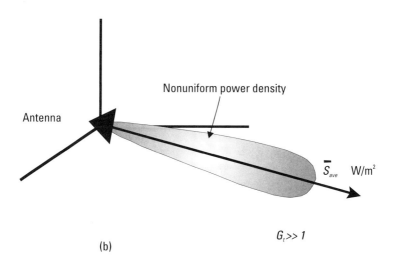

(b)

Figure 11.2 Antenna gain: (a) isotropic antenna, and (b) directive antenna.

$$P_r(R) = \overline{S}_{ave}(R) \cdot \hat{r} A_{er} \qquad (11.9)$$

Upon substitution of (11.7), the received power can be expressed as

$$P_r(R) = \frac{P_{ant} G_t A_{er}}{4 \pi R^2} \qquad (11.10)$$

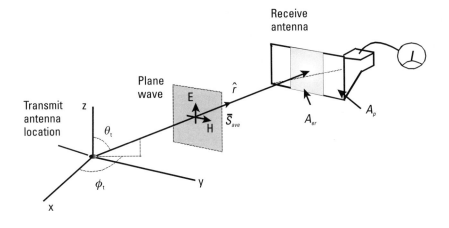

Figure 11.3 Effective aperture.

For a constant power, P_{ant}, at a fixed range R, the received power can be increased by: (1) increasing the antenna gain, G_r, shown in Figure 11.1, or (2) increasing the effective aperture A_{er} in (11.9). Since both factors impact the received signal in the same manner, it is concluded that gain is proportional to effective aperture according to the relationship [3, p. 294].

$$G_r = \frac{4\pi A_{er}}{\lambda_o^2} \qquad (11.11)$$

Therefore, gain of the receive antenna is proportional to its *effective electrical area* A_{er}/λ_o^2. Thus, gain depends upon the physical size of the structure as well as wavelength, or frequency. The free space wavelength of an RF carrier is determined as the quotient of the speed of light, $c = 2.998 \times 10^8$ m/s and the frequency, f, in hertz.

$$\lambda_o = \frac{c}{f} \qquad (11.12)$$

Power at the terminals of the receive antenna in terms of the power available from the source is calculated by solving (11.11) for A_{er}, then substituting the derived expression with (11.1) and (11.4) into (11.10) to provide

$$P_r(R) = \left[\frac{P_s(1 - |\Gamma_t|^2)G_t}{L_{tc}} \right] \frac{G_r\lambda_0^2}{(4\pi R)^2} \qquad (11.13)$$

This form of the Friis transmission formula accounts for cable attenuation and mismatch losses on the transmit end. The expression within the brackets is the effective isotropic radiated power; however, to explicitly show the dependence upon dissipative losses and impedance mismatch the variable *EIRP* is not substituted into the form shown in (11.13).

Equation (11.13) calculates the power at the terminals of the receive antenna, a convenient reference plane. Sometimes it is necessary to compute the power at the terminals of the receiver, P_{rec}, which can be reduced from P_r by impedance mismatch and feed line attenuation. The power, P_{line}, delivered to the transmission line of characteristic impedance, Z_0, has the same form as (11.4). In this case, P_{line} is the difference between the power available at the terminals of the receive antenna, P_r, and that reradiated by the antenna, P_{rerad} (the reflected power depicted on the receive end in Figure 11.1).

$$P_{line} = P_r(1 - |\Gamma_r|^2) \qquad (11.14)$$

The voltage reflection coefficient looking from the receive antenna into the transmission line is defined as

$$\Gamma_r = \frac{Z_0 - Z_{ant}}{Z_0 + Z_{ant}} \qquad (11.15)$$

In practice, this is measured looking from the receive cable into the terminals of the antenna, a measurement of $-\Gamma_r$. Since (11.14) only depends upon the magnitude of Γ_r, the sign of the reflection coefficient doesn't impact the result. Dissipation in the receive cable is accounted for by its loss factor, L_{rc}, given as

$$L_{rc} = \frac{P_{line}}{P_{rec}} \qquad (11.16)$$

The attenuation, A_{rc} in dB, in terms of the receive cable loss factor is given by

$$A_{rc} = 10\log_{10}(L_{rc}) \qquad (11.17)$$

Substituting (11.14) and (11.16) into (11.13), the power at the terminals of the receiver is determined as

$$P_{rec}(R) = \frac{P_s(1 - |\Gamma_t|^2)}{L_{tc}} \frac{G_t G_r \lambda_o^2}{(4\pi R)^2} \frac{(1 - |\Gamma_r|^2)}{L_{rc}} \qquad (11.18)$$

In addition to free space operation, assumptions implicit to (11.18) are: (1) The transmitter and receiver are well matched to their respective cables; (2) the peak of the transmit beam is pointed toward the receiver and vice versa; and (3) the polarization of the two antennas is matched. The first assumption is acceptable for most telemetry applications but the remaining two must be quantified to obtain a realistic measure of system performance.

11.3 Antenna Fundamentals

The objective of this section is twofold: (1) to provide the reader with necessary formulas to quantify the performance of a real-world telemetry link, and (2) to introduce the Fourier transform relationship between the current distribution along an antenna and the resulting radiation patterns. In telemetry applications, separation between the transmitter and receiver is generally large enough so that the two antennas are considered to be in each other's far field. The discussion below is limited to this case to allow adoption of the simplifying assumption of plane wave propagation.

11.3.1 Basic Antenna Theory

Antennas fall into two basic classes: (1) wires with electric currents, and (2) apertures with magnetic currents. The former are configurations such as monopoles, spikes, or blades commonly used in aircraft applications. The latter are primarily the horn and reflector antennas used in the ground station segment. Antenna analysis generally begins with the solution of Maxwell's equations, which are based upon the laws governing electromagnetic wave propagation and radiation. A complete treatment of the associated analysis methods, the magnetic and electric vector potential techniques, is presented in a variety of textbooks [2, 3, 6, 7]. The derivation and solution of these methods is beyond the scope of this presentation; however, key concepts can be introduced by considering a simplified form. For the purposes of this discussion it is sufficient to say that a time-varying current produces propagating electromagnetic fields.

To illustrate the relationship between gain introduced in Section 11.2.1 and physical dimensions, consider the linear wire antenna of length, L, shown in Figure 11.4. It is an electrically thin wire with a radius much less than a free space wavelength, $a \ll \lambda_0$. In this case the three-dimensional problem reduces to a one-dimensional one because the circumferential and radial currents are zero. The fundamental problem is to find the transverse electric and magnetic fields, $\overline{E}$ and $\overline{H}$, respectively, given the electric current along the wire, I.

The fields $\overline{E}$ and $\overline{H}$ are obtained in a two-step process, by first computing the magnetic vector potential, $\overline{A}$, and then the fields. The simplified form of the magnetic vector potential, for a z-directed current, $I(z')$, has the following Fourier form in the far field; that is, when $r \geq r_{ff}$ [3, p. 27]

$$\overline{A} = \mu_0 \frac{e^{-j\beta_0 r}}{4\pi r} \int_{-L/2}^{L/2} \hat{z} I(z') e^{-j\beta_0 z' \cos\theta} dz' \tag{11.19}$$

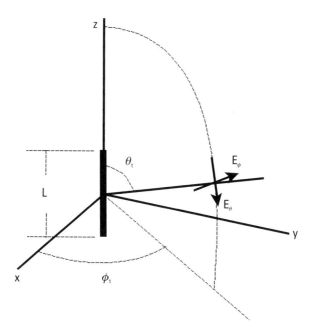

Figure 11.4 Short dipole.

The permeability of free space is $\mu_o = 4\pi \times 10^{-7}$ H/m and $j = \sqrt{-1}$. The phase coefficient, β_o, has units of rad/m and in the free space medium surrounding the antenna it is given by

$$\beta_o = \frac{2\pi}{\lambda_o} \tag{11.20}$$

The far-field distance is defined as the minimum antenna separation beyond which the phase front of the propagating wave appears to be planar; that is, produces an electromagnetic plane wave to some level of approximation. There are three criteria that are used to determine r_{ff}. It is taken to be the largest of the three quantities [3, p. 30]:

$$r_{ff} \geq 1.6\lambda_o \tag{11.21a}$$

$$r_{ff} \geq 5L \tag{11.21b}$$

$$r_{ff} \geq \frac{2L^2}{\lambda_o} \tag{11.21c}$$

If $L/\lambda_o < 0.32$, then equation (11.21a) is the appropriate criterion, whereas if $L/\lambda_o > 2.5$, (11.21c) is valid. Equation (11.21b) is selected for all values such that $0.32 \leq L/\lambda_o \leq 2.5$. For a linear antenna, such as the dipole, the dimension L to use in (11.21) is obvious; it is the length of the thin wire. For other two- and three-dimensional structures, the value of L to be used in (11.21) is the maximum dimension of the antenna.

The transverse electric field components of interest, E_θ and E_ϕ, are found by transforming the vector potential into radial and transverse components. In the far field, the radial component is negligible and discarded. Electric fields transverse to the direction of propagation are depicted in Figure 11.4 and are found in terms of the corresponding components of $\overline{A}$ and the radian frequency, $\omega = 2\pi f$.

$$E_\theta = -j\omega A_\theta \tag{11.22a}$$

$$E_\phi = -j\omega A_\phi \tag{11.22b}$$

In the far field, the electric and magnetic fields are related via the impedance of free space, η_o, which is computed from the permeability and permittivity of free space. The former was defined above, the permittivity of free space is $\epsilon_o = 8.854 \times 10^{-12}$ F/m.

$$\eta_o = \sqrt{\frac{\mu_o}{\epsilon_o}} = 120\pi \ \Omega \tag{11.23}$$

The transverse magnetic fields may then be written in terms of the electric fields.

$$H_\theta = \frac{-E_\phi}{\eta_o} \tag{11.24a}$$

$$H_\phi = \frac{E_\theta}{\eta_o} \tag{11.24b}$$

Since H_θ and H_ϕ are known via (11.24), it is only necessary to find the transverse components of $\overline{E}$ in the far field. To do this, the transverse components of $\overline{A}$ are required. The transformation matrix between the rectangular and spherical forms is given by:

$$\begin{Bmatrix} A_r \\ A_\theta \\ A_\phi \end{Bmatrix} = \begin{bmatrix} \sin\theta\cos\phi & \sin\theta\sin\phi & \cos\theta \\ \cos\theta\cos\phi & \cos\theta\sin\phi & -\sin\theta \\ -\sin\phi & \cos\phi & 0 \end{bmatrix} \begin{Bmatrix} A_x \\ A_y \\ A_z \end{Bmatrix} \tag{11.25}$$

The radial component, A_r, is negligible in the far field and discarded [3, p. 31].

Equation (11.19) transforms between the current distribution along the wire and the far-field pattern (in angle space). The angular dependence of the field components, E_θ and E_ϕ, is a function of the angular variation of the Fourier transform as well as the projection of the result onto the transverse directions through (11.25). To reinforce antenna concepts, analogies based upon the properties of Fourier transforms are shown in Table 11.1.

Example 11.1: Short Dipole

Example 11.1 computes the far fields using the magnetic vector potential technique. Here the length of the source is small enough to approximate a delta function. In this case, the angular dependence of the patterns is due solely to the projection of $\overline{A}$ onto the transverse plane wave components, E_θ and E_ϕ. Although it is concluded that the pattern shape for an electrically small antenna is relatively insensitive to its size, this is not the case for most telemetry antennas.

Table 11.1
Antenna/Signal Analogies

Current/Pattern Relationship	Time/Frequency Relationship
For a fixed-shape current distribution, $I(z)$, the angular extent of the mainbeam will be inversely proportional to the length of the source, L. A longer source will have a narrower beam pattern.	For a fixed-pulse shape, bandwidth will be inversely proportional to pulse width. A longer duration pulse will have a smaller bandwidth.
For a fixed-length source, a tapered distribution will produce a wider mainbeam and reduced-level secondary lobes.	For a fixed-pulse length, a rounded pulse without sharp transitions reduces the level of the high frequency components.

When the length of the z-directed dipole in Figure 11.4 is short, that is, when $L \ll \lambda_0$, A_x and A_y are zero, and the exponential term is approximately unity since $\beta_0 L \ll 1$. This results in the following expression for the magnetic vector potential.

$$\overline{A} = \hat{z}A_z = \hat{z}\mu_0 \frac{I_0 L e^{-j\beta_0 r}}{4\pi r}$$

This is analogous to the Fourier transform of a delta function; the source of infinitesimal extent produces an A_z with no angular variation. Using (11.25) in conjunction with (11.22) through (11.24) the far fields are obtained as

$$E_\theta = j\omega\mu_0 \frac{I_0 L e^{-j\beta_0 r}}{4\pi r} \sin\theta$$

$$H_\phi = \frac{j\omega\mu_0}{\eta_0} \frac{I_0 L e^{-j\beta_0 r}}{4\pi r} \sin\theta$$

For the z-directed dipole, both E_ϕ and H_θ are zero. The distinct dependencies upon range and angle are evident. In the far field, when $r \geq r_{ff}$ and (11.19) is valid, the radial dependence will always be that shown in this example.

11.3.2 Radiation Patterns

The radiation pattern of an antenna is important because it describes the angular variation of the power density of the signal throughout space. Understanding the radiation characteristics facilitates proper selection of the location and orientation of transmit and receive antennas to maximize the received signal. To quantify the performance of the link it is necessary to determine the power density impinging upon the receive antenna. In the far field of the transmit antenna the average Poynting vector is given by [3, p. 37]

$$\overline{S}_{ave}(r,\ \theta,\ \phi) = \frac{1}{2}\operatorname{Re}\{\overline{E} \times \overline{H}^*\} = \hat{r}\frac{1}{2}\frac{|\overline{E}|^2}{\eta_o} \qquad (11.26)$$

Section 11.3.1 highlighted the inverse dependence of the electric and magnetic fields upon range, which supports the inverse square law relationship for the Poynting vector postulated in (11.6) and (11.7) for $R = r$. That is, (11.22) in conjunction with (11.19) will always yield an inverse square law relationship for the far-field power density when substituted into (11.26). This is well known, and for simplicity it is convenient to describe the angular dependence only. This is accomplished by introducing the radiation intensity, U, which is measured in units of watts/steradian (W/sr). It is defined as

$$U(\theta,\ \phi) = \overline{S}_{ave}(r,\ \theta,\ \phi) \cdot \hat{r}r^2 \qquad (11.27)$$

In (11.27) the dependence upon r is eliminated. Although a subtle difference, this is important because a complete description is obtained having only an angular dependence for any range $r > r_{ff}$. Furthermore, if the angular characteristics of the antenna were measured for $r > r_{ff}$ the results would hold at any distance in the far field. It is evident that the radiation intensity has a maximum value in the direction of greatest power density, so it is commonly expressed as U_{max} multiplied by a normalized function describing the angular variation. $F(\theta,\ \phi)$ is termed the field pattern function.

$$U(\theta,\ \phi) = U_{max}|F(\theta,\ \phi)|^2 \qquad (11.28)$$

At microwave and radio frequencies, power rather than electric field intensity is usually measured. Thus, the normalized power pattern is defined as

$$P(\theta,\ \phi) = |F(\theta,\ \phi)|^2 \qquad (11.29)$$

Such patterns are often plotted on a logarithmic scale, which is facilitated by defining the power pattern in dB as

$$P_{dB}(\theta, \phi) = 10 \log_{10} |F(\theta, \phi)|^2 \qquad (11.30)$$

A representative pattern is shown in Figure 11.5. The primary lobe is termed the mainbeam and is characterized by its half-power beamwidth, *HPBW*, that is, the points that are 3 dB lower than the peak value. The pattern also displays secondary lobes or sidelobes, which have amplitude less than the mainbeam peak. The sidelobe level can be characterized in two ways: (1) by a single value, the worst-case value anywhere outside the mainbeam, or

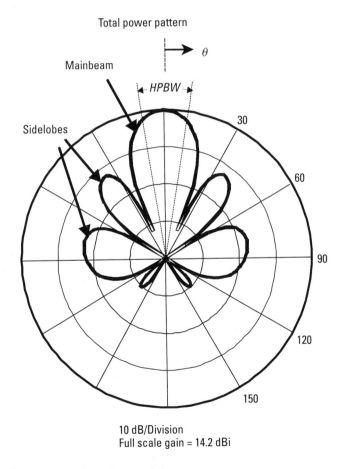

Figure 11.5 Antenna pattern characteristics.

(2) as an envelope; that is, the maximum level within specific angular sectors displaced from the mainbeam.

The normalized power pattern for the short dipole of Example 11.1 is given by

$$P(\theta, \phi) = \sin^2 \theta \qquad (11.31)$$

Figure 11.6 plots the normalized power pattern, in dB, for this case. The gain of the antenna is determined by taking the normalized value at the peak of the beam (negative dB), wherever it occurs on the plot, and adding it to the specified full scale gain (the value at the outer edge of the plot). For this case, the peak of the beam is 0 dB down from the edge,

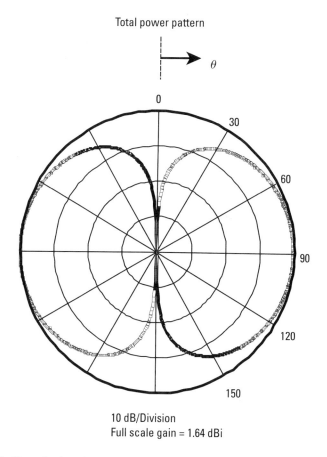

10 dB/Division
Full scale gain = 1.64 dBi

Figure 11.6 Short dipole polar pattern.

which has a specified full-scale value of 1.64 dBi, so that the gain is given by $G = 1.64 - 0 = 1.64$ dBi. Gain will be discussed in detail in Section 11.3.3. The half-power beamwidth for any plane containing the z-axis is $HPBW = 90°$. An absence of sidelobes is typical for this type of electrically short, wide-beam antenna; this is a direct consequence of the Fourier transform of (11.19) for the current distribution along an electrically short antenna.

The form shown in Figure 11.6 displays the angular dependence of the antenna in one plane only and is acceptable for antennas that are ϕ-symmetric such as the dipole. For antennas having a single main beam but not symmetry, a pattern set (such as $\phi = 0°$ and $\phi = 90°$ cuts) is often adequate. A more complete representation, a radiation distribution plot, or RDP, is shown in Figure 11.7. The radiation distribution plot may be

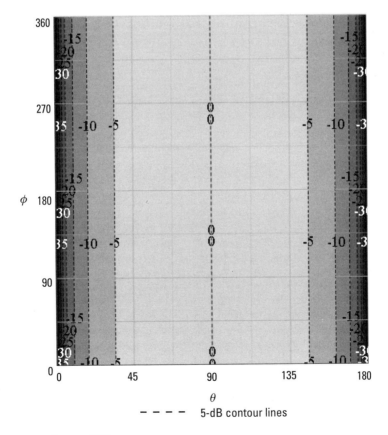

Figure 11.7 Contour RDP.

presented as a contour or alternatively as a surface plot. The former is more useful when a quantitative description is required; the latter is purely qualitative. In this example, the ϕ-symmetry of the pattern is evident because the contour lines are parallel to the ϕ axis.

An optional form is the three-dimensional surface plot shown in Figure 11.8. The distance of the surface from the origin is proportional to the radiation intensity in that direction. This form is useful when a qualitative representation showing the directions of the mainbeam and sidelobes of the pattern is required.

11.3.3 Directivity and Gain

Section 11.2 stated that an antenna achieves gain by directing energy to particular portions of space that would otherwise radiate equally in all directions as from a point source, and is thus always relative to an isotropic antenna. Directivity, D, of an antenna, which is the maximum radiation intensity divided by the average radiation intensity (averaged over angular space) [3, p. 39], is directly related to gain through the radiation efficiency introduced previously.

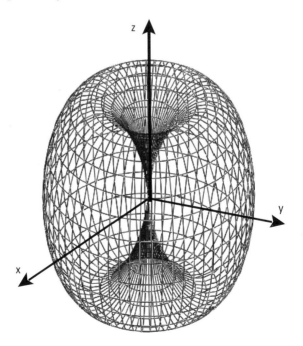

Figure 11.8 Three-dimensional pattern plot.

$$D = \frac{U_{max}}{U_{ave}} \tag{11.32}$$

The average radiation intensity is simply the total power radiated by the antenna P_{rad}, divided by the solid angle of a sphere, 4π sr. It may be computed by

$$U_{ave} = \frac{1}{4\pi} \int_0^{2\pi} \int_0^{\pi} U_{max} |F(\theta, \phi)|^2 \sin\theta \, d\theta d\phi \tag{11.33}$$

Therefore, the directivity can be written in terms of the beam solid angle, Ω_A, which is expressed in steradians, or square radians.

$$D = \frac{4\pi}{\Omega_A} \tag{11.34}$$

where

$$\Omega_A = \int_0^{2\pi} \int_0^{\pi} |F(\theta, \phi)|^2 \sin\theta \, d\theta d\phi \tag{11.35}$$

The power pattern, $|F(\theta, \phi)|^2$, must be normalized to unity to achieve the correct result.

The directivity of an antenna is important because it quantifies the directional properties. How does it relate to the antenna gain introduced in Section 11.2.1? Since the average radiation intensity is simply the total power radiated into 4π steradians, the directivity may also be written as

$$D = \frac{4\pi U_{max}}{P_{rad}} \tag{11.36}$$

The definition of antenna gain, G, is [3, p. 41]

$$G = \frac{4\pi U_{max}}{P_{ant}} \tag{11.37}$$

Substitution of (11.5) into (11.37) demonstrates that gain is proportional to directivity.

$$G = e_{rad}D \qquad (11.38)$$

A mathematical derivation of the relationship between gain and the physical antenna dimensions is well beyond the scope of this chapter; however, perusal of (11.19), (11.22), (11.26), through (11.28), (11.34), (11.35), and (11.38) demonstrates the dependence of patterns and gain upon the Fourier transform of the antenna current distribution. For antennas of planar extent, the resulting two-dimensional Fourier transform is always proportional to antenna area. Properties of Fourier transforms and the use of Parseval's theorem allow the proportionality constant to be derived as $4\pi/\lambda_o^2$, supporting the result of (11.11).

$$G = \frac{4\pi A_e}{\lambda_o^2} \qquad (11.39)$$

This expression is equally applicable to antennas operating in either transmit or receive mode. The effective aperture, A_e, is related to the physical area, A_p, through the aperture efficiency, ϵ_{ap} [3, p. 295].

$$A_e = \epsilon_{ap}A_p \qquad (11.40)$$

When the physical area of the antenna is on the order of a square wavelength or greater, A_e is typically some fraction of A_p resulting in values of aperture efficiency ranging as $0.25 < \epsilon_{ap} < 0.9$. For electrically small antennas that are less than about a half wavelength in extent, the effective aperture may be larger than its physical area. For example, the effective aperture of a thin, half-wavelength long dipole is usually greater than its cross-sectional area.

The gain of the antenna, G_{dB} in decibels, or dB, is expressed as

$$G_{dB} = 10 \log_{10}(G) \qquad (11.41)$$

The result is in dBi. That is, the gain in dB is relative to that of a lossless, isotropic antenna ($G = 1$ or equivalently $G_{dB} = 0$ dBi). For simplicity, the subscript dB will often be implied. That is, if it is given as a number, it is G, otherwise if it is specified in dB it is G_{dB}. For example, it is common to state the gain as $G = 20$ dBi (with an implicit dB subscript); this would correspond to a gain of $G = 100$ (dimensionless).

Example 11.2: Directivity and Gain of a Short Dipole

Example 11.2 illustrates the calculation of directivity and gain for a short dipole. The length is $L = 0.05$m and it operates at 400 MHz. Find the directivity, D, and gain, G, of this antenna given that its radiation efficiency is 63%, $e_{rad} = 0.63$. What is the far-field distance associated with this antenna?

The wavelength at 400 MHz is $\lambda_0 = 0.75$m, thus $L \ll \lambda_0$. It can be assumed that this is an electrically short dipole. From the result of Example 11.1 and (11.35) the beam solid angle can be found by integrating the normalized power pattern.

$$\Omega_A = \int_0^{2\pi} \int_0^\pi |\sin\theta|^2 \sin\theta \, d\theta d\phi = \frac{8\pi}{3} \quad \text{sr}$$

The directivity is

$$D = \frac{4\pi}{\Omega_A} = \frac{4\pi}{\dfrac{8\pi}{3}} = \frac{3}{2}$$

$$G = e_{rad}D = (0.63)\frac{3}{2} = 0.945$$

$$G_{dB} = 10\log_{10}(0.945) = -0.245 \text{ dBi}$$

It is interesting to note that the gain of an antenna can be less than $G_{dB} = 0$ dBi. This is particularly true in the case of electrically small antennas.
Compute the quotient $L/\lambda_0 = 0.066$ and determine if (11.21a) applies. The far-field distance for this antenna is $r_{ff} = 1.6 \, (0.75) = 1.2$m.

11.3.4 Antenna Impedance

The input, or driving point, impedance of an antenna is important to quantify to ensure that it functions properly in a system of specified characteristic impedance, usually $Z_0 = 50 \, \Omega$. An antenna is attached to its interconnecting transmission line as shown in Figure 11.9. A reflection results when the antenna is mismatched to the line; that is, when $Z_{ant} = R_{ant} + jX_{ant} \neq Z_0$. Determination of the reflection coefficient is dependent upon knowing Z_{ant} given Z_0. The antenna impedance is determined by relating the real and reactive power associated with fields $\overline{E}$ and $\overline{H}$ to circuit elements such as

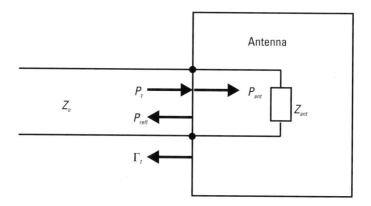

Figure 11.9 Antenna reflection coefficient.

resistors, capacitors, and inductors. The determination of impedance is specific to the exact antenna geometry and is beyond the scope of this treatment. Typically, the telemetry engineer measures the antenna impedance in the laboratory.

The antenna impedance is modeled with three circuit elements: (1) R_Ω, representing the power dissipated in the antenna structure, (2) R_{rad}, an equivalent resistance that absorbs an amount of power equal to that actually radiated by the antenna, and (3) X_{ant}, which represents the energy stored in the region close to the antenna, see Figure 11.10. In most antenna applications, the latter term, X_{ant}, is only important because it impacts the ability to impedance match the transmit antenna to the transmitter.

This model clearly shows that it is desirable to have R_Ω as small as possible. In this case, the real power radiated by the antenna is maximized. The radiation efficiency, introduced in Section 11.2.1, is given by

$$e_{rad} = \frac{P_{rad}}{P_{ant}} = \frac{R_{rad}}{R_\Omega + R_{rad}} \qquad (11.42)$$

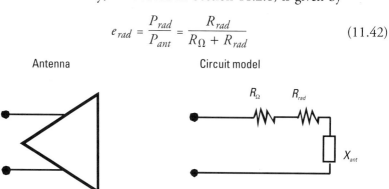

Figure 11.10 Antenna impedance circuit model.

The radiation efficiency varies between zero and unity, 1 being the desired value.

In order to maximize the power delivered to a transmit antenna, its impedance must be the conjugate of the source impedance, $Z_{ant} = Z_s^*$. Since nearly all transmitters and receivers have a characteristic impedance of $Z_s = Z_0 = 50 + j0\Omega$, the task is reduced to a simple impedance matching problem. When the antenna impedance is not matched, a reflection results. Restating (11.3), the voltage reflection coefficient looking into the terminals of either a transmit or receive antenna is

$$\Gamma = \frac{Z_{ant} - Z_0}{Z_{ant} + Z_0} \tag{11.43}$$

The requirement for a perfect match is $Z_{ant} = Z_0$. When the practical assumption of a real-valued characteristic impedance is made, the best reflection coefficient also results in maximum power transfer. In practice, the reflection coefficient is measured with an instrument such as a network analyzer and is most often displayed in one of three formats: (1) the magnitude and phase of the reflection coefficient on the complex plane, (2) voltage standing wave ratio, *VSWR*, or (3) the power reflection coefficient in dB. The first is simply a plot of Γ as a function of frequency on the complex reflection coefficient plane. The voltage standing wave ratio, or *VSWR*, is defined as

$$VSWR = \frac{1 + |\Gamma|}{1 - |\Gamma|} \tag{11.44}$$

It is usually displayed as a function of frequency on a rectangular plot. In most telemetry applications, a $VSWR \leq 2$ is generally acceptable. The associated value of reflection coefficient is $|\Gamma| = 1/3$. The power reflection coefficient is usually referred to as S_{11} in dB.

$$S_{11} = 10 \log_{10} |\Gamma|^2 \tag{11.45}$$

The corresponding value of this quantity for a $VSWR = 2$ is $S_{11} = -9.54$ dB.

To quantify the reduction in received power due to mismatch on both ends of the system, the total loss factor, q_{tot}, is defined.

$$q_{tot} = q_t \, q_r \qquad \qquad (11.46)$$

On the transmit end and receive ends the individual factors are given by

$$q_t = 1 - |\Gamma_t|^2 \qquad \qquad (11.47)$$

and

$$q_r = 1 - |\Gamma_r|^2 \qquad \qquad (11.48)$$

The mismatch loss is often stated in decibels; it is defined as

$$MML_{tot} = -10 \log_{10}(q_{tot}) \qquad \qquad (11.49)$$

The bandwidth of an antenna is determined by the variation of three factors as a function of frequency: (1) impedance, (2) pattern or gain, and (3) polarization. Most often, the bandwidth of an antenna is taken to be that for which its reflection coefficient is acceptably low. This is illustrated in Example 11.3 of Section 11.3.4.1. It is important to note that in some cases the bandwidth is limited by the gain and/or polarization characteristics, not the impedance bandwidth.

Example 11.3: Antenna *VSWR*

The input impedance of an antenna is measured using a network analyzer. The values R_{ant} and X_{ant} are plotted in Figures 11.11 and 11.12, respectively.

Using (11.43) and (11.44), the *VSWR* is computed and plotted as a function of frequency, as shown in Figure 11.13. The bandwidth for *VSWR* ≤ 2 is from 1,338 to 1,488 MHz, a total of 150 MHz. If satisfactory system performance dictated a better match, say *VSWR* ≤ 1.5, the bandwidth would decrease to 78 MHz.

11.3.5 Antenna and Wave Polarization

In order for the telemetry system to function properly, the polarization of the transmit and receive antennas must be properly selected. The concepts of wave and antenna polarization are introduced in this section. As indicated in Figure 11.14, the fields associated with waves emanating from an antenna are transverse electromagnetic; that is, E_θ and E_ϕ are orthogonal to the direction of propagation, $\hat{r}$.

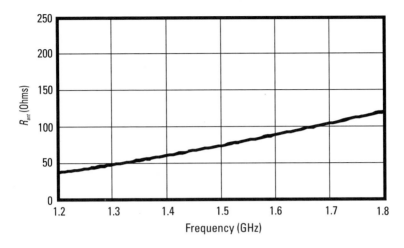

Figure 11.11 Measured R_{ant} versus frequency.

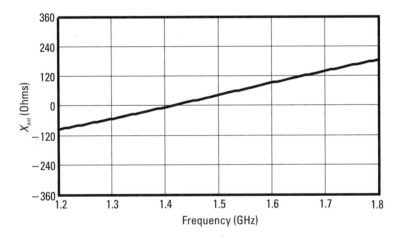

Figure 11.12 Measured X_{ant} versus frequency.

Wave polarization is defined in terms of the electric field variation as a function of time at a fixed location. The polarization of an antenna is taken to be that of the wave it radiates. The polarization of a reciprocal receive antenna is that of the wave it would radiate if transmitting. The polarization state generally varies over angular space; however, if the predominate polarization is linear it is termed a linearly polarized antenna. In a similar fashion, if an antenna primarily radiates circularly polarized waves over its mainbeam it's usually considered to be circularly polarized even if the radiation outside the mainbeam degrades to linear. Since telemetry

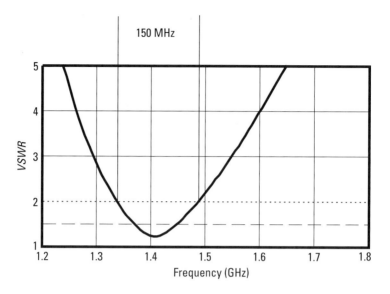

Figure 11.13 *VSWR* versus frequency.

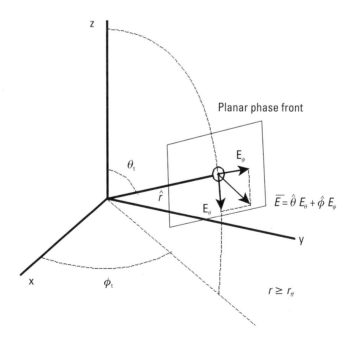

Figure 11.14 Transverse electric far fields.

antennas often have wide coverage areas, it is important to note that the polarization state can vary considerably throughout space. Consider the following mathematical representation for the transverse fields propagating in some known (θ, ϕ) direction.

$$E_\theta(r, t) = E_1 \cos(\omega t - \beta_o r) \tag{11.50}$$

$$E_\phi(r, t) = E_2 \cos(\omega t - \beta_o r + \delta) \tag{11.51}$$

The amplitude of the two components, E_1 and E_2, are real-valued. The phase angle by which E_ϕ leads E_θ is defined to be δ. The range on the phase is $-180° \leq \delta \leq 180°$. To describe the relative contributions of the two components, the quantity γ is defined as

$$\gamma = \tan^{-1} \frac{E_2}{E_1} \tag{11.52}$$

This quantity varies over the range $0 \leq \gamma \leq 90°$. Choosing $\beta_o r = n2\pi$ ($n = 0, \pm1, \pm2 \ldots$) for convenience, and plotting the total electric field as a function of time, the polarization ellipse depicted in Figure 11.15 is

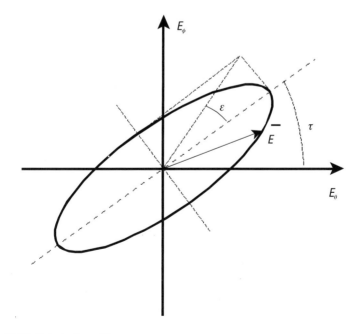

Figure 11.15 Polarization ellipse.

obtained. In this figure it is assumed that the wave is propagating out of the page.

Two important quantities depicted in Figure 11.15 are the angles τ and ϵ. The tilt angle, τ, indicates the orientation of the major axis of the ellipse relative to the reference direction and is easily measured on an antenna range. By convention, the reference direction from which τ is measured is the positive E_θ axis. The angle ϵ describes the shape of the ellipse and a related quantity, the axial ratio. The instantaneous electric field in Figure 11.15 may be rotating in the clockwise or counterclockwise direction. The IEEE standard is that the sense is given by the right-hand rule; that is, if one puts the thumb of his or her right hand in the direction of propagation and the electric field in the fixed plane rotates counterclockwise, the sense is termed right hand [8, p. 77]. Of course, if the field rotates clockwise it is called left hand. For convenience ϵ is signed to account for the sense, positive for left hand, negative for right hand. The magnitude of axial ratio is defined as the ratio of the electric field along the semimajor axis to the electric field along the semiminor axis of the ellipse, by convention it is often given a sign to determine the sense.

$$AR = \cos(-\epsilon) \tag{11.53}$$

The axial ratio is specified in decibels as

$$AR_{dB} = 20 \log_{10}|AR| \tag{11.54}$$

In accordance with (11.53), the sign of the axial ratio is opposite that given to ϵ; that is, the IEEE standard is that right-hand elliptical or circular polarization has a positive value of axial ratio; left-hand is negative.

In general, an antenna generates elliptical polarization. There are two special cases of interest: (1) when the ellipse collapses to a line, linear polarization results; and (2) when the major and minor axes are equal, circular polarization is produced. These limiting cases are depicted in Figures 11.16 and 11.17, respectively.

To achieve circular polarization requires that $E_1 = E_2$ and that the two components be in phase quadrature. Thus, the conditions for circular polarization, in terms of the mathematical description, are $\gamma = 45°$ and $\delta = \pm 90°$. Through (11.50) and (11.51), it is observed that positive values of δ yield left-hand sense and vice versa.

In the limit as the polarization ellipse becomes a line, it is clear that ϵ approaches zero and the axial ratio asymptotically approaches ∞. It is

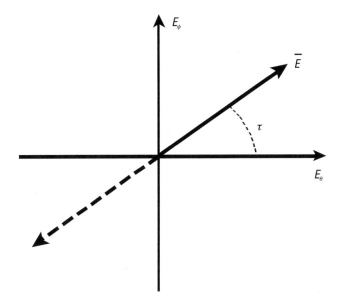

Figure 11.16 Linear polarization.

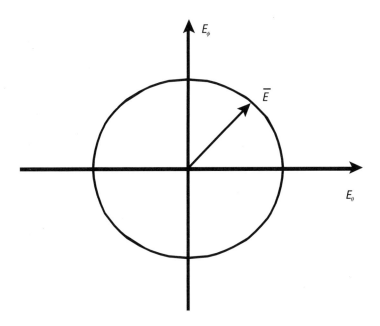

Figure 11.17 Circular polarization.

apparent that when the polarization ellipse becomes a circle, the major and minor axes are equal and $|\epsilon| = 45°$. The resulting axial ratio is $|AR| = 1$. The pair (γ, δ) is a convenient mathematical description; often quantities easily related to (ϵ, τ) are measured on an antenna range. The relationship between the two representations is determined by spherical trigonometry.

$$\tan 2\tau = \tan 2\gamma \cos \delta \qquad (11.55)$$

$$\sin 2\epsilon = \sin 2\gamma \sin 2\delta \qquad (11.56)$$

To compute the polarization mismatch between a radiated wave and a receive antenna it is more convenient to convert the measured quantities from (ϵ, τ) to (γ, δ)

$$\cos 2\gamma = \cos 2\epsilon \cos 2\tau \qquad (11.57)$$

$$\tan \delta = \frac{\tan 2\epsilon}{\sin 2\tau} \qquad (11.58)$$

Polarization mismatch is quantified in terms of the quantity p [3, p. 398].

$$p = |\hat{e} \cdot \hat{h}^*|^2 \qquad (11.59)$$

In order to maximize coupling between the wave and receiving antenna, the semimajor axes of the respective polarization ellipses must be aligned; they must have the same axial ratio in magnitude and sign. That is, if the sense of the incoming wave is right hand, the receive antenna's sense must be right hand as well. The vector $\hat{e}$ represents the polarization of the wave emanating from the transmit antenna. The polarization state of the receive antenna, $\hat{h}$, is that of the wave it would radiate if transmitting. The conjugate in (11.59) is necessary in the receive case. The wave polarization may be described by [3, p. 399]

$$\hat{e} = \cos \gamma \hat{\theta} + \sin \gamma e^{j\delta} \hat{\phi} \qquad (11.60)$$

The receive antenna polarization in terms of the pair (γ_r, δ_r) is

$$\hat{h} = \cos \gamma_r \hat{\theta} + \sin \gamma_r e^{j\delta_r} \hat{\phi} \qquad (11.61)$$

Equations (11.55) through (11.58) apply to the transformations between (γ_r, δ_r) and (ϵ_r, τ_r) as well. The polarization mismatch is often expressed in decibels.

$$PML = -10 \log_{10}(p) \tag{11.62}$$

This quantity is the number of decibels down the received power is relative to in the polarization-matched case.

Table 11.2 summarizes the range of the parameters (γ, δ) and (ϵ, τ) for various polarization states. In practice the designations linear and circular depend upon the application. In many applications an antenna with $|AR| > 10$ (or $AR_{dB} > 20$ dB) is considered linearly polarized. In other cases, an antenna with $|AR| < 2$ (or $AR_{dB} < 6$ dB) might be acceptable in a system requiring a circular polarization.

To further illustrate the concept of polarization mismatch loss, consider a telemetry system that uses right-hand circular polarization. If the ground station has essentially perfect RHCP but the transmit antenna is nonideal, the polarization mismatch loss depends upon the axial ratio as shown in Figure 11.18. This plot demonstrates that if the polarization of the ground station is good quality circular, the polarization mismatch loss is quite low even for relatively large values of transmit antenna axial ratio. In many applications an axial ratio of $AR_{dB} = 6$ dB, which corresponds to a polarization mismatch loss of about $PML = 0.45$ dB, may be acceptable.

Example 11.4: Polarization Mismatch

A left-hand, circularly polarized transmit antenna is found to have an axial ratio of 3 dB upon measurement at an antenna range. The tilt angle is

Table 11.2
Polarization Summary

Polarization	Parameters
Linear ($AR = \infty$)	$\delta = 0, 180°; \forall \gamma$ or $\epsilon = 0°; \forall \tau$
Right-hand elliptical ($1 < AR < \infty$)	$-180° < \delta < 0°; 0 < \gamma < 90°$ or $0° > \epsilon > -45°; \forall \tau$
Left-hand elliptical ($-\infty < AR < -1$)	$0° > \delta > 180°; 0 < \gamma < 90°$ or $0° < \epsilon < 45°; \forall \tau$
Right-hand circular ($AR = 1$)	$\gamma = 45°; \delta = -90°$ or $\epsilon = -45°$
Left-hand circular ($AR = -1$)	$\gamma = 45°; \delta = 90°$ or $\epsilon = 45°$

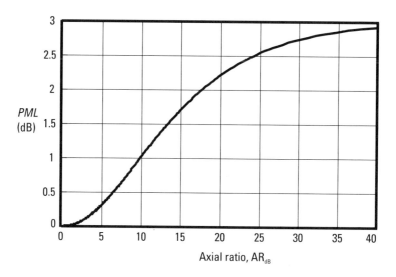

Figure 11.18 Polarization mismatch loss versus axial ratio.

measured to be $\tau = 45°$. The receive antenna is linearly polarized but it has not been installed properly and its tilt angle is $\tau = 135°$. Find the polarization mismatch loss, *PML*, in decibels.

First, the parameters (γ, δ) are determined for the incident wave. The tilt angle was given, and the axial ratio is specified to be 3 dB; find *AR*. Since the sense is specified to be left-hand, ϵ and δ will be positive; *AR* will be negative.

$$|AR| = 10^{3/20} = 1.413 \quad AR = -1.413 \quad \epsilon = \cot^{-1}(-AR) = 35.3°$$

Convert from (ϵ, τ) to find (γ, δ). The tilt angle was specified to be $\tau = 45°$.

$$\gamma = \frac{1}{2}\cos^{-1}(\cos(2\epsilon)\cos(2\tau)) = 45°$$

$$\delta = \tan^{-1}\left(\frac{\tan(2\epsilon)}{\sin(2\tau)}\right) = 70.6°$$

The polarization vector for the wave is

$$\hat{e} = \cos 45°\hat{\theta} + \sin 45°e^{j70.6°} = \frac{1}{\sqrt{2}}(\hat{\theta} + e^{j70.6°}\hat{\phi})$$

The receive antenna is linearly polarized so that $\epsilon_r = 0°$, $\tau = 135°$ was specified in the problem statement. Converting from (ϵ_r, τ_r) to (γ_r, δ_r)

$$\gamma_r = \frac{1}{2}\cos^{-1}(\cos(2\epsilon_r)\cos(2\tau_r)) = 45°$$

$$\delta_r = \tan^{-1}\left(\frac{\tan(2\epsilon_r)}{\sin(2\tau_r)}\right) = 180°$$

The polarization vector for the receive antenna is

$$\hat{h} = \cos 45°\hat{\theta} + \sin 45°e^{j180°} = \frac{1}{\sqrt{2}}(\hat{\theta} - \hat{\phi})$$

The polarization mismatch factor is computed from (11.56)

$$p = \left|\frac{1}{\sqrt{2}}(\hat{\theta} + e^{j70.6°}\hat{\phi})\cdot\frac{1}{\sqrt{2}}(\hat{\theta} + \hat{\phi})\right|^2 = 0.334$$

The resulting polarization mismatch loss is

$$PML = -10\log_{10}(p) = 4.76 \text{ dB}$$

That is, the received signal is 4.76 dB lower than it would have been if the polarization states had been matched.

11.3.6 Common Practical Antennas

This section summarizes the performance and characteristics of several well-known, practical antennas. Antennas particularly suited to telemetry applications, specifically those transmitting from the test article, are discussed in Chapters 12 and 14. The properties of a short dipole were discussed in the previous sections of this chapter; more practical forms of this type of antenna are the half-wavelength dipole and the quarter-wavelength monopole. In addition, this section presents the nominal gain and radiation patterns for antennas such as spirals, helices, horns, and paraboloidal reflectors.

The most common form of the dipole antenna is the half-wavelength version, $L = \lambda_o/2$, shown in Figure 11.19.

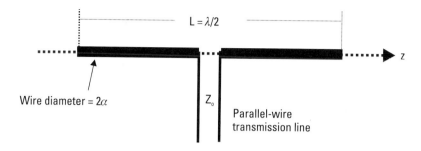

Figure 11.19 Half-wavelength dipole.

The radiation patterns are computed using the method presented earlier; the primary difference is that the current along the half-wavelength long dipole is sinusoidal rather than the uniform distribution assumed in Example 11.1. The distribution that meets the boundary condition of zero current at the ends of the wire is given by

$$I(z') = \begin{Bmatrix} I_0 \sin\left[\beta_o \left(\frac{L}{2} - z' \right) \right] & 0 \le z' \le \frac{L}{2} \\ I_0 \sin\left[\beta_o \left(\frac{L}{2} + z' \right) \right] & -\frac{L}{2} \le z' \le 0 \end{Bmatrix} \tag{11.63}$$

Substituting into (11.19) and applying the method of the remainder of Section 11.3.1, the normalized field pattern of an arbitrary-length dipole with the z-directed sinusoidal current of 11.63 is

$$F(\theta, \phi) = \frac{\left[\cos\left(\beta_o \frac{L}{2} \cos\theta \right) - \cos\left(\beta_o \frac{L}{2} \right) \right]}{\left[1 - \cos\left(\beta_o \frac{L}{2} \right) \right] \sin\theta} \tag{11.64}$$

Equation (11.64) is normalized and there is a single mainbeam when the length is restricted to $L \le 1.25\lambda_o$. For $L = \lambda_o/2$ the result becomes

$$F(\theta, \phi) = \frac{\left[\cos\left(\frac{\pi}{2} \cos\theta \right) \right]}{\sin\theta} \tag{11.65}$$

A plot of the radiation pattern is given in Figure 11.20. The half-power beamwidth of the half-wavelength dipole is about $HPBW = 77°$ in any plane containing the dipole; in the orthogonal direction the pattern is uniform. When (11.65) is integrated to find the beam solid angle, the resulting directivity is $D = 1.64$ (2.15 dBi). If good conductors are used and the wire diameter is large enough the radiation efficiency is high; the corresponding gain is within a couple of tenths of a decibel of the maximum value when $2a > \lambda_o/1,000$ for the half-wavelength long dipole.

The input impedance of the dipole at the center frequency is about $Z_{ant} = 70 + j0\Omega$. Over the bandwidth for which $VSWR \leq 2$ the resistive part varies between about 55 and 80 Ω and the reactive portion can be the same order of magnitude depending upon the diameter of the wire used in construction.

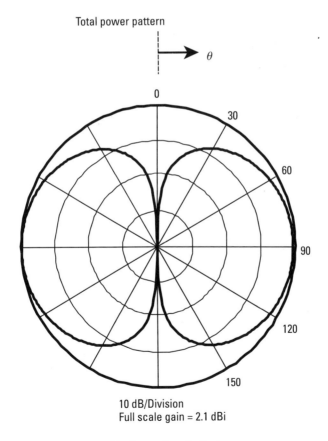

Figure 11.20 Radiation pattern of half-wavelength dipole.

The monopole antenna shown in Figure 11.21 is a modified form of the half-wavelength dipole. It consists of a quarter-wavelength long section driven against a ground plane.

For a z-directed monopole, the current distribution is the top half of (11.63). When the ground plane is infinite in extent, the pattern is given by (11.65), but the pattern is only valid in the upper half space; that is, (11.65) produces the correct result over the range $0 \le \theta \le \pi/2$ while the fields are zero for $\pi/2 \le \theta \le \pi$.

$$F(\theta, \phi) = \begin{cases} \dfrac{\left[\cos\left(\dfrac{\pi}{2} \cos\theta \right) \right]}{\sin\theta} & 0 \le \theta \le \dfrac{\pi}{2} \\[4mm] 0 & \dfrac{\pi}{2} < \theta \le \pi \end{cases} \qquad (11.66)$$

As in the case of the dipole, the radiation efficiency is high when the wire diameter is sufficiently large and the gain is approximately $G = 5.1$ dBi.

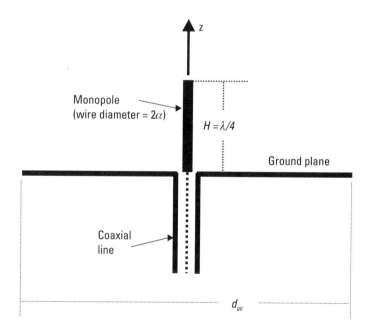

Figure 11.21 Quarter-wavelength monopole.

In practice, monopoles are implemented on ground planes of finite extent. In this case the pattern can be approximated by

$$F(\theta, \phi) = \frac{\left[\cos\left(\dfrac{\pi}{2}\cos\theta\right)\right]}{F_{max}\sin\theta}\left(\frac{1+\cos\theta}{2}\right) \qquad 0 \leq \theta \leq \pi \quad (11.67)$$

For the quarter-wavelength-long version, the value of F_{max} is set to 0.618 to normalize the pattern. The direction of maximum radiation is above the x-y plane; that is, the peak of the beam occurs at about $\theta \approx 65°$ rather than 90° as in the case of (11.66). The half-power beamwidth of the approximate pattern given by (11.67) is $HPBW = 62°$; the pattern is uniform in the orthogonal direction. For large wire diameter, the radiation efficiency is high and the gain is almost $G_{dB} = 3.3$ dBi. For the finite ground plane case, the pattern is given in Figure 11.22.

The input impedance to the monopole at its center frequency is about half that of the dipole; that is, $Z_{ant} \approx 35 + j0\Omega$.

Dipoles and monopoles have relatively narrow impedance bandwidth; that is, the frequency band over which the $VSWR$ is less than or equal to 2 is typically about 15% to 20%. (Their radiation patterns and polarization maintain desirable properties over a larger bandwidth.)

The impedance bandwidth limitation of wire antennas such as $\lambda/2$ dipoles and $\lambda/4$ monopoles is overcome with the use of antennas such as helices and spirals. Axial mode helices and equiangular spirals are circularly polarized; there are other antennas such as log periodic structures that are linearly polarized. This discussion will be restricted to approximate patterns and gain expressions for the helix and spiral. The helix antenna is depicted in Figure 11.23.

The frequency range over which the helix produces a single beam along the positive z-axis is $0.75f_c$ to $1.33f_c$ when the circumference of the helix, $2\pi d_h$, is one wavelength at the center frequency, f_c. The spacing between turns, S_{turn}, is usually taken to be between 0.2 and $0.25\lambda_o$. An approximate pattern of the helix is given by (11.68).

$$F(\theta, \phi) = \frac{\sin\left[\dfrac{N_{turn}}{2}\beta S_{turn}\left(\cos\theta - 1 - \dfrac{\pi}{N_{turn}}\right)\right]}{N_{turn}\sin\left[\dfrac{1}{2}\beta S_{turn}\left(\cos\theta - 1 - \dfrac{\pi}{N_{turn}}\right)\right]}\cos\theta$$

$$(11.68)$$

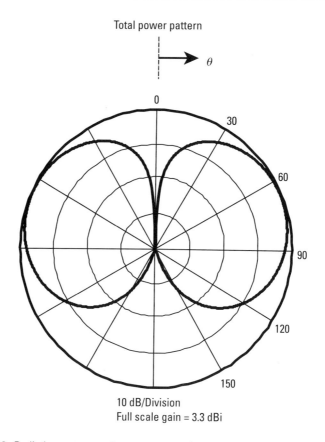

Figure 11.22 Radiation patterns of a quarter-wavelength monopole.

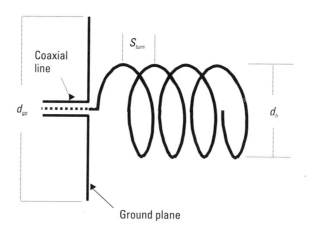

Figure 11.23 Helix antenna.

$$G \approx G_o \left(\frac{2\pi d_h}{\lambda_o} \right)^2 \frac{N_{turn} S_{turn}}{\lambda_o} \qquad (11.69)$$

Values of G_o ranging from 6.2 to 15 are found in the literature. A nominal value of $G_o = 10$ can be used as an approximation. The gain, G_{dB}, is found by applying (11.41) to (11.69) and the result is expressed in dBic, that is, dB relative to isotropic circular polarization. Typically, the number of turns used is between 3 and 15 producing gain values in the range $G_{dB} = 7$ to 15 dBic. When the number of turns, N_{turn}, exceeds about 15, the gain does not actually continue to increase as indicated by (11.69). The beamwidth in degrees can be approximated by

$$HPBW^\circ = \frac{52^\circ}{\dfrac{2\pi d_h}{\lambda_o} \sqrt{\dfrac{N_{turn} S_{turn}}{\lambda_o}}} \qquad (11.70)$$

The radiation pattern of a 10-turn helix at $f = 2{,}251.5$ MHz is shown in Figure 11.24. The diameter and spacing are $d_h = 2.12$ cm and $S_{turn} = 2.93$ cm, respectively. The gain of this unit is $G_{dB} = 13$ dBic.

For a helix operating near its center frequency, when the circumference is one wavelength, the gain versus the number of turns can be computed. At $f = 2{,}251.5$ MHz the gain versus N_{turn} for $d_h = 2.12$ cm and $S_{turn} = 2.93$ cm is plotted in Figure 11.25.

The impedance of the helix is given by

$$Z_{ant} = 140 \frac{2\pi d_h}{\lambda_o} + j0 \ \Omega \qquad (11.71)$$

In practice, a shunt element can be added within the first quarter turn from the input connector, or the helix shape itself can be deformed, to produce a good match to $Z_o = 50 \ \Omega$.

The equiangular spiral antenna is depicted in Figure 11.26. It is an approximation to a self-complementary structure that scales its active region according to the frequency of operation (for a spiral of finite extent over some large but limited bandwidth). The most popular form of this antenna is backed by a cavity to provide unidirectional radiation. It is also circularly polarized.

Since the active region that produces radiation is about one wavelength in circumference, the spiral produces a relatively wide pattern. An approximate, empirical form is given by

Total power pattern

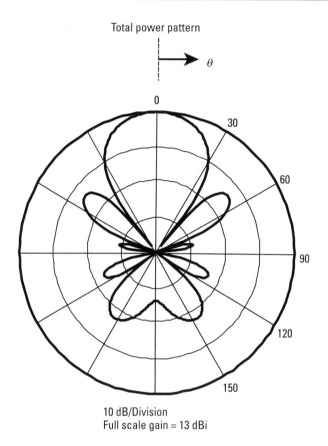

10 dB/Division
Full scale gain = 13 dBi

Figure 11.24 Radiation pattern for helix.

$$F(\theta, \phi) = \cos\left(\frac{\theta}{2}\right)\left(\frac{1 + \cos\theta}{2}\right)^2 \qquad (11.72)$$

The resulting pattern is plotted in Figure 11.27. The gain is approximately $G_{dB} = 5$ dBic and the beamwidth is about $HPBW = 75°$ to $80°$. A balun is used to drive the arms of the spiral with equal amplitude currents and provide an impedance match to a 50-Ω system. The pattern, polarization and impedance characteristics are often achieved over a 3-octave bandwidth.

Pyramidal horns and paraboloidal reflectors or dish antennas are widely used. These structures fall under the heading of aperture antennas, and the details of analysis are beyond the scope of this rudimentary treatment. Antennas such as these are used when more directive, higher-gain patterns are required. This section simply outlines the general characteristics of these two

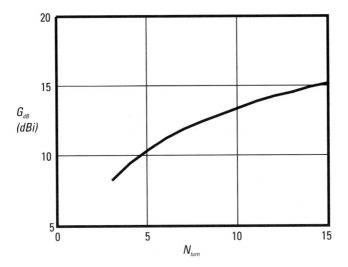

Figure 11.25 *G* versus N_{turn}.

antennas. For example, the equations presented can be used to determine the nominal size of a reflector antenna to be used in a ground station application to develop a baseline design of the telemetry system.

Pyramidal horns are useful in applications where gain values ranging between about 10 and 24 dBi are required. For gain values larger than about G_{dB} = 24 dBi, the dish antenna usually provides a more compact design. Horn antennas can be either linearly or circularly polarized depending upon the exact nature of their connecting waveguide. The details of horn analysis are beyond the scope of this discussion. A rectangular pyramidal horn is shown in Figure 11.28.

Similar results can be achieved with a conical horn, but the nominal characteristics given below are for a linearly polarized rectangular horn fed by a waveguide operating in the dominant TE_{10} mode. The aperture efficiency of a typical horn is about 50%, ϵ_{ap} = 0.5. Thus, using (11.39) and (11.40), the gain can be approximated by

$$G \approx \frac{2\pi A_p}{\lambda_o^2} \tag{11.73}$$

Again, the gain in dB, G_{dB}, is determined by applying (11.41) to (11.73). The physical area is $A_p = W \times H$. One usually takes $H = 0.68\,W$ to achieve equal beamwidths in the $\phi = 0°$ and $90°$ planes. An approximate expression for the half-power beamwidth in degrees is

Top view

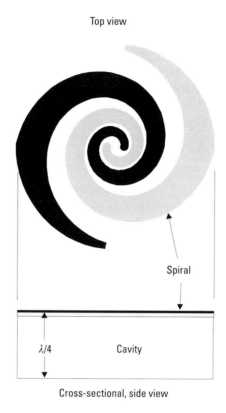

Spiral

$\lambda/4$ Cavity

Cross-sectional, side view

Figure 11.26 Cavity-backed spiral antenna.

$$HPBW \approx 65° \frac{\lambda_0}{W} \qquad (11.74)$$

Typical patterns for a horn at f = 2,251.5 MHz is shown in Figure 11.29. For the linearly polarized version depicted in Figure 11.28, the half-power beamwidths are approximately equal in the x-z (ϕ = 0°) and y-z (ϕ = 90°) planes; the major difference between the patterns is due to the excitation fields produced by the TE_{10} mode waveguide and the phase errors associated with the structure.

The most common reflector geometry is a parabola of revolution, the paraboloidal form. The version considered here is most commonly implemented as a prime-focus fed device; that is, a feed antenna is placed at the focal point (a distance f_p from the vertex) of the parabola, (see Figure 11.30). The radiation patterns, sidelobe levels, and gain are a function of many parameters such as f_p/d_{dish}, taper efficiency, spillover efficiency, aperture

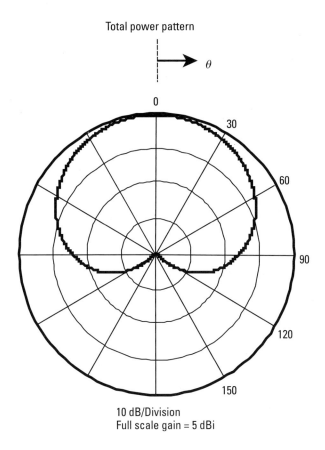

Figure 11.27 Approximate spiral radiation pattern.

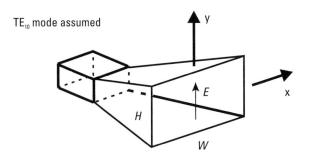

Figure 11.28 Pyramidal horn.

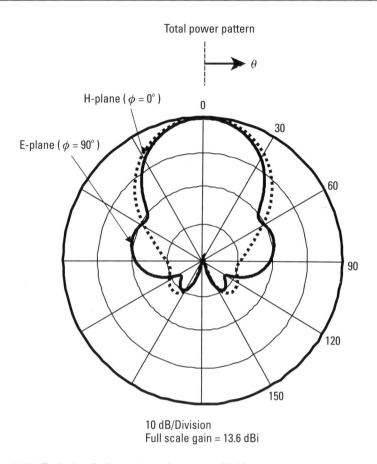

Total power pattern

θ

H-plane ($\phi = 0^{\circ}$)

E-plane ($\phi = 90^{\circ}$)

0

30

60

90

120

150

10 dB/Division
Full scale gain = 13.6 dBi

Figure 11.29 Typical radiation patterns for a pyramidal horn.

blockage, and so on. A typical value of aperture efficiency for a dish antenna is about 50%. Thus, the gain can be *approximated* with a form similar to that in (11.73). In terms of the diameter of the dish, d_{dish}, the gain is given by

$$ G \approx \frac{1}{2} \left(\frac{\pi d_{dish}}{\lambda_o} \right)^2 \tag{11.75} $$

The gain of the reflector antenna in decibels, G_{dB}, is calculated by using (11.41) in conjunction with (11.75). The required reflector electrical diameter to obtain a specified value of gain is plotted in Figure 11.31. The

Paraboloidal reflector

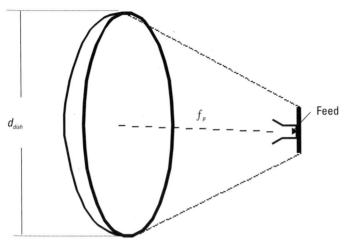

Figure 11.30 Paraboloidal reflector geometry.

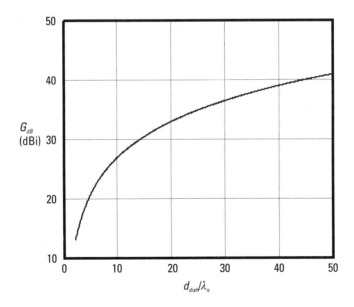

Figure 11.31 G versus d_{dish}.

nominal beamwidth of the reflector antenna system can be approximated by (11.74) by replacing W with d_{dish}.

11.4 Calculation of Carrier Power

It is necessary to calculate the signal power to quantify the performance of the system, that is, to determine the signal-to-noise ratio. Here, calculation of the signal power is addressed; characterization of system noise is the topic of Chapter 13. The link equation (11.18) incorporated some but not all real-world factors affecting the received signal level. The objective of this section is to set forth the form of the Friis transmission formula applicable in telemetry applications. Two situations are considered: (1) fixed transmitter and receiver sites, and (2) a moving transmitter and fixed ground station site. It is still assumed that the transmitter and receiver operate in a free space environment; that is, the Earth is not present to cause reflections, and so on.

For fixed transmitter and receiver sites, the system would be configured such that the antennas would be pointed directly at each other as shown in Figure 11.32.

Combining (11.59), (11.46) and (11.18) results in an expression applicable when the peak of the mainbeam of the receive antenna is pointed at the test article and vice versa.

$$P_{rec} = \frac{P_s G_t G_r}{L_{tc} L_{rc}} \left(\frac{\lambda_o}{4\pi R} \right)^2 pq_{tot} \tag{11.76}$$

The propagation loss can be due to factors other than the free space, or spreading loss, which is the term in the parentheses in (11.76). That is, at some frequencies the attenuation due to propagation through the atmosphere can be significant and must be considered. For some geometries, free space operation cannot be presumed, and the signal level can be seriously

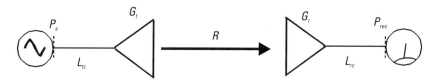

Figure 11.32 Fixed transmit and receive sites.

degraded by reflections from the ground as well as nearby objects. Thus, the term in the parentheses of (11.76) can expanded to accommodate these effects.

$$L_{path} = L_{fsl} L_{atm} L_{mpl} \tag{11.77}$$

The first term, the free space loss denoted as L_{fsl}, is written as

$$L_{fsl} = \left(\frac{4\pi R}{\lambda_o} \right)^2 \tag{11.78}$$

In the far field, when $R \geq r_{ff}$, the free space loss will be rather large; that is, $L_{fsl} \gg 1$. The second term, L_{atm}, can normally be assumed to be unity in typical telemetry applications, particularly for S-band operation; however, at higher frequencies this term must be considered. The third term is the multipath loss, L_{mpl}, and is the topic of Section 11.5. For some geometrical configurations it can be less than one and range up to ∞. When there is a single specular reflection, it varies as $1/2 \leq L_{mpl} \leq \infty$. Rewriting (11.76) in more compact form

$$P_{rec} = \frac{P_s G_t G_r}{L_{tc} L_{rc} L_{path}} pq_{tot} \tag{11.79}$$

The most commonly used version of (11.79) is the logarithmic form. To derive the desired form, define the path attenuation in dB as

$$A_{path} = 10 \log_{10}(L_{path}) \tag{11.80}$$

Using the results of (11.2), (11.17), (11.62), and (11.80)

$$\begin{aligned} P_{rec} \text{ (dBm)} = {} & P_s \text{ (dBm)} + G_t \text{ (dBi)} + G_r \text{ (dBi)} - A_{tc} \text{ (dB)} \\ & - A_{rc} \text{ (dB)} - A_{path} \text{ (dB)} - PML \text{ (dB)} \\ & - MML_{tot} \text{ (dB)} \end{aligned} \tag{11.81}$$

As demonstrated by (11.1) and (11.2) as well as (11.16) and (11.17), the attenuation in dB is 10 times the logarithm of the ratio of powers. This is also true of the *PML* and *MML* factors in dB. The antenna gain values are specified in dBi, gain relative to an isotropic antenna (or a gain of

$G = 1$). The usual definition of dBm is used in this text; it is defined as 10 times the logarithm of the ratio of the power in milliwatts to 1 mW. P_s and P_{rec} must be in the same format; as presented in (11.81) they are both to be specified in dBm. Alternatively they could both be specified in dBW, dB with respect to 1W.

Of course, if the power delivered to a $Z_0 = 50\ \Omega$ line attached to the terminals of the receive antenna is desired, rather than that delivered to the receiver, this equation is modified accordingly.

$$P_{line}\ (\text{dBm}) = P_s\ (\text{dBm}) + G_t\ (\text{dBi}) + G_r\ (\text{dBi}) - A_{tc}\ (\text{dB})$$
$$- A_{path}\ (\text{dB}) - PML\ (\text{dB}) - MML_{tot}\ (\text{dB}) \quad (11.82)$$

The free space loss term, L_{fsl}, is often referred to as spreading loss. Since it is usually the primary contributor to L_{path}, and therefore the A_{path} term, it is convenient to define it in terms of dB as in (11.83). A plot of the free space loss, FSL in dB, versus range at $f = 2{,}251.5$ MHz is given in Figure 11.33.

$$FSL = 20 \log_{10}\left(\frac{4\pi R}{\lambda_o}\right) \quad (11.83)$$

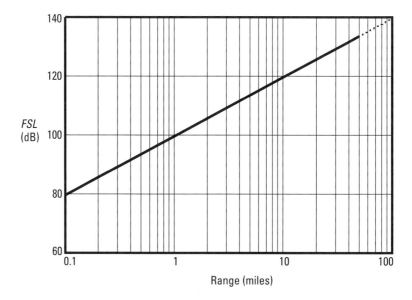

Figure 11.33 Free space loss versus range at $f = 2{,}251.5$ MHz.

In many telemetry applications the orientation of the transmitter does not allow the peak of the transmit beam to be pointed toward the ground station, and receive antenna tracking errors may also result in reduced signal level. In addition, the distance between the transmitter and receiver as well as the aspect angle usually vary with time. These considerations are depicted in Figure 11.34. Also, the polarization loss factor may change as the test article moves along the trajectory. Thus, in addition to the range varying, the possibility exists for the gain and polarization to vary too.

Accounting for these effects, the received power for the dynamic-link case becomes

$$P_{rec}(\theta_t, \phi_t; \theta_r, \phi_r) = \frac{P_S G_t(\theta_t, \phi_t) G_r(\theta_r, \phi_r)}{L_{tc} L_{rc} L_{path}} p(\theta_t, \phi_t; \theta_r, \phi_r) q$$

$$(11.84)$$

The angles (θ_t, ϕ_t) are with respect to the transmit antenna system; (θ_r, ϕ_r) are with respect to the ground station antenna. The quantities

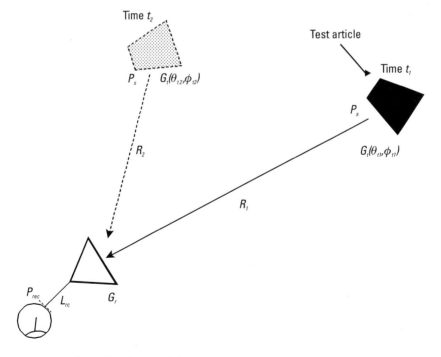

Figure 11.34 Dynamic telemetry link.

$G_t(\theta_t, \phi_t)$ and $G_r(\theta_r, \phi_r)$ are referred to as the directive gain and are given by (11.85) and (11.86), respectively.

$$G_t(\theta_t, \phi_t) = G_t|F_t(\theta_t, \phi_t)|^2 \qquad (11.85)$$

$$G_r(\theta_r, \phi_r) = G_r|F_r(\theta_r, \phi_r)|^2 \qquad (11.86)$$

If the tracking errors are small, $G_r(\theta_r, \phi_r)$ is essentially equal to G_r, or at least within an acceptable quantifiable error, say 0.5 dB. On the other hand, the directive gain of the transmit antenna in the direction of the ground station can vary by 10 dB or more.

The impedance mismatch factor is usually insensitive to the location and orientation of the transmitter and receiver; an exception is if one or both of the systems has to operate within an enclosed region for a portion of the test. For example, a test article that has to radiate while inside a launch canister may have a significantly different *VSWR* than outside the enclosure.

$$
\begin{aligned}
P_{rec}(\theta_t, \phi_t; \theta_r, \phi_r) \text{ (dBm)} = {} & P_s \text{ (dBm)} + G_t(\theta_t, \phi_t) \text{ (dBi)} \\
& + G_r(\theta_r, \phi_r) \text{ (dBi)} \\
& - A_{tc} \text{ (dB)} - A_{rc} \text{ (dB)} - A_{path} \text{ (dB)} \\
& - PML(\theta_t, \phi_t; \theta_r, \phi_r) \text{ (dB)} \\
& - MML_{tot} \text{ (dB)} \qquad (11.87)
\end{aligned}
$$

If the power delivered to a $Z_0 = 50\ \Omega$ line attached to the terminals of the receive antenna is desired, this equation is modified to be

$$
\begin{aligned}
P_{line}(\theta_t, \phi_t; \theta_r, \phi_r) \text{ (dBm)} = {} & P_s \text{ (dBm)} + G_t(\theta_t, \phi_t) \text{ (dBi)} \\
& + G_r(\theta_r, \phi_r) \text{ (dBi)} \\
& - A_{tc} \text{ (dB)} - A_{path} \text{ (dB)} \\
& - PML(\theta_t, \phi_t; \theta_r, \phi_r) \text{ (dB)} \\
& - MML_{tot} \text{ (dB)} \qquad (11.88)
\end{aligned}
$$

Example 11.5: Fixed-Link Geometry

A 20-dBm transmitter, operating at $f = 2{,}251.5$ MHz, is connected through a cable with a loss of 0.5 dB to an antenna of gain, $G_t = 2$ dBi. The worst case *VSWR* produced by the transmit antenna over a ±5-MHz bandwidth about the center frequency is 1.8. The transmit antenna is considered to be left-hand circularly polarized but has an axial ratio of 3 dB. It has been installed such that the major axis of its polarization ellipse is 45° from vertical, $\tau = 45°$. The receive antenna is well matched to its cable and receiver; the resulting *VSWR* is 1.2. The attenuation of the receive cable is 1.1 dB. The receive antenna gain is 23.7 dBi. The receive antenna has an axial ratio of 42 dB, and due to an error in installation is mounted such that $\tau = 135°$. The distance between the transmitter and receiver is 10 km. The system impedance is $Z_0 = 50\ \Omega$.

The task is to compute the power at the terminals of the receiver as well as the input to a 50 Ω transmission line. This is easily accomplished using (11.81) and (11.82) in the following steps:

1. Convert the impedance mismatch information from *VSWR*.
2. Compute q.
3. Compute the impedance mismatch loss in dB, MML_{tot}.
4. Calculate the polarization mismatch loss in dB, PML.
5. Compute the *FSL* in dB.

$$q_t = 1 - |\Gamma_t|^2 = 1 - \left(\frac{1.8 - 1}{1.8 + 1}\right)^2 = 0.918$$

$$q_r = 1 - |\Gamma_r|^2 = 1 - \left(\frac{1.2 - 1}{1.2 + 1}\right)^2 = 0.992$$

$$MML_{tot} = -10\log_{10}(q_t q_r) = -10\log_{10}(0.918 \cdot 0.992) = 0.41 \text{ dB}$$

The results of Example 11.4 can be used because the axial ratio of the receive antenna is large enough to ignore; that is, in an engineering sense it is linearly polarized. The polarization mismatch loss factor in dB was

$$PML = -10\log_{10}(0.334) = 4.76 \text{ dB}$$

Specify the range in meters.

$$R = 10^4 \text{m}$$

Compute the wavelength in meters.

$$\lambda_o = (2.998 \times 10^8)/(2.2515 \times 10^9) = 0.133\mathrm{m}$$

Compute the free space loss in decibels

$$FSL = -20 \log_{10}\left(\frac{4\pi 10^4}{0.133}\right) = 119.5 \text{ dB}$$

In this example it will be assumed that propagation is in free space. Thus, atmospheric and multipath loss will be ignored. The path attenuation in decibels is then given as

$$A_{path} = FSL = 119.5 \text{ dB}$$

Using (11.81)

$$P_{rec} \text{ (dBm)} = P_s \text{ (dBm)} + G_t \text{ (dBi)} + G_r \text{ (dBi)} - A_{tc} \text{ (dB)} - A_{rc} \text{ (dB)}$$
$$- A_{path} \text{ (dB)} - PML \text{ (dB)} - MML_{tot} \text{ (dB)}$$

The result is given in Table 11.3.

As a side note it is observed that the output power of the transmitter is 100 mW.

Using (11.82) the power delivered to the line is

$$P_{line} \text{ (dBm)} = P_s \text{ (dBm)} + G_t \text{ (dBi)} + G_r \text{ (dBi)} - A_{tc} \text{ (dB)} - A_{path} \text{ (dB)}$$
$$- PML \text{ (dB)} - MML_{tot} \text{ (dB)}$$

The result is given in Table 11.4.

Table 11.3
Power Delivered to Receiver in Example 11.5

Source power	P_s	20.0	(dBm)
Transmit antenna gain	G_t	+2.0	(dBi)
Receive antenna gain	G_r	+23.7	(dBi)
Transmit cable attenuation	$-A_{tc}$	−0.5	(dB)
Receive cable attenuation	$-A_{rc}$	−1.1	(dB)
Path attenuation	$-A_{path}$	−119.5	(dB)
Polarization mismatch	$-PML$	−4.76	(dB)
Impedance mismatch	$-MML_{tot}$	−0.41	(dB)
Power at receiver terminals	$P_{rec} =$	−80.6	(dBm)

Table 11.4
Power Delivered to Transmission Line in Example 11.5

Source power	P_s	20.0	(dBm)
Transmit antenna gain	G_t	+2.0	(dBi)
Receive antenna gain	G_r	+23.7	(dBi)
Transmit cable attenuation	$-A_{tc}$	−0.5	(dB)
Path attenuation	$-A_{path}$	−119.5	(dB)
Polarization mismatch	$-PML$	−4.76	(dB)
Impedance mismatch	$-MML_{tot}$	−0.41	(dB)
Power delivered to 50Ω line	$P_{line} =$	−79.5	(dBm)

Example 11.6: Dynamic-Link Geometry

When the link geometry varies continuously throughout the trajectory, there are two basic approaches using (11.87) or (11.88): (1) Compute the carrier power at a large number of discrete points, and (2) Estimate the signal level for the best and worst case situations. To accomplish the former, the location and orientation of the object must be known at all points so that the directive gain and polarization mismatch factors can be computed. In the latter, it is sometimes difficult to identify the best and worst case configurations. Generally speaking, the power will be reduced at maximum range and this condition should be analyzed. The range dependence, explicitly shown in (11.78) and (11.83), dictates that the power should increase at reduced distance; however, other factors such as directive gain and polarization mismatch may affect the signal level and must be estimated. To illustrate these effects, the received power will only be computed at two times, t_1 and t_2. To simplify this example, it will be assumed that the ground station can track the target so that the gain and polarization of the receive antenna in the direction of the test article does not vary. All other system parameters will remain the same as Example 11.5.

The telemetry system designer has selected a transmit antenna to provide wide coverage. At time t_1 the maximum range is $R = 20$ km, and the gain of the transmit antenna toward the ground station is $G_t(\theta_{t1}, \phi_{t1}) = 2$ dBi; the axial ratio and tilt angle at time t_1 are $AR_{dB} = 3$ dB and $\tau = 45°$. Later in the flight the test article begins to tumble, and the worst-case directive gain is $G_t(\theta_{t2}, \phi_{t2}) = -6$ dBi; the axial ratio and tilt angle at time t_2 are $AR_{dB} = 6$ dB and $\tau = 90°$. The range at time t_2 is 5 km.

Using the results of Example 11.5 the impedance mismatch loss is

$$MML_{tot} = 0.41 \text{ dB}$$

The range at time t_1 is 20 km, the free space loss is

$$FSL = 20 \log_{10}\left(\frac{4\pi(2 \times 10^4)}{0.133}\right) = 125.5 \text{ dB}$$

Determine the polarization parameters at time t_1.

$$\epsilon_1 = \cot^{-1}(10^{3/20}) = 35.3° \quad \tau = 45°$$

The corresponding values of (γ, δ) at times t_1 and t_2 are

$$\gamma_1 = \frac{1}{2} \cos^{-1}(\cos(2 \cdot 35.3°)\cos(2 \cdot 45°)) = 45°$$

$$\delta_1 = \tan^{-1}\left(\frac{\tan(2 \cdot 35.3°)}{\sin(2 \cdot 45°)}\right) = 35.3°$$

The ground station antenna in this case has the same gain as in Example 11.5, but is configured to operate as LHCP. The axial ratio is instrumentation quality, $AR_{dB} = 0.2$ dB. The tilt angle is $\tau = 0°$. To simplify the example, make an engineering approximation; assume the receive antenna's axial ratio is perfect, 0 dB. In this case, $\gamma_r = 45°$ and $\delta_r = +90°$. The corresponding values of polarization mismatch factor are

$$p_1 = \left|\cos\gamma_1 \cos\gamma_r + \sin\gamma_1 \sin\gamma_r e^{j(\delta-\delta_r)}\right|^2 = 0.789$$

The polarization mismatch loss in dB is then given by

$$PML = -10 \log_{10}(0.789) = 1.03 \text{ dB}$$

The result is given in Table 11.5.

Of course, if the power delivered to the 50-Ω line interconnecting the antenna and receiver is desired, the receive cable attenuation is eliminated. In that case, $P_{line} = -81.7$ dBm.

The transmit gain at time t_2 is $G_{t2} = -6.0$ dBi and the range has decreased to 5 km, resulting in a lower free space loss.

$$FSL = 20 \log_{10}\left(\frac{4\pi(5 \times 10^3)}{0.133}\right) = 113.5 \text{ dB}$$

Table 11.5
Power Delivered to Receiver in Example 11.6

Source power	P_s	20.0	(dBm)
Transmit antenna gain	G_t	+2.0	(dBi)
Receive antenna gain	G_r	+23.7	(dBi)
Transmit cable attenuation	$-A_{tc}$	−0.5	(dB)
Receive cable attenuation	$-A_{rc}$	−1.1	(dB)
Path attenuation	$-A_{path}$	−125.5	(dB)
Polarization mismatch	$-PML$	−1.03	(dB)
Impedance mismatch	$-MML_{tot}$	−0.41	(dB)
Power at receiver terminals	$P_{rec} =$	−82.8	(dBm)

The polarization of the transmit antenna at time t_2 is found to be

$$\epsilon_2 = \cot^{-1}(10^{3/20}) = 26.6° \quad \tau = 90°$$

Converting to (γ, δ)

$$\gamma_2 = \frac{1}{2}\cos^{-1}(\cos(2 \cdot 26.6°)\cos(2 \cdot 90°)) = 63.4°$$

$$\delta_2 = \tan^{-1}\left(\frac{\tan(2 \cdot 26.6°)}{\sin(2 \cdot 90°)}\right) = 90°$$

The corresponding value of polarization mismatch at time t_2 is

$$p_2 = \left|\cos\gamma_2\cos\gamma_r + \sin\gamma_2\sin\gamma_r e^{j(\delta-\delta_r)}\right|^2 = 0.9$$

Converting to dB

$$PML = -10\log_{10}(0.9) = 0.46 \text{ dB}$$

The result is given in Table 11.6.

Again, if the power delivered to the 50-Ω line interconnecting the antenna and receiver is desired the power increases to $P_{line} = -77.2$ dBm.

11.5 Multipath Effects

Analysis and characterization of antennas is usually conducted in a free space environment. However, the presence of the Earth as well as obstructing

Table 11.6
Power Delivered to Transmission Line in Example 11.6

Source power	P_s	20.0	(dBm)
Transmit antenna gain	G_t	−6.0	(dBi)
Receive antenna gain	G_r	+23.7	(dBi)
Transmit cable attenuation	$-A_{tc}$	−0.5	(dB)
Receive cable attenuation	$-A_{rc}$	−1.1	(dB)
Path attenuation	$-A_{path}$	−113.5	(dB)
Polarization mismatch	$-PML$	−0.46	(dB)
Impedance mismatch	$-MML_{tot}$	−0.41	(dB)
Power at receiver terminals	$P_{rec} =$	20.0	(dBm)

objects can have a deleterious effect on telemetry system performance. The computations of Section 11.4 are based upon a single path, the direct one between the transmitter and receiver. In practice, there can be multiple propagation paths; the complex phasor addition of these signals can cause destructive interference resulting in multipath fading. The concept is introduced in this section; it is discussed in practical terms in Chapter 14.

Although multipath fading can be due to any number of interfering signals, the concept is illustrated in this section using only two paths between transmit and receive locations. The direct and reflected paths are depicted in Figure 11.35. The direct path is the one that would exist in free space.

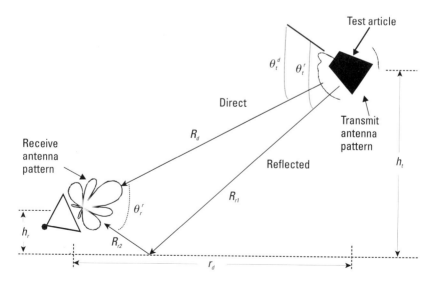

Figure 11.35 Multipath problem.

The reflected signal can be due to ground reflections or scattering from nearby objects. The superposition of the direct and reflected signals can result in constructive or destructive interference.

Figure 11.35 depicts typical transmit and receive antenna patterns. The total voltage at the terminals of the receive antenna, V_{TOT}, will be the superposition of the direct and reflected signals. The first component is weighted by transmit and receive directive gains along the direct path. The second term is weighted by the value of these two quantities along the reflected path.

$$V_{TOT} \propto \sqrt{G_r(\theta_r^d)G_t(\theta_t^d)} \, \frac{e^{-j\beta_o R_d}}{R_d} + \sqrt{G_r(\theta_r^r)G_t(\theta_t^r)} \, \Gamma_g \frac{e^{-j\beta_o R_r}}{R_r}$$

$$(11.89)$$

The ground reflection coefficient, Γ_g, depends upon the electrical parameters of the ground, the polarization, the surface roughness, and frequency. The angles θ_t^d and θ_t^r are with respect to the transmit antenna along the direct and reflected paths, respectively. The angles θ_r^d and θ_r^r are with respect to the receive antenna along the direct and reflected paths, respectively. The sketch depicted in Figure 11.35 assumes no pointing error; therefore, $\theta_r^d = 0°$ and the directive gain is equal to its peak value $G_r(\theta_r^d) = G_r$.

When r_d is large compared to h_t and h_r, the distance of the direct path can be approximated by

$$R_d = r_d + \frac{(h_t - h_r)^2}{2r_d}$$

$$(11.90)$$

The length of the reflected path can be approximated in a similar fashion

$$R_r = R_{r1} + R_{r2} = r_d + \frac{(h_2 + h_1)^2}{2r_d}$$

$$(11.91)$$

Depending upon the path length difference and the magnitude/phase of the ground reflection coefficient, Γ_g, the two terms in (11.89), which are weighted by the directive gains, can either add or interfere destructively, causing multipath fading. The effect of the second term can be reduced if Γ_g and/or $G_r(\theta_r^r)G_t(\theta_r^r)$ along the reflected path are small.

The worst-case situation usually occurs when both antennas are near the ground, causing the direct and reflected signals to be weighted equally

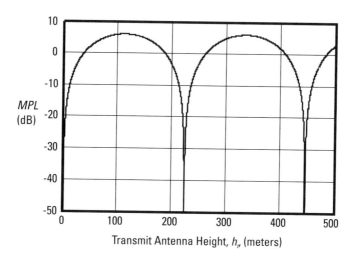

Figure 11.36 Path loss factor versus transmit antenna height.

by the transmit and receive antenna patterns. For near grazing incidence, the magnitude of the reflection coefficient is approximately unity with a phase of 180°. In this case, $R_d \approx R_r$ and the expression for the total voltage becomes

$$V_{TOT} \propto \sqrt{G_r G_t(\theta_d)} \, \frac{e^{-j\beta_o R_d}}{R_d} \left\{ \left| 2 \sin\left(\frac{\beta_o h_1 h_2}{r_d} \right) \right| \right\} \quad (11.92)$$

The power at the terminals of the receive antenna in the presence of multipath, P_r^M, can be expressed in terms of that available in the free space case, P_r^o, the latter quantity being the result obtained using the equations in Section 11.4.

$$P_r^M \text{ (dBm)} = P_r^o \text{ (dBm)} + MPL \text{ (dB)} \quad (11.93)$$

MPL in decibels is given in terms of the multipath loss factor, L_{mpl}, as

$$MPL = 20 \log_{10} |L_{mpl}| \quad (11.94)$$

For the special case of (11.92) the multipath loss factor is given by

$$L_{mpl} = \left| 2 \sin\left(\frac{\beta_o h_t h_r}{r_d} \right) \right| \quad (11.95)$$

It is observed that constructive interference can occur; the received voltage can double (a 6-dB change in power). More importantly it is observed that deep fades can occur as well, and this often occurs in practice when the antennas are located too close to the ground. Figure 11.36 shows the multipath loss factor, *MPL*, as a function of transmit antenna height, h_t, for operation at $f = 2,251.5$ MHz for a range $R = 10$ km when the receive antenna is fixed at a height of $h_r = 3$m.

Problems

Problem 11.1

A low-loss cable is used for an S-band telemetry application. The vendor-supplied data specifies the attenuation of the $Z_o = 50$-Ω cable as 0.197 and 0.207 dB/m at 2.2 and 2.4 GHz, respectively. If a 30-cm-long piece is used to connect the transmitter to the antenna, find: (a) A_{tc} (dB) at 2.2 and 2.4 GHz, and (b) L_{tc} at 2,251.5 MHz.

Problem 11.2

A 10-W transmitter is connected to an antenna with a driving point impedance of $Z_{ant} = 65 - j10$ Ω through the cable specified in Problem 1.1. Find: (a) the *EIRP* in watts at $f = 2,251.5$ MHz if the gain of the transmit antenna is $G_t = 3$ dBi, and (b) the power density, $\overline{S}_{ave}$, in W/m^2 at a distance of $R = 10$ km.

Problem 11.3

A 5-W transmitter delivers $P_{ant} = 4.8$W to an antenna whose maximum power density is $\overline{S}_{ave} = 10$ nW/m^2 at $R = 5$ km. Find: (a) *VSWR*, and (b) the gain, G, of the antenna in dBi. State any necessary assumptions.

Problem 11.4

The *VSWR* of the transmit and receive antennas in a telemetry system are 1.7 and 2.3, respectively. Find the mismatch loss for the system, MML_{tot}, in dB.

Problem 11.5

The input impedance of a transmit antenna is $Z_{ant} = 73 + j42\ \Omega$. If the characteristic impedance of the system is $Z_0 = 50$, find: (a) the complex reflection coefficient, Γ_t, (b) the VSWR on the line connecting the antenna to the source, and (c) the corresponding value of mismatch loss in dB, MML.

Problem 11.6

Given the current distribution of (11.63), show that the normalized field pattern of an arbitrary-length dipole is given by (11.64). Show that the result reduces to (11.65) for $L = \lambda_0/2$.

Problem 11.7

A 5/8-wavelength long dipole operates at 421 MHz; determine the far-field distance, r_{ff}, for this antenna.

Problem 11.8

Determine the far-field distance, r_{ff}, for a 10-m diameter dish operating at $f = 2,251.5$ MHz.

Problem 11.9

A helix operates across the 1.5 to 2.67 GHz band. The diameter is 4.58 cm and it consists of 12 turns of 8 AWG wire. The spacing between turns is $S = 0.25\lambda_0$. Find the far-field distance for this band of operation.

Problem 11.10

Find the far fields, $\overline{E}$ and $\overline{H}$, respectively, for a z-directed, triangular current distribution. That is, the current is given as

$$\hat{z}I(z') = \begin{cases} I_0\left(1 - \left|\dfrac{2z'}{L}\right|\right) & 0 \leq |z'| \leq \dfrac{L}{2} \\ 0 & \text{otherwise} \end{cases}$$

Derive the expression for the fields for an antenna of arbitrary length; do not assume $L << \lambda_0$.

Problem 11.11

If the wire antenna of Problem 11.10 is $L = 1$-m long and operates at $f = 100$ MHz, what is the far-field distance, r_{ff}, in meters?

Problem 11.12

The input impedance to a transmit antenna is measured using a network analyzer. The result is displayed and also produces a file of frequency versus S_{11} data over the range 2.2 to 2.4 GHz. A curve fit to the amplitude data over this frequency band produces a function approximating S_{11} as

$$S_{11} = \frac{1}{2} \left\{ \left[(100 + f_{GHz}) \left(1 - \frac{f_{GHz}}{2.3} \right) + 1 \right]^2 \right\} - 15 \text{ (in dB)}$$

Determine: (a) the center frequency, (b) the bandwidth such that $VSWR \leq 2$, and (c) the bandwidth such that $VSWR \leq 1.5$.

Problem 11.13

If the maximum allowable mismatch loss on the transmit end is 1 dB, what is the corresponding value of $VSWR$ produced by the transmit antenna?

Problem 11.14

Show that the directivity of a thin, half-wavelength long dipole is $D = 2.15$ dBi.

Problem 11.15

The normalized power pattern of a thin, linear, z-directed antenna is given by

$$P(\theta, \phi) = \left(\frac{\sin \left(\beta_o \frac{L}{2} \cos \theta \right)}{\beta_o \frac{L}{2} \cos \theta} \right)^4$$

If the frequency of operation is $f = 1.5$ GHz and the length is 5 cm, find: (a) the directivity, D, in dBi, and (b) the gain, G, in dBi if the radiation efficiency is $e_{rad} = 0.85$. Hint: Perform the integration numerically.

Problem 11.16

Plot the normalized pattern specified in Problem 11.15 over a 40-dB dynamic range on a polar plot. Find the half-power beamwidth, *HPBW*, in degrees.

Problem 11.17

Plot a radiation distribution plot of the pattern of Problem 11.15 over a 40-dB dynamic range in contour format.

Problem 11.18

Determine the directivity, D, in dBi, of the practical monopole through numerical integration of (11.67).

Problem 11.19

A 12-ft diameter reflector antenna has a measured gain of $G = 35$ dBi at $f = 2,222.2$ MHz; find the aperture efficiency, ϵ_{ap}.

Problem 11.20

The radiation patterns of a helix are measured on an antenna range. At the center frequency of operation the axial ratio is found to be $AR_{dB} = 1.8$ dB, the tilt angle is $\tau = 30°$ and its sense is right-hand. Find: (a) ϵ, (b) the polarization mismatch factor, p, and (c) the polarization mismatch loss in dB, *PML* if a same-sense antenna at the opposite end of the link has an axial ratio of $AR_{dB} = 0.5$ dB and the tilt angle is $\tau = 210°$.

Problem 11.21

A transmit antenna has an axial ratio of $AR_{dB} = 2$ dB and a tilt angle of $\tau = 0°$. If the receive antenna is a dipole, plot the polarization mismatch loss, *PML*, versus tilt angle τ_r.

Problem 11.22

Design a helix antenna to have $G = 12$ dBi gain minimum over the frequency band 2,200 to 2,400 MHz. Assume the efficiency is $e_{rad} = 0.90$. Analyze the antenna at midband and the endpoints and tabulate the gain, G, in dBi, and *HPBW*.

Problem 11.23

A helix with a 3-dB beamwidth of $HPBW = 40°$ is required in a particular telemetry application. If the frequency is $f = 2,295.5$ MHz and the spacing between turns is selected to be $0.2\lambda_o$, find the number of turns necessary to produce the desired beamwidth. What is the corresponding gain?

Problem 11.24

A 2.5-m diameter dish is to be used to receive data at $f = 2,215.5$ MHz. (a) What is the gain, G, in dBi if the aperture efficiency is $\epsilon_{ap} = 0.5$? (b) What is the directivity if the radiation efficiency is $e_{rad} = 90.9\%$?

Problem 11.25

Repeat Example 11.5 at $f = 2,235.5$ MHz, if the receive antenna axial ratio is only $AR_{dB} = 18$ dB. The receive antenna is also left-hand sense.

Problem 11.26

A 2-W transmitter is connected to an antenna whose $VSWR$ is 1.7. The peak gain of the single-lobed pattern is $G_t = 6.1$ dBi and it has a half-power beamwidth of $HPBW = 75°$. The transmit antenna is LHCP and has a worst-case axial ratio of $AR_{dB} = 6$ dB over its half-power beamwidth. The system operates at $f = 2.225$ GHz and has a characteristic impedance of $Z_o = 50$ Ω. The attenuation values for the transmit and receive cables are 0.8 and 1.2 dB, respectively. The receive antenna is a 2.5-m dish with a half-power beamwidth of about $HPBW = 3.6°$. The tracking system keeps the dish pointed with 0.5° so that the gain of the LHCP receive antenna is within 0.2 dB of its peak gain. Find the power in dBm at the terminals of the receiver. State any and all assumptions you make.

Problem 11.27

It is critical to receive telemetry data from an article under test at a range of $r_d = 5$ km when it is between 100 and 200m above the Earth; that is, when $100 \leq h_t \leq 200$m. The frequency of operation is $f = 2,251.5$ MHz and the intervening terrain is electrically smooth. Determine the receive antenna height to minimize the signal variation.

References

[1]   Collin, R. E., *Foundations for Microwave Engineering*, New York: McGraw-Hill, 1966.

[2]   Kraus, J. D., *Antennas*, New York: McGraw-Hill, 1950.

[3]   Stutzman, W. L., and G. A. Thiele, *Antenna Theory and Design*, 2nd Edition, New York: Wiley, 1998.

[4]   Pozar, D. M., *Microwave Engineering*, 2nd Edition, New York: Wiley, 1998.

[5]   Elliot, R. S., *An Introduction to Guided Waves and Microwave Engineering*, Englewood Cliffs, NJ: Prentice-Hall, 1993.

[6]   Balanis, C. A., *Antenna Theory—Analysis and Design*, 2nd Edition, New York: Wiley, 1997.

[7]   Milligan, T. A., *Modern Antenna Design*, New York: McGraw-Hill, 1985.

[8]   IEEE Antenna Standards Committee, *IEEE Standard Test Procedures for Antennas*, IEEE STD 149-1979, New York: Wiley, 1979.

12

Conformal Antennas

Chapter 12 introduces conformal antennas, which are particularly suited to telemetry applications, and discusses their terminal properties. Simple formulas for candidate ground station antennas such as reflectors and horns were provided in Chapter 11; this chapter concentrates on antennas mounted to the article under test. The general pattern characteristics of several different antennas are introduced, and a simple model for the microstrip patch antenna is presented.

12.1 Learning Objectives

This chapter focuses on transmit antennas mounted on the article under test. Upon completing this chapter the reader should understand the following:

- Radiating elements used in telemetry applications;
- Operation of a microstrip patch antenna;
- Simple patch design techniques;
- Characteristics of conformal arrays.

12.2 Single Element Antennas

A variety of different antenna types have been used in telemetry applications over the years. In the past, some projectiles were excited as dipoles (at low operating frequencies) to produce broad patterns with correspondingly wide coverage. In the current S-band telemetry channels, individual radiating elements are a practical size and can provide satisfactory coverage on stabilized platforms. Some individual elements, which are well suited to telemetry applications where aerodynamic drag can affect vehicle performance, are low-profile blades and conformal antennas such as cavity-backed slots and microstrip patch antennas. In the sections to follow, the general characteristics of the blade and stripline slot are given; then a transmission-line model for the microstrip patch is presented. This simple model affords the telemetry engineer a method by which to design and quantify the nominal performance of the patch antenna.

12.2.1   Blades

Although not truly conformal antennas, blades are presented in this chapter because they are widely used in telemetry applications. The blade shown in Figure 12.1 is essentially a quarter-wavelength-long monopole, fabricated in a relatively aerodynamic form, with added mechanical strength to increase physical integrity. Furthermore, methods of feeding have been developed to ensure that the entire structure is connected to ground to provide lightning protection and eliminate static charge buildup [1, Chapter 37]. This blade has characteristics similar to the quarter-wavelength monopole discussed in

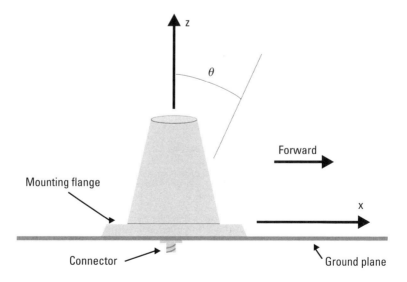

Figure 12.1   Blade antenna.

Section 11.3.6. The gain is G_{dB} = 3 to 5 dBi, depending upon the exact mounting location and the size of the ground plane. The radiation pattern on a ground plane about two wavelengths in extent is approximately that shown in Figure 11.22; the size of the ground plane mainly affects the radiation behind the antenna and the direction of peak radiation. The impedance bandwidth is usually better than the thin monopole due to the larger electrical diameter of the blade.

To improve the aerodynamic properties for use on high-performance aircraft and projectiles, the blade may be tilted as shown in Figure 12.2. This modifies the pattern somewhat; the null, directly above the antenna and clearly evident in Figure 11.22, fills, providing wider coverage than the standard $\lambda_0/4$ monopole. The resulting patterns are dependent upon the shape of the object to which the blade is mounted as well as the location of the radiator relative to edges, and so on. The nominal total power pattern envelope of the modified blade is sketched in Figure 12.3. The gain, when mounted to an object of moderate electrical extent, varies between G_{dB} = 0 and 5 dBi. Again, the impedance bandwidth of this blade is generally greater than that of the thin monopole; a good match is not difficult to obtain over the entire range of S-band telemetry frequencies.

The quadraloop is a lower profile variant of the quarter-wavelength monopole. The blade was derived from the monopole by adjusting the angle for reduced aerodynamic drag. The quadraloop goes one step further by bending the monopole over and driving it against a parallel ground plane in close proximity. A sketch of a quadraloop antenna is shown in Figure

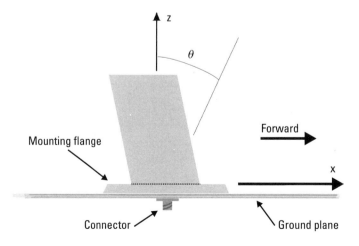

Figure 12.2 Modified blade antenna.

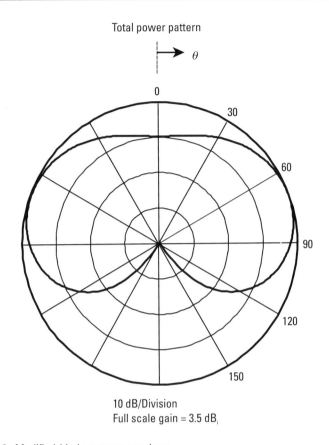

Total power pattern

10 dB/Division
Full scale gain = 3.5 dB

Figure 12.3 Modified blade pattern envelope.

12.4; the dielectric supports the structure. The length, L, is somewhat less than $\lambda_o/4$ due to the dielectric loading as well as fringing off the tail end. The general characteristics of the total power pattern are sketched in Figure 12.5 (the forward direction is along $\theta = 90°$ on the plot). The pattern coverage is similar to that shown in Figure 12.3, but the dip along the $\theta = 0°$ direction is not as pronounced. Furthermore, the pattern is less symmetric than that of the blade; the peak of the beam is typically in the aft direction. The gain of a single element mounted on a flat surface varies between $G_{dB} = 0$ and about 3 dBi, depending upon the ground plane size. The gain of a unit mounted axially on a curved surface with reasonable radius of curvature is about the same as that cited above. The input impedance can be adjusted by varying the location of the coaxial feed point; that is, by adjusting the distance d_f. When the connector is near the front and d_f is close to zero, the impedance is correspondingly small; toward the tail it

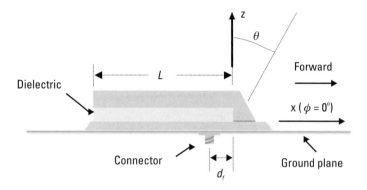

Figure 12.4 Quadraloop antenna.

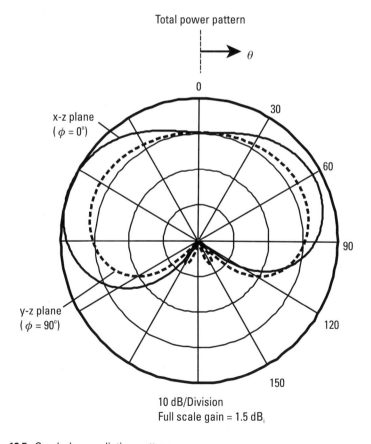

Figure 12.5 Quadraloop radiation pattern.

becomes rather large. The location producing Z_{ant} = 50 Ω is in between, usually closer to the front. The frequency band for which $VSWR \leq 2$ is considerably smaller than that of the blades discussed above; typically the impedance bandwidth for $VSWR \leq 2$ is on the order of 2%.

12.2.2 Printed Circuit Antennas

The remaining two elements discussed in this section, which are truly conformal antennas, are the cavity-backed slot and microstrip patch. Although conformal versions of the former can be constructed by a variety of means, the one presented here is typically referred to as a stripline slot. One difference between the two conformal antennas is that the stripline slot requires that multiple printed circuit board layers be bonded to form the antenna while the microstrip patch can be monolithic. However, in many applications, to survive the environment to which it is exposed, the microstrip patch must have a protective radome layer attached, and the same bonding process as that utilized for stripline is required.

12.2.2.1 Stripline Slot

This antenna is simply a slot in a ground plane excited by a strip transmission line, or stripline. Since one circuit layer and two ground planes are required to fabricate the stripline, it is convenient to back the slot with a cavity created by plated edges, or vias. The presence of the lower ground plane ensures that the pattern will be unidirectional. These antennas are fabricated by etching a slot in the upper ground plane and a feed line on the opposite side of the same board. The circuit is completed by laminating this board to one with no copper on one side and a ground plane on the other as depicted in Figure 12.6. For simplicity, the plated edges are not shown. The dimensions of the cavity are nominally a half-wavelength.

The transmission line can be electromagnetically coupled to the slot by having it extend a quarter-wavelength past the slot. The impedance can be controlled by the width of the feed line and its location relative to the centerline of the slot [2, 3]. Of course, multiple slots can be printed on a board to create an array similar to the ones discussed in Section 12.3.

Stripline slots can be used in applications where significant thermodynamic heating occurs. A metallic heat shield with holes cut out to expose only the radiating slots can be used in high-temperature applications.

The radiation pattern has the general form of that shown in Figure 12.7. The gain of the element, mounted on a flat ground plane a couple of wavelengths in extent, is about G_{dB} = 2 to 4 dBi. The frequency band for

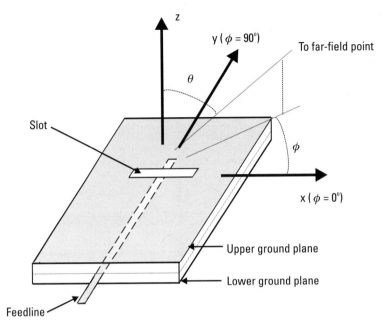

Figure 12.6 Stripline slot.

which $VSWR \leq 2$ is dependent upon the size of the slot and the characteristics of the cavity; typical values for thin structures range from about 2% up to 5%.

12.2.2.2 Microstrip Patch

Microstrip patch antennas have come into wide usage over the last several decades. These antennas are low profile and can be produced at relatively low cost using simple photolithographic techniques. In their most common form, microstrip patches are etched on one side of a printed circuit board; the other side is a copper ground plane. In applications requiring thermal protection, a dielectric superstrate may be bonded on the exterior surface, creating an integrated radome. A linearly polarized patch is shown in Figure 12.8. The width of the patch is W, the height is H. The thickness of the dielectric substrate supporting the patch is h while that of the copper is t. The version shown depicts a feed point representing the connection of a coaxial line a distance d_f from the edge of the patch to achieve $Z_{ant} = 50 \ \Omega$. To ensure proper excitation of the antenna to achieve linear polarization, the feed point is on the centerline of the patch (midpoint of the antenna along the x direction). Alternatively, the patch could have been fed by a

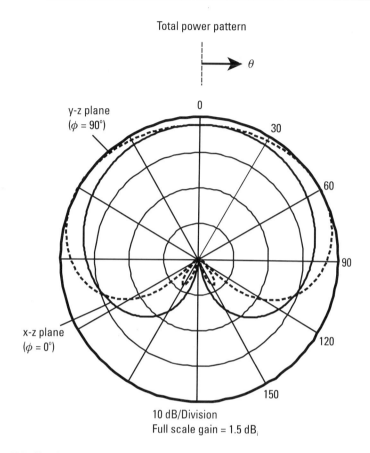

Figure 12.7 Nominal pattern characteristics for stripline slot.

microstrip transmission line connected to the edge of the patch along the dashed line shown in Figure 12.8.

Circularly polarized versions may be achieved by the two configurations shown in Figure 12.9: (a) the nearly square patch, and (b) the square, trimmed-corners microstrip antenna. These two configurations produce good circular polarization over a very limited bandwidth; less than 1% for AR_{dB} ≤ 5 dB is typical at S-band telemetry frequencies on standard, thin substrates. Good axial ratio, say AR_{dB} ≤ 2 dB, can be achieved over a wider frequency range by driving a square patch from two orthogonal sides with a 90° hybrid.

The intent of the remainder of this section is to provide the reader with a simple means to design individual patch antennas, not to provide detailed theoretical descriptions that are available elsewhere [4–6]. To this end, a simple transmission line model, representing the microstrip antenna

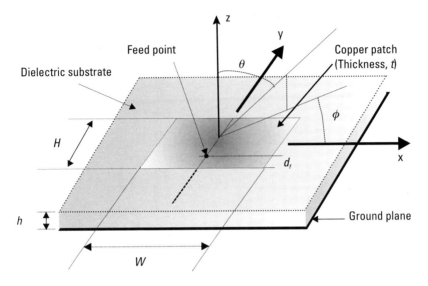

Figure 12.8 Linearly polarized microstrip patch antenna.

as a pair of radiating slots, is presented. The analysis technique presented is suitable to determine the nominal size of microstrip radiating structures and nominally quantify their performance. This rudimentary model, which was later improved, is applicable for operation at the dominant resonance of the structure [7, 8]. The equivalent slots at the top and bottom of Figure 12.10 are interconnected with an intervening transmission line as shown in Figure 12.11.

Although the metallic patch and the ground plane are on opposite sides of the dielectric substrate layer, the patch antenna is modeled in terms of the radiation from a slot in a ground plane. The size of each equivalent slot is the width of the patch, W, by the ground plane spacing, h. The resonant dimension of the antenna, which determines its operating frequency, is primarily dependent upon the dimension H. The frequency at which H is slightly less than a half-wavelength in the dielectric determines the center frequency. The input impedance, Z_{ant}, and *VSWR* versus frequency can be computed using the simple transmission line model presented below; this allows the nominal size of the antenna to be estimated and the impedance bandwidth to be calculated. In addition, expressions for the radiation patterns are given.

The analysis begins by applying standard transmission line theory after the admittance of the equivalent slot is determined. For the slots at the two edges of the patch separated by the dimension H (along the y direction) the admittance, Y_{slot}^{y}, can be approximated by [9, p. 183]

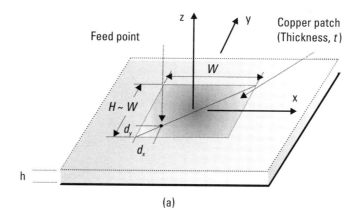

(a)

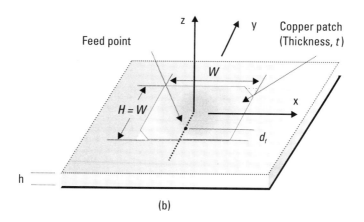

(b)

Figure 12.9 Circularly polarized microstrip patch antennas: (a) nearly square patch, and (b) square, trimmed-corners patch.

$$Y_{slot}^{y} = \left(\frac{\pi W}{\eta_o \lambda_o}\right)\left[\left(1 - \frac{(\beta_o h)^2}{24}\right) + j(1 - 0.636\ln(\beta_o h))\right] \quad (12.1)$$

where β_o is the free space phase constant given by (11.20) and η_o is the intrinsic impedance of free space. The real part of Y_{slot}^{y} represents the radiation conductance of the slot, the imaginary part the susceptance. The latter quantity accounts for fringing off the top and bottom edges of the patch. The distance from the top slot in Figure 12.10 to the feed point is $(H - d_y)$, the distance from the bottom slot to that same point is d_y. Transforming the slot admittances to the feed point through the dielectric-loaded microstrip transmission line yields

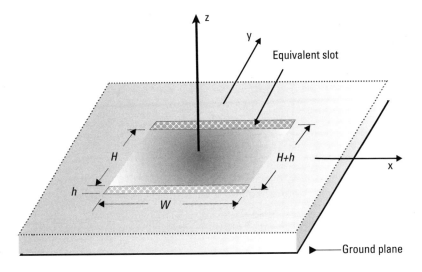

Figure 12.10 Equivalent slots of linearly polarized patch.

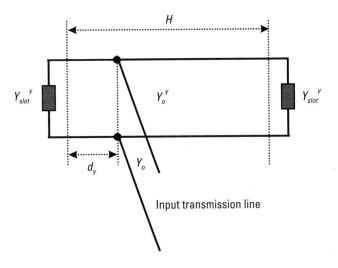

Figure 12.11 Transmission-line model of linearly polarized patch.

$$Y_{iny1} = Y_o^y \frac{Y_{slot}^y + jY_o^y \tan(\beta_y(H - d_y))}{Y_o^y + jY_{slot}^y \tan(\beta_y(H - d_y))} \tag{12.2}$$

and

$$Y_{iny2} = Y_o^y \frac{Y_{slot}^y + jY_o^y \tan(\beta_y d_y)}{Y_o^y + jY_{slot}^y \tan(\beta_y d_y)} \qquad (12.3)$$

where

$$\beta_y = \frac{2\pi}{\lambda_o} \sqrt{\epsilon_{reff}^y} \qquad (12.4)$$

The effective relative permittivity of the microstrip line interconnecting the two equivalent slots is obtained in terms of the width, W, substrate height, h, and copper thickness, t [10]. In this context, ϵ_r is the relative permittivity of the microwave substrate material, not the angle defined in Section 11.3.5.

$$\epsilon_{reff}^y = \frac{\epsilon_r + 1}{2} + \frac{\epsilon_r - 1}{2}\left(1 + 10\frac{h}{W}\right)^{1/2} - (\epsilon_r - 1)\frac{\frac{t}{h}}{4.6\sqrt{\frac{W}{h}}} \qquad (12.5)$$

The characteristic admittance of the intervening transmission line is given by

$$Y_o^y = \frac{\sqrt{\epsilon_{reff}^y}\, W}{120\pi h} \qquad (12.6)$$

The input admittances add in parallel at the feed point.

$$Y_{iny} = Y_{iny1} + Y_{iny2} \qquad (12.7)$$

Convert to find the antenna impedance

$$Z_{ant} = \frac{1}{Y_{iny}} \qquad (12.8)$$

Since the antenna impedance is a function of β_o and β_y, which depend upon the frequency, f, the reflection coefficient Γ, $VSWR$, and/or S_{11} can be computed across the band of interest using the results of Section 11.4. To ensure linear polarization and provide an acceptable range of impedances,

W is usually taken to be on the order of half a free-space wavelength. Typical dimensions for a linearly polarized patch are

$$W \approx \frac{\lambda_o}{2} \tag{12.9}$$

$$H \approx 0.49 \frac{\lambda_o}{\sqrt{\epsilon_{reff}^{y}}} \tag{12.10}$$

For a given W, Equations (12.1) through (12.8) can be used to find H and d_f to provide a good impedance match at the desired frequency. When $\beta_o h$ is small, which it typically is, the patterns of the linearly polarized patch due to the resonance along the y dimension can be shown to be (ignoring terms that are essentially unity)

$$F_\theta^y = \sin\phi \left[\frac{\sin\left(\frac{\beta_o W}{2} \sin\theta\cos\phi\right)}{\left(\frac{\beta_o W}{2} \sin\theta\cos\phi\right)} \cos\left(\frac{\beta_o(H+h)}{2} \sin\theta\sin\phi\right) \right]$$

$$\times \left(\frac{1+\cos\theta}{2}\right) \tag{12.11}$$

$$F_\phi^y = \cos\phi \left[\frac{\sin\left(\frac{\beta_o W}{2} \sin\theta\cos\phi\right)}{\left(\frac{\beta_o W}{2} \sin\theta\cos\phi\right)} \cos\left(\frac{\beta_o(H+h)}{2} \sin\theta\sin\phi\right) \right]$$

$$\times \left(\frac{1+\cos\theta}{2}\right) \tag{12.12}$$

The linearly polarized microstrip patch, oriented as shown in Figure 12.8 produces only F_θ^y in the $\phi = 90°$ plane, and only F_ϕ^y in the = 0° plane. In any plane exclusive of the two principal ones (x-z or y-z), the two components both exist but since they are in-phase they will always produce linear polarization. Although W and H are not usually taken to be the same value for a linearly polarized patch, they are both less than or equal to $\lambda_o/2$ for the dominant resonance; this produces patterns with a single, broad, main lobe. In this case, the two terms in the square brackets of (12.11) and

(12.12) can be simply *approximated* by the expression $(1 + \cos \theta/2)$. Thus, the total power pattern in any plane containing the z-axis can be then be written in its most simple approximate form as

$$F_{ap} = \left(\frac{1 + \cos \theta}{2} \right)^2 \qquad (12.13)$$

It is understood that in the $\phi = 90°$ plane the result is F_θ^y and F_ϕ^y in the $\phi = 0°$ plane.

Typical microstrip patch configurations etched on thin, low-loss microwave substrates have gain values between $G_{dB} = 4.5$ and 7 dBi and a bandwidth of about 2% to 3% for $VSWR \leq 2$.

An interesting configuration occurs when $W \approx H \approx \lambda_d/2$; that is, when both dimensions are about a half-wavelength in the dielectric. In this case, two orthogonal modes can be excited on the structure producing circular polarization if the nearly square patch is fed along the diagonal. When the antenna is fed along the lower left diagonal as shown in Figure 12.9(a) it produces right-hand circular polarization (RHCP) along the $\theta = 0°$ direction for $H > W$, left-hand circular polarization (LHCP) for $H < W$. The ratio of the larger dimension to that of the smaller is generally about 1.01 to 1.04 to achieve circular polarization with acceptable axial ratio.

The equivalent slots for the nearly square patch of Figure 12.9(a) are shown in Figure 12.12. The slots on the left- and right-hand sides of Figure 12.12 are interconnected with a microstrip transmission line analogous to Figure 12.11.

For the nearly square patch the analysis for the resonance along the y direction is the same as that presented in (12.1) through (12.8), the model must simply be amended to accommodate a resonance along the x direction as well. The input impedance due to the mode along the y direction was obtained in (12.8); rewrite it as shown below to explicitly associate it with the mode along the y direction.

$$Z_{iny} = \frac{1}{Y_{iny}} \qquad (12.14)$$

The model for the resonance along the x direction is essentially the same as that presented above, but the slot dimension is H by h and the distance between the two slots is W. The admittance of the equivalent slots along the left and right edges in Figure 12.12 will be

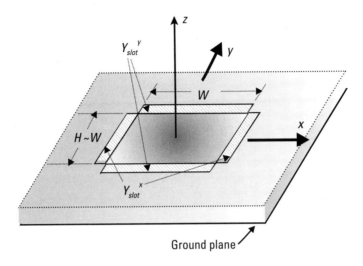

Figure 12.12 Equivalent slots for nearly square patch.

$$Y_{slot}^x = \left(\frac{\pi H}{\eta_o \lambda_o}\right)\left[\left(1 - \frac{(\beta_o h)^2}{24}\right) + j(1 - 0.636\ln(\beta_o h))\right]$$

(12.15)

The input admittances at the feedpoint due to the mode across the x direction are

$$Y_{inx1} = Y_o^x \frac{Y_{slot}^x + jY_o^x \tan(\beta_x(W - d_x))}{Y_o^x + jY_{slot}^x \tan(\beta_x(W - d_x))}$$

(12.16)

and

$$Y_{inx2} = Y_o^x \frac{Y_{slot}^x + jY_o^x \tan(\beta_x d_x)}{Y_o^x + jY_{slot}^x \tan(\beta_x d_x)}$$

(12.17)

where

$$\beta_x = \frac{2\pi}{\lambda_o}\sqrt{\epsilon_{reff}^x}$$

(12.18)

The effective relative permittivity of the microstrip line interconnecting the two equivalent slots along this direction is analogous to (12.5).

$$\epsilon_{reff}^{x} = \frac{\epsilon_r + 1}{2} + \frac{\epsilon_r - 1}{2} \left(1 + 10 \frac{h}{H} \right)^{1/2} - (\epsilon_r - 1) \frac{\frac{t}{h}}{4.6 \sqrt{\frac{H}{h}}}$$

(12.19)

The characteristic admittance of the transmission line connecting the slots along the x direction is given by

$$Y_o^x = \frac{\sqrt{\epsilon_{reff}^{x}} \, H}{120 \pi h}$$

(12.20)

The input admittances add in parallel at the feed point.

$$Y_{inx} = Y_{inx1} + Y_{inx2}$$

(12.21)

The input impedance for the mode along the x direction is

$$Z_{inx} = \frac{1}{Y_{inx}}$$

(12.22)

The antenna impedance of the nearly square patch is obtained by the series combination of the results due to the two orthogonal modes.

$$Z_{ant} = Z_{inx} + Z_{iny}$$

(12.23)

Given the characteristic impedance, Z_o, of the line connected to the feed point of the patch, the reflection coefficient Γ, $VSWR$ or S_{11} can be computed across the band of interest using (12.1) through (12.7) in conjunction with (12.14) through (12.23).

The patterns due to the mode along the y direction are still given by (12.11) and (12.12); for the mode along the x direction they can be approximated in a similar fashion

$$F_\theta^x = \cos\phi \left[\frac{\sin\left(\dfrac{\beta_o H}{2}\sin\theta\sin\phi\right)}{\left(\dfrac{\beta_o H}{2}\sin\theta\sin\phi\right)} \cos\left(\dfrac{\beta_o(W+h)}{2}\sin\theta\cos\phi\right) \right]$$

$$\times \left(\frac{1+\cos\theta}{2}\right) \tag{12.24}$$

$$F_\phi^x = \sin\phi \left[\frac{\sin\left(\dfrac{\beta_o H}{2}\sin\theta\sin\phi\right)}{\left(\dfrac{\beta_o H}{2}\sin\theta\sin\phi\right)} \cos\left(\dfrac{\beta_o(H+h)}{2}\sin\theta\cos\phi\right) \right]$$

$$\times \left(\frac{1+\cos\theta}{2}\right) \tag{12.25}$$

The polarization state of the wave propagating along the bore sight direction, the z-axis or $\theta = 0°$, is found in terms of (γ, δ) as discussed in Section 11.3.5. Approach the $\theta = 0°$ direction in the $\phi = 0°$ plane. In this case, F_θ^y and F_ϕ^x are both zero. The ratio of the amplitudes of the two field components required to compute γ through (11.52) will be proportional to the voltage associated with the modes along the x and y directions. Since the modal impedances add in series, this is then obtained in terms of the input impedances for the two directions

$$\gamma = \tan^{-1}\left|\frac{Z_{iny}}{Z_{inx}}\right| \tag{12.26}$$

The phase is determined in a similar fashion.

$$\delta = \tan^{-1}\left(\frac{\mathrm{Im}\left(\dfrac{Z_{iny}}{Z_{inx}}\right)}{\mathrm{Re}\left(\dfrac{Z_{iny}}{Z_{inx}}\right)}\right) \tag{12.27}$$

Using Equation (11.56)

$$\epsilon = \frac{\sin^{-1}[\sin(2\gamma)\sin(\delta)]}{2} \tag{12.28}$$

Restating (11.53), the corresponding value of axial ratio is

$$AR = \cot(-\epsilon) \qquad (12.29)$$

Since the impedance values are functions of frequency [through the wavelength dependence in (12.4) and (12.18)], the axial ratio can also be computed in terms of frequency. The polarization bandwidth, the frequency range for which the axial ratio is below some acceptable value, is usually the limiting factor for circularly polarized microstrip patches.

Example 12.1: Linearly Polarized Microstrip Patch

Design a linearly polarized patch for operation at $f = 2{,}251.5$ MHz. The antenna is to be etched on a microwave substrate material with the following parameters: $\epsilon_r = 2.94$, $h = 0.1524$ cm and 1 oz copper, or $t = 0.003556$ cm. The design is approached in the following steps:

1. Determine the free space wavelength.
2. Select the width, W.
3. Calculate the effective relative permittivity, ϵ_{reff}^y.
4. Determine an initial value of H using (12.10).
5. Analyze the structure to find the resonant frequency for the current values of W and H.
6. Adjust H to achieve the desired frequency by scaling by the ratio of the current resonant frequency to that desired.

$$H^{new} = H^{old} \frac{f_{current}}{f_{desired}}$$

7. Adjust d_y to obtain a good impedance match.

The free space wavelength at the center frequency is

$$\lambda_0 = 13.324 \text{ cm}$$

Select the width to be that specified by (12.9), $W = \lambda_0/2 = 6.662$ cm. Using (12.5), the effectively relative permittivity for this width is

$$\epsilon_{reff}^y = 2.843$$

The initial value of H can now be specified from (12.10); upon substituting the effective relative permittivity, the value $H \approx 3.872$ cm is obtained. Analysis of this configuration yields a resonant frequency of 2,216 MHz and the impedance for $d_y = 0$ is $Z_{ant} = 122 + j0\ \Omega$. Using (12.1) through (12.7) the dimensions H and d_y can be modified until a good match at the desired frequency is achieved. By scaling the antenna a new value for H is obtained.

$$H = 3.872 \frac{2,216}{2,251.5} = 3.811 \text{ cm}$$

This increases the resonant frequency to 2,250.5 MHz; the impedance is about the same, approximately 120 Ω. To reduce the resonant resistance to 50 Ω, the coaxial probe must be moved toward the center of the patch. Theoretically, the impedance at the center should be 0 Ω; trying a value halfway in between, $d_y = 0.95$ cm, the impedance reduces to $Z_{ant} = 57 + j0\ \Omega$. Further minor adjustments bring the patch dimensions to $H = 3.808$ cm and $d_y = 1.02$ cm. The input *VSWR* versus frequency is computed using (12.1) through (12.8) in conjunction with (11.43) and (11.44); the result is plotted in Figure 12.13. The bandwidth for *VSWR* ≤ 2 is 44 MHz.

The patterns are computed using (12.11) and (12.12) and are compared to the approximate result of (12.13) in Figure 12.14. Integration of these patterns to find the beam solid angle using (11.35) results in a directivity of $D = 4.9$ (6.9 dBi). Using a typical efficiency of about 90% for a low-loss substrate results in a peak gain of $G_{dB} = 6.5$ dBi.

The nominal half-power beamwidth of this design is 85°; the value for the approximate pattern slightly larger at *HPBW* = 92°. When microstrip

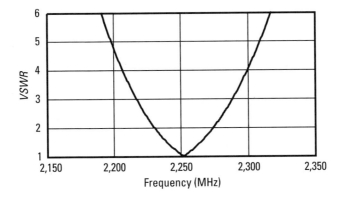

Figure 12.13 Linearly polarized patch input *VSWR* versus frequency.

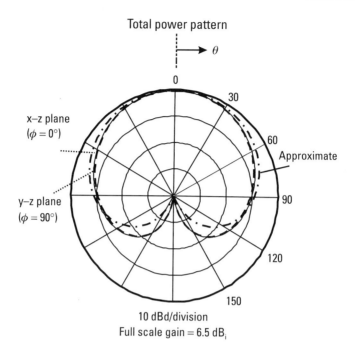

Figure 12.14 Radiation patterns of linearly polarized patch.

patches are fabricated on higher relative permittivity substrates and the ground plane size is less than a wavelength, the pattern can be made even wider (approaching 105°). However, materials with higher values of relative permittivity are often avoided because their use usually reduces the impedance bandwidth and their dissipative losses are generally greater.

Example 12.2: Circularly Polarized Microstrip Patch

Design a nearly square microstrip patch for operation at a frequency of $f = 2,251.5$ MHz using the same substrate material as in Example 12.1. The polarization is to be RHCP. The design of the circularly polarized patch requires four values to be specified: W, H, d_x, and d_y. To obtain the correct center frequency and good circular polarization characteristics using the nearly square patch requires that $W \approx H$ and $d_x \approx d_y$. To achieve RHCP when the patch is fed near the diagonal in the lower left-hand corner of Figure 12.9(a) requires that H be slightly greater than W. The overall dimensions of the patch are a nominal half-wavelength in the dielectric; one approach is to:

1. Compute the free space wavelength, λ_o.

2. Estimate an initial value for the effective relative permittivity using the approximation

$$\epsilon_{reff} \approx \epsilon_r$$

3. Select an initial dimension for the patch as

$$dim \approx 0.49 \frac{\lambda_o}{\sqrt{\epsilon_r}} = 3.808 \text{ cm}$$

The variable *dim* controls the nominal resonant frequency of the patch.

4. Select an initial value for the aspect ratio, Δ. The variable Δ is manipulated to obtain an acceptable value of axial ratio. ($\Delta < 0$ for RHCP, $\Delta > 0$ for LHCP. A typical initial value to achieve RHCP on standard microwave substrates is $\Delta \approx -0.02$.) Using $\Delta = -0.02$,

$$H = dim(1 - \Delta) = 3.884 \text{ cm}$$
$$W = dim(1 + \Delta) = 3.731 \text{ cm}$$

5. Compute the effective relative permittivity, ϵ^y_{reff}, based upon this value of W.

$$\epsilon^y_{reff} = 2.401$$

6. Compute the effective relative permittivity, ϵ^x_{reff}, based upon this value of H.

$$\epsilon^x_{reff} = 2.394$$

7. Set d_x and d_y equal to zero.

8. Analyze the patch using equations (12.1) through (12.29). The axial ratio is found to be $AR = 1.33$ ($AR_{dB} = 2.48$ dB) at a resonant frequency of 2,276 MHz. The *VSWR* at this frequency is 3. The

nominal patch size, *dim*, must be increased to reduce the center frequency; the aspect ratio, Δ, must be modified to improve the axial ratio; and the feed point must be moved from the corner to improve the impedance match.

9. Adjust d_x and d_y to obtain a better match maintaining the same ratio as the overall dimensions of the patch. This is most convenient if a single parameter, d, is defined and d_x and d_y are calculated from it. Set $d = 1.26$ cm and find d_x and d_y as

$$d_y = d(1 - \Delta) = 1.289 \text{ cm}$$
$$d_x = d(1 + \Delta) = 1.251 \text{ cm}$$

This improves the *VSWR* at 2,276 MHz to *VSWR* = 1.4.

10. Adjust the aspect ratio variable, Δ, to obtain an acceptable value of axial ratio at the current center frequency. Changing the aspect ratio variable to $\Delta = -0.015$ results in improved circular polarization; the axial ratio is reduced to $AR = 1.01$ ($AR_{dB} = 0.09$ dB) at 2,274 MHz.

11. Adjust the nominal size of the patch to obtain acceptable axial ratio and impedance match at the desired frequency. It is most straightforward to do a simple frequency scaling

$$dim^{new} = dim^{old} \frac{f_{current}}{f_{desired}} = 3.808 \frac{2,274}{2,251.5} = 3.846 \text{ cm}$$

12. Reanalyze the configuration and perform minor adjustments to W, H, d_x and d_y. The final dimensions are

$W = 3.904$ cm;

$H = 3.788$ cm;

$d_x = 1.251$ cm;

$d_y = 1.289$ cm.

The complex input reflection coefficient is computed and plotted versus frequency on the Smith chart in Figure 12.15. The plot of the antenna impedance, Z_{ant}, versus frequency on the complex plane clearly demonstrates the cusp at $f = 2,251.5$ MHz associated with the orthogonal-mode excitation required to obtain circular polarization.

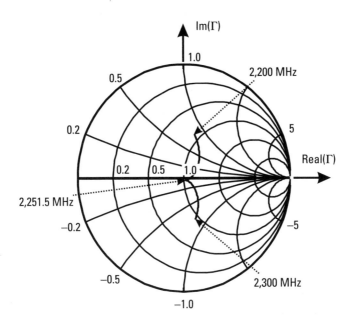

Figure 12.15 Circularly polarized patch input impedance versus frequency.

The *VSWR* is plotted versus frequency in Figure 12.16. The bandwidth such that *VSWR* ≤ 2 is from 2,206 to 2,295 MHz, or a total of 89 MHz.

The patterns are similar in shape to those observed in Figure 12.14 and won't be plotted here. Instead, the on-axis ($\theta = 0°$) axial ratio versus frequency is computed using (12.29) and plotted in Figure 12.17. The frequency range over which this model predicts $AR \leq 2$ ($AR_{dB} \leq 6$ dB) is from 2,230 to 2,275 MHz, or a bandwidth of 45 MHz.

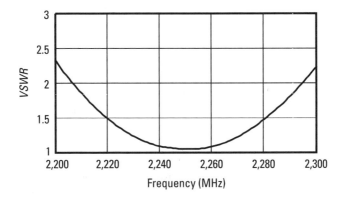

Figure 12.16 Circularly polarized patch input *VSWR* versus frequency.

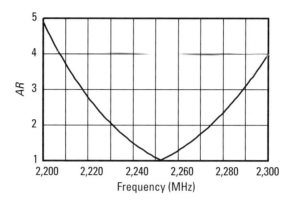

Figure 12.17 Bore sight axial ratio versus frequency.

12.3 Conformal Arrays

A set of antennas rather than a single radiator are often used to achieve desired pattern or beam scanning characteristics. These configurations are referred to as array antennas. The characteristics of linear and planar arrays are well documented in a variety of texts [11–13]. Conformal antennas, such as microstrip arrays, have been addressed in [14]. General cylindrical arrays are also presented in the literature as well as texts; this section focuses on antennas that conform to the exterior surface of the test article [1, Chapter 7], [7]. These conformal configurations are often termed wrap-around antennas. The total field produced by such an array is obtained by superposition via a method similar to that outlined in Section 11.3.1. If the array consists of N_a identical, equispaced elements, the procedure is somewhat simplified. The rudimentary model presented below does not include the effects of wings or tail, et cetera. As such, it is useful to estimate the general pattern characteristics of a wrap-around antenna configuration but not specific pattern details. However, this does not limit the use of the model for the telemetry engineers developing a baseline system design, particularly when the antenna-mounting locations are offset from wing structures (so that the blockage is a small portion of the radiation sphere as viewed from the radiating element).

In this discussion it is assumed that the fields produced by a single radiating element are known and that mutual coupling between the individual antennas is small enough to be ignored. Unidirectional elements such as microstrip patch antennas are generally used in telemetry applications because they have wide beamwidths to provide relatively comprehensive pattern coverage. A typical pattern for such a microstrip antenna was depicted in Figure 12.14.

A wrap-around array can be used to increase the coverage achievable with a single element; that is, to approximate omnidirectional coverage. A possible microstrip array configuration is shown in Figure 12.18. The $\theta = 0°$ direction is taken to be along the forward axis in accordance with IRIG 253-93 [15]. The roll plane is the *x-y* plane. Since many cylindrical test articles are not spin-stabilized, the choice of yaw and pitch planes are somewhat arbitrary. In this discussion both the yaw and pitch planes will be referred to as elevation plane patterns with appropriate reference to the ϕ angle of the reference spherical coordinate system. It consists of N_a radiating elements that are interconnected by microstrip transmission lines printed on the same surface. To simplify the drawing, the feed network is not shown. The radius of the vehicle is r_v.

The overall pattern and polarization of the conformal array depends upon the beam shape of the individual radiators as well as their location and complex excitation (amplitude and phase). The location of each element is specified in spherical coordinates as (r_v, θ', ϕ'). For a cylindrical array

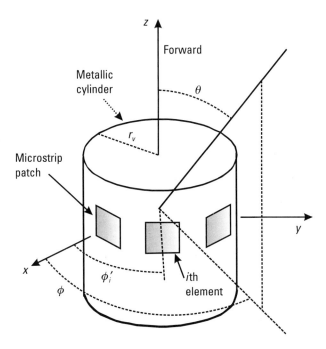

Figure 12.18 Conformal microstrip patch array.

it is convenient to take the centers of the radiating elements in the x-y plane, in this case $\theta' = 90°$. Let the complex value for the ith element be

$$I^i = I_o e^{j\Phi^i} \tag{12.30}$$

That is, this discussion is limited to uniform amplitude excitation; the phase can vary around the circumference of the array. The transverse field patterns of the entire array are obtained by superposition in terms of the component patterns of an individual element, $F_\theta^i(\theta, \phi, \theta_i', \phi_i')$ and $F_\phi^i(\theta, \phi, \theta_i', \phi_i')$.

$$F_\theta^{Tot}(\theta, \phi) = \sum_{i=1}^{N} I^i F_\theta^i(\theta, \phi; \theta_i', \phi_i') e^{j\beta \hat{r} \cdot \bar{r}_i'} \tag{12.31}$$

and

$$F_\phi^{Tot}(\theta, \phi) = \sum_{i=1}^{N} I^i F_\phi^i(\theta, \phi; \theta_i', \phi_i') e^{j\beta \hat{r} \cdot \bar{r}_i'} \tag{12.32}$$

The radial vector to the source point, $\bar{r}'$, has a reduced form when the analysis is restricted to a circumferential ring in the x-y plane, that is, when $\theta_i' = 90°$, $\forall i$.

$$\bar{r}_i' = \hat{x} r_v \cos \phi_i' + \hat{y} r_v \sin \phi_i' \tag{12.33}$$

The unit vector in the (θ, ϕ) direction to the far-field point, $\hat{r}$, is given by

$$\hat{r} = \hat{x} \sin \theta \cos \phi + \hat{y} \sin \theta \sin \phi + \hat{z} \cos \theta \tag{12.34}$$

For cylindrical arrays operating at a frequency such that $r_v > \lambda_o$, simplified expressions for the element pattern similar to (12.13) can be used. At position $(r_v, \theta_i' = 90°, \phi_i')$ the component patterns of the ith element are

$$\left\{\begin{matrix} F_\theta^i(\theta,\ \phi,\ \phi_i') \\ F_\phi^i(\theta,\ \phi,\ \phi_i') \end{matrix}\right\} = \begin{bmatrix} \cos\theta\cos\phi & \cos\theta\sin\phi & -\sin\theta \\ -\sin\phi & \cos\phi & 0 \end{bmatrix} \tag{12.35}$$

$$\times \begin{bmatrix} \cos\phi_i' & -\sin\phi_i' & 0 \\ \sin\phi_i' & \cos\phi_i' & 0 \\ 0 & 0 & 1 \end{bmatrix} \left\{\begin{matrix} F_x^i \\ F_y^i \\ F_z^i \end{matrix}\right\}$$

To determine F_x^i, F_y^i and F_z^i, it is convenient to express the patterns in a local coordinate system referenced to the location of the ith element, ϕ_i', see Figure 12.19.

The relationships between the local and global coordinate systems are determined through spherical trigonometry via the angles, θ, ϕ, ϕ_i', σ_i, and ζ_i.

$$\cos\sigma_i = \sin\theta\cos(\phi - \phi_i') \tag{12.36}$$

and

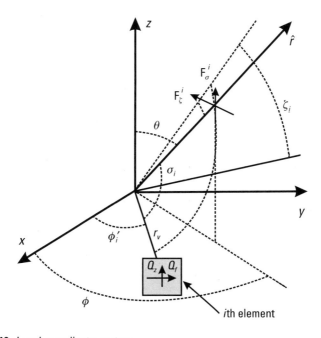

Figure 12.19 Local coordinate system.

$$\tan \zeta_i = \frac{\cos \theta}{\sin \theta \cos \left(\frac{\pi}{2} - \phi + \phi_i' \right)} \tag{12.37}$$

Equations (12.36) and (12.37) can be solved for σ and ζ and used to find the vector projections of F_σ^i and F_ζ^i into F_x^i, F_y^i, and F_z^i. The vector on the right-hand side of (12.35) can then be specified.

$$\begin{Bmatrix} F_x^i \\ F_y^i \\ F_z^i \end{Bmatrix} = \begin{bmatrix} -\sin \sigma_i & 0 \\ \cos \sigma_i \cos \zeta_i & -\sin \zeta_i \\ \cos \sigma_i \sin \zeta_i & \cos \zeta_i \end{bmatrix} \begin{Bmatrix} F_\sigma^i \\ F_\zeta^i \end{Bmatrix} \tag{12.38}$$

F_σ^i and F_ζ^i depend upon the precise form of the radiating element. For microstrip antennas displaying nominal pattern characteristics similar to those in Figure 12.14, an approximate form based upon the constants Q_ϕ and Q_z is

$$\begin{Bmatrix} F_\sigma^i \\ F_\zeta^i \end{Bmatrix} = g_a(\theta, \phi, \phi_i') \begin{bmatrix} \cos \zeta_i & \sin \zeta_i \\ -\sin \zeta_i & \cos \zeta_i \end{bmatrix} \begin{Bmatrix} Q_\phi \\ Q_z \end{Bmatrix} \tag{12.39}$$

The pattern of a microstrip patch mounted on a flat ground plane residing in the x-y plane was shown in Figure 12.14. Transforming the approximate expression in (12.13) to cylindrical coordinates, the pattern of a patch mounted on the cylindrical surface shown in Figure 12.18 may be approximated by

$$g_a(\theta, \phi, \phi_i') = \left(\frac{1 + \sin \theta \cos(\phi - \phi_i')}{2} \right)^{N_e} \tag{12.40}$$

N_e is a variable used to adjust for the electrical diameter of the vehicle. A reasonable value is $N_e = 1 \rightarrow 2$. On cylinders of large radius of curvature, approaching a flat ground plane, the value would be close to 2; on a cylinder with a radius of curvature on the order of a wavelength it is likely to be closer to 1. For the approximation of (12.40) to be valid requires the length and diameter of the vehicle to be at least $2\lambda_0$. The polarization of the individual radiating elements is controlled by the complex constants Q_ϕ and Q_z listed in Table 12.1.

Table 12.1

Parameters to Control Element Polarization

Element polarization	a_ϕ	a_z
Linear (aligned with axis)	0	1
Linear (circumferential)	1	0
LHCP	1	j
RHCP	1	$-j$

An interesting feature of the cylindrical array is that even if linearly polarized elements are used, circular polarization can be achieved in the regions forward and aft of the vehicle by selecting $\Phi^i = \pm\phi_i'$. For equispaced elements this becomes $\Phi^i = \pm(i - 1)2\pi/N_a$. If the forward portion of the vehicle is along the positive z-axis, the positive sign will produce LHCP in that direction and vice versa.

The normalized power pattern of the conformal array is then obtained as

$$P_{Tot}(\theta, \phi) = \frac{\left|F_\theta^{Tot}(\theta, \phi)\right|^2 + \left|F_\phi^{Tot}(\theta, \phi)\right|^2}{\left|F_\theta^{Tot}(\theta_o, \phi_o)\right|^2 + \left|F_\phi^{Tot}(\theta_o, \phi_o)\right|^2} \qquad (12.41)$$

The direction of maximum radiation is (θ_o, ϕ_o).

The number of individual antennas required, N_a, depends upon the electrical size of the vehicle. A rule of thumb sets the spacing to be a half-wavelength. Dividing the circumference by a half wavelength yields

$$N_a = \frac{4\pi r_v}{\lambda_o} \qquad (12.42)$$

Examples 12.3 and 12.4 contrast the patterns, or coverage, available from two similar configurations with different excitations. Both of the designs are often used in telemetry applications. The results do not include the effect of wings and fins, but do show the overall pattern characteristics.

Example 12.3: Conformal Array Fed with Equal Amplitude and Phase

A conformal array is required for a sounding rocket; the radius of the vehicle is $r_v = 0.219$m. The frequency of operation is $f = 2,251.5$ MHz

(λ_o = 0.1332m). It is desired for the pattern to be as isotropic as possible and required that the roll plane variation be minimized; that is, the variation in the θ = 90° plane must be less than 2 dB. Such a pattern is often termed omnidirectional. To meet this pattern requirement, the estimated number of elements is determined using (12.42). The result must be rounded to the nearest integer.

$$N_a = \frac{4\pi(0.225\text{m})}{0.1332\text{m}} = 21$$

One candidate would be a microstrip patch antenna printed on a microwave substrate material. The individual microstrip elements selected for this array produce linear polarization along the axis of the vehicle; thus, the element polarization parameters are set as Q_ϕ = 0 and Q_z = 1. In this application, the individual radiators will be fed with a transmission line network to produce equal amplitude and phase. The number of elements will be limited to N_a = 16 to facilitate fabrication using two-way power dividers only. To approximate the pattern of a microstrip antenna mounted on a surface with this radius of curvature, select N_e = 1.

The patterns were computed using (12.31) and (12.32) and the total power patterns are depicted in Figures 12.20 and 12.21. The first shows an elevation plane pattern, a cut through the x-z, or ϕ = 0°, plane. The patterns for other values of ϕ were quite similar and were not plotted. The second is the roll plane pattern, or θ = 90° plane cut. This array produces linear polarization everywhere in space.

Do the patterns meet the stated objective of an omnidirectional pattern? The roll plane pattern is relatively uniform, the variation is much less than 1 dB and indistinguishable on the plot in Figure 12.21; however, the elevation plane result shown in Figure 12.20 displays large nulls along the positive and negative z-axis. Elimination of these nulls is addressed in Example 12.4. A radiation distribution plot was computed and integrated numerically to find the beam solid angle. The resulting value of directivity was D = 1.7 (2.3 dBi). For the sake of illustration, the feed network and element losses are assumed to 1 dB (e_{rad} = 0.794) resulting in a peak gain value of G_{dB} = 1.3 dBi.

Example 12.4: Conformal Array Fed with Equal Amplitude and Phase Progression

To fill the nulls along the positive and negative axes, this example considers the same geometric configuration as in Example 12.3; however, the excitation

Total power pattern

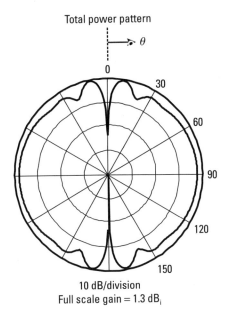

10 dB/division
Full scale gain = 1.3 dB$_i$

Figure 12.20 Elevation plane pattern (equal amplitude and phase excitation).

Total power pattern

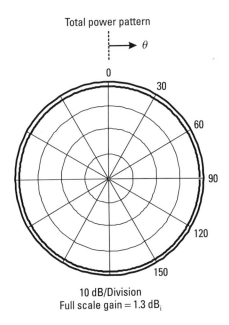

10 dB/Division
Full scale gain = 1.3 dB$_i$

Figure 12.21 Roll plane pattern (equal amplitude and phase excitation).

is selected to be equal amplitude with a phase progression around the circumference. The angular spacing between adjacent patches is set to be $\Delta\phi = 2\pi/N_a$. To achieve phase progression, the phase of the ith patch is set to be $\Phi^i = +\phi_i' = (i-1)\Delta\phi$. This will produce LHCP along the positive z-axis of the vehicle.

The resulting elevation ($\phi = 0°$) and roll ($\theta = 0°$) patterns are shown in Figures 12.22 and 12.23, respectively. Again, the elevation patterns for different ϕ values demonstrated only minor differences and are not plotted. The elevation patterns improved, the nulls fore and aft filled, greatly reducing the variation in that plane. This improvement was at the expense of the roll pattern; the variation in this plane increased slightly. A radiation distribution plot was computed and the resulting directivity was found to $D = 2.344$ (3.7 dBi). Assuming the feed network losses to be nominally the same 1 dB as the previous example, the gain of this configuration is $G_{dB} = 2.7$ dBi.

For the chosen excitation, the polarization in the regions broadside to the vehicle, in the x-y plane, is linear. Forward in the vicinity of the positive z-axis, the polarization is LHCP while that behind the vehicle is RHCP.

In many applications the radiation pattern of Figure 12.22 would be more desirable than that shown in Figure 12.20; however, to be useful the ground station must have the ability to receive multiple polarizations. That

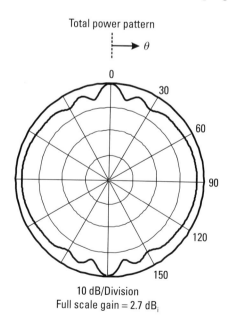

10 dB/Division
Full scale gain = 2.7 dB$_i$

Figure 12.22 Elevation plane pattern (equal amplitude with progressive phase excitation).

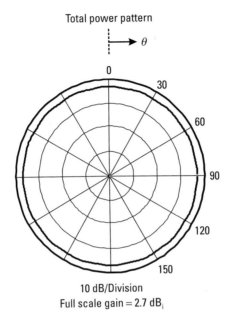

Total power pattern

10 dB/Division
Full scale gain = 2.7 dB$_i$

Figure 12.23 Roll plane pattern (equal amplitude with progressive phase excitation).

is, the receiving equipment would have to have the ability to receive LHCP when the link is established between the ground station and the forward portion of the vehicle and vice versa for the aft end. The results presented here will be used to make comparisons between candidate airborne telemetry antennas in Section 14.5.2.

Problems

Problem 12.1

Perform a survey of blade antennas that operate over any portion of the 500-MHz to 3-GHz band. Tabulate the results including gain, $VSWR = 2$ bandwidth, power handling capability, and size.

Problem 12.2

The principal plane ($\phi = 0°$ and $90°$) patterns of a stripline slot antenna are measured and the following functions approximate its performance. Assume that the two components are in-phase everywhere in space.

$$F_\theta(\theta, \phi) = \sin\phi \left[\left(\frac{\cos\left(\frac{\pi}{2}\cos\theta\right)}{1.1018\sin\theta} \right)^2 | \cos\theta \right] (1 + \cos\theta)$$

$$F_\phi(\theta, \phi) = \cos\phi \left(\frac{1 + \cos\theta}{2} \right)$$

If the radiation efficiency is $e_{rad} = 0.8$, find the gain, G_{dB} in dBi.

Problem 12.3

Estimate the nominal dimensions of a linear-polarized microstrip patch operating at $f = 2,272.5$ MHz using (12.9) and (12.10). The substrate material to be used has a relative permittivity of $\epsilon_r = 2.2$ and a thickness of $h = 0.1524$ cm.

Problem 12.4

A linearly polarized microstrip patch is to operate at $f = 2,295.5$ MHz. The width, W, is selected to be a half of a free space wavelength at this frequency. Compute and plot the nominal value of the dimension, H, using (12.10) for relative permittivity values ranging as $1 \le \epsilon_r \le 10$.

Problem 12.5

Perform a survey of the microwave substrate materials available that would be suitable for S-band telemetry applications. Include the relative permittivity (often referred to as dielectric constant), loss characteristics, the range of thickness values available, and the temperature range over which they can be used.

Problem 12.6

Rework Example 12.1 if the antenna is to be etched on a microwave substrate with a dielectric constant, $\epsilon_r = 9.8$. The thickness of the dielectric substrate remains $h = 0.1524$ cm, and it is covered with 1 oz copper, $t = 0.003556$ cm. If $Z_0 = 50$ Ω, what is the bandwidth for which the $VSWR \le 2$?

Problem 12.7

For the final dimensions W, H, h, and t specified in Example 12.1, compute and plot the input *VSWR* for feed point offset distances varying as $0 \leq d_f \leq H/2$. The system characteristic impedance is $Z_0 = 50 \ \Omega$.

Problem 12.8

Rework Example 12.2 if the center frequency of operation is required to be $f = 2,295.5$ MHz and the substrate thickness is doubled. What is the bandwidth for: (a) $VSWR \leq 2$ if $Z_0 = 50 \ \Omega$, and (b) an axial ratio $AR_{dB} \leq 5$ dB?

Problem 12.9

For the final dimensions W, H, h, and t specified in Example 12.1, set $d_y/d_x = H/W$ and compute; then plot the *VSWR* for feed point offset distances varying as $0 \leq d \leq dim/2$. The system characteristic impedance is $Z_0 = 50 \ \Omega$.

Problem 12.10

A sounding rocket is 56 cm in diameter and has been assigned a telemetry channel at $f = 2,254.5$ MHz. Part (a), estimate the number of radiating elements required. Part (b), if circularly polarized patches etched on a substrate with $\epsilon_r = 2.2$ are used, and if the number of patches determined in Part (a) is used, will they fit around the circumference of the rocket?

Problem 12.11

Design a microstrip wrap-around array to achieve a roll plane variation less than 1.5 dB for a 50-cm-diameter vehicle. The frequency is $f = 2,215.5$ MHz. It is desired to have as uniform coverage as possible. A groove 0.3048-cm deep is available for mounting the conformal antenna. The maximum axial length of the configuration, including the microstrip feed network, is restricted to be 15 cm; however, the transmission lines printed the same layer as the patches will consume 8 cm of the available length. Furthermore, the distance from any patch to the edge of the board must be at least 1.5 cm.

References

[1]   Johnson, R. C., and H. Jasik, *Antenna Engineering Handbook*, 2nd Edition, New York: McGraw-Hill, 1984.

[2]   Rao, J. S., and B. N. Das, "Impedance of Off-Centered Stripline Fed Series Slot," *IEEE Transactions on Antennas and Propagation*, Vol. AP-26, No. 6, Nov. 1978, pp. 893–895.

[3]   Das, B. N., and K. V. S. V. R. Prasad, "Impedance of a Transverse Slot in the Ground Plane of an Offset Stripline," *IEEE Transactions on Antennas and Propagation*, Vol. AP-32, No. 11, Nov. 1984, pp. 1245–1248.

[4]   Carver, K. R., and J. W. Mink, "Microstrip Antenna Technology," *IEEE Transactions on Antennas and Propagation*, Vol. AP-29, No. 1, Jan. 1981, pp. 2–24.

[5]   Lo, Y. T., D. Solomon, and W. F. Richards, "Theory and Experiment on Microstrip Antennas," *IEEE Transactions on Antennas and Propagation*, Vol. AP-27, Jan. 1978, pp. 137–45.

[6]   Richards, W. F., Y. T. Lo, and D. Harrison, "An Improved Theory for Microstrip Antennas and Applications," *IEEE Transactions on Antennas and Propagation*, Vol. AP-29, No. 1, Jan. 1981, pp. 38–46.

[7]   Munson, R. E., "Conformal Microstrip Antennas and Microstrip Phased Arrays," *IEEE Transactions on Antennas and Propagation*, Vol. AP-22, No. 1, Jan. 1974, pp. 74–78.

[8]   Van de Capelle, A., and H. Pues, "Accurate Transmission-Line Model for the Rectangular Microstrip Antenna," *IEE Proc.* Vol. 131, Pt. H, 1984, pp. 334–340.

[9]   Harrington, R. F., *Time-Harmonic Electromagnetic Fields*, New York: McGraw-Hill, 1961, p. 183.

[10]   Bahl, I. J., and R. Garg, "Simple and Accurate Formulas for Microstrip with Finite Strip Thckness," *Proceedings of the IEEE*, Vol. 65, Nov. 1977, pp. 1611–12.

[11]   Stutzman, W. L., and G. A. Thiele, *Antenna Theory and Design*, 2nd Edition, New York: Wiley, 1998.

[12]   Balanis, C. A., *Antenna Theory—Analysis and Design*, 2nd Edition, New York: Wiley, 1997.

[13]   Lo, Y. T., and S. W. Lee, *Antenna Handbook—Theory, Applications and Design*, New York: Van Nostrand Reinhold, 1988.

[14]   Mailloux, R. J., J. F. McIlvenna, and N. P. Kernweis, "Microstrip Array Technology," *IEEE Transactions on Antennas and Propagation*, Vol. AP-29, No. 1, Jan. 1981, pp. 25–37.

[15]   Range Commanders Council, Electronic Trajectory Measurements Group, IRIG STANDARD 253-93, U.S. Army White Sands Missile Range, NM, Aug. 1993.

13

Receiving Systems

The intent of this chapter is to provide the reader a means by which to quantify the performance of a given design in terms of noise figure, noise temperature, and system figure of merit, G/T. The end result is an ability to compute the carrier-to-noise ratio, C/N, and ultimately the link margin.

13.1 Learning Objectives

This chapter focuses on the end-to-end noise characterization of the telemetry system. Upon completing this chapter the reader should understand the following concepts:

- Equivalent noise temperature;
- Receiver noise figure;
- Performance of a cascaded system of components;
- Antenna temperature;
- System noise temperature;
- System noise figure;
- Gain-to-noise temperature ratio (G/T);
- Carrier-to-noise ratio (C/N);
- Bit energy to spectral noise density (E_b/N_o);
- Link margin.

13.2 Receiver Noise Figure

A basic introduction to thermal noise will be presented and then a figure of merit for receiving systems will be defined. In this chapter, noise is represented in terms of the power available from a resistor at a specified temperature. Noise associated with all telemetry system components can then be modeled in terms of an equivalent temperature, which can be related to the figure of merit.

13.2.1 Thermal Noise and Equivalent Noise Temperature

Chapter 11 introduced fundamental antenna concepts and provided the reader with techniques to determine signal level at the receiver input in terms of transmitter output power and the remaining system parameters. The noise associated with the system determines if received power is adequate; that is, the signal-to-noise ratio must exceed some specified value in order to maintain data integrity. The thermal noise to be quantified is governed by Planck's blackbody radiation law [1]. In the microwave region, it is convenient to use the Rayleigh-Jeans approximation that states the relationship between the rms noise voltage, v_N, produced by a resistor, R, and its physical temperature, T_p [2].

$$v_N = \sqrt{4kT_p BR} \tag{13.1}$$

where T_p is the temperature in kelvins, k is Boltzman's constant and is equal to 1.38×10^{-23} J/K. B is the bandwidth in hertz.

The noise power, P_N, delivered to a matched load resistor connected to the source in Figure 13.1 is determined by solving for the current and substituting to find

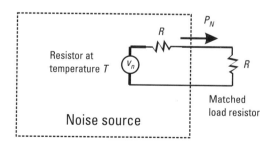

Figure 13.1 Noise power.

$$P_N = \left(\frac{v_N}{2R}\right)^2 R = kT_p B \qquad (13.2)$$

The noise added by any component, receiver, or system is considered to be that of an equivalent resistor, limited by the bandwidth. The noise power available from the equivalent resistor is proportional to its temperature; hence the concept of equivalent temperature is used. The relationship between the noise power delivered to a matched load, N, and equivalent temperature, T_e for any component, subsystem, or system is expressed as

$$N = kT_e B \qquad (13.3)$$

Henceforth the symbol N, rather than P_N, will be used when referring to noise power. It is apparent from (13.2) and (13.3) that the bandwidth must be minimized to optimize noise performance.

The sources of noise within the system are depicted in Figure 13.2. The external background consists of celestial and terrestrial sources and will be discussed further in Section 13.5. In addition, nonideal system components such as the antenna and receiver produce noise as well. For example, the total noise impinging upon the receiver terminals is due to two factors, the noise received from external sources such as the Earth and sky and that generated within the antenna structure itself. Noise appearing at the output of the receiver is a combination of the input noise amplified by the gain of the receiver and that added by individual components within the system. The undesired interference signals also depicted in Figure 13.2 can seriously

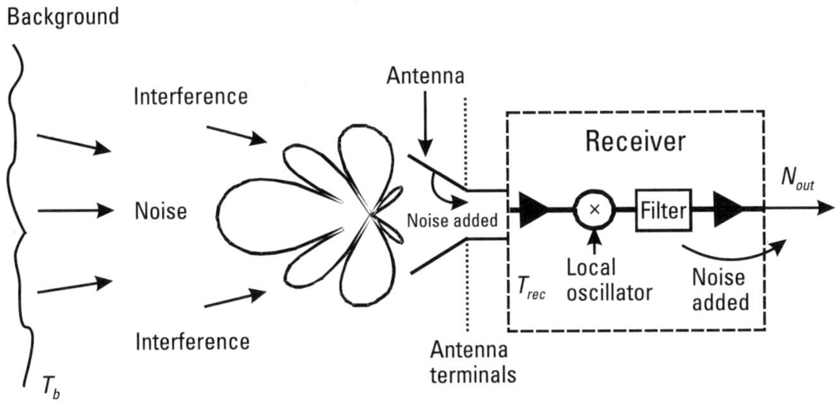

Figure 13.2 Noise sources.

degrade the performance of a telemetry system. Interference will not be addressed here since the factors affecting it are particularly configuration and location dependent. The telemetry engineer should be aware of possible performance degradations and be prepared to take corrective action.

Since noise is added by components at different points within the system, a method to determine the overall, end-to-end performance must be identified. The ultimate performance depends upon the signal-to-noise ratio (S/N) at the output of the system. The S/N throughout the system is affected not only by noise added by nonideal components but also the bandwidth at each point. Thus filters are selected to restrict the noise bandwidth as much as possible while passing the required information.

For convenience in quantifying the noise performance, a reference plane is selected. The terminals of the receive antenna are typically chosen and the signal amplitude is compared to an equivalent noise level at that point. The equivalent noise is defined in such a manner that it relates to the S/N at the output of the system. The first step is to relate noise added at the output of a component or subsystem to an equivalent input value. Once this is accomplished it is possible to refer all noise to a common point in the system. For the purpose of illustration, the relationship between noise power and equivalent temperature for an amplifier is considered in Figure 13.3.

In Figure 13.3(a), a resistor at a physical temperature of 0K is attached to the input of a real-world, noise-producing amplifier with gain G. In this

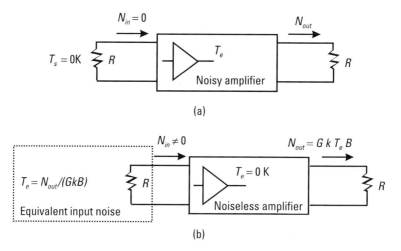

Figure 13.3 Equivalent noise temperature: (a) real-world amplifier with no input noise, and (b) ideal amplifier with equivalent noise at input.

case, it is evident from (13.3) that the input noise will be zero. However, since the amplifier is nonideal, the components within the circuit will generate noise appearing at the output; the resultant output noise power is N_{out}. In this chapter, an equivalent input noise power that produces the same noise power at the output of the device will be used. That is, an equivalent input noise is defined such that the same noise power appears at the output of an ideal, noiseless amplifier as shown in Figure 13.3(b). The noise generated by the actual amplifier, which is modeled as an equivalent resistor connected across the input terminals of an ideal unit, is at a temperature

$$T_e = \frac{N_{out}}{kGB} \tag{13.4}$$

Since $N_{out} = GN_{in}$ in Figure 13.3(b), the equivalent input noise power is

$$N_{in} = kT_e B \tag{13.5}$$

It is desirable to have a low equivalent temperature since it is proportional to noise power. A figure of merit is established in the next section to provide a basis of comparison between candidate components and systems.

13.2.2 Noise Figure Definition

The total noise power at the output of a system is due to prior components in the chain. A figure of merit, or noise figure, for each device is specified in terms of the degradation in signal-to-noise ratio between its input and output ports. Since the noise level depends upon the bandwidth, which is not necessarily defined at the input to a given device, a reference noise is applied to the input terminals so that the signal-to-noise ratio at that point can be determined; this can then be compared to the result at the output. The input and output signal-to-noise ratios are given as S_{in}/N_{in} and S_{out}/N_{out}, respectively. The signal level at the input terminals of the receive antenna is calculated using the Friis transmission formula, which was discussed in Section 11.4. The input noise at this point is that received from external sources plus that generated internally in the imperfect antenna. As defined here, the noise input to the actual receiver, N_{in}, is limited to these two contributions. As the signals propagate through the system, the physical components of the receiver will add to the noise appearing at the output port. Thus, S_{out}/N_{out} will be less than or equal to S_{in}/N_{in} (assuming a constant bandwidth throughout the system). The noise figure, or figure of merit, is defined as

$$F = \frac{\left(\dfrac{S_{in}}{N_{in}}\right)}{\left(\dfrac{S_{out}}{N_{out}}\right)} \tag{13.6}$$

The noise figure is greater than or equal to unity, $F \geq 1$. Since the noise is characterized in terms of an equivalent temperature, F can be related to T_e.

Figure 13.4 depicts a nonideal network with two inputs: (1) a signal, S_{in}, and (2) some reference input noise, N_{ref}.

The input noise, N_{ref}, will be at the corresponding reference temperature, T_o. The value of the reference temperature is taken as $T_o = 290K$ (relatively close to ambient).

$$N_{ref} = kT_o B \tag{13.7}$$

In addition to the signal and noise applied to the input, the receiver will produce additional noise that appears at the output. The pertinent form of (13.3) for the amplifier of Figure 13.4 is

$$N_{amp} = kT_{amp} B \tag{13.8}$$

This is the equivalent noise applied to the input of a noiseless amplifier. To streamline the notation, the equivalent temperature of the amplifier will be designated T_{amp}, not T_{eamp}; the fact that it is an equivalent temperature is implicit. This convention will be followed throughout the remainder of the text. Since the internally generated noise is referenced to the input, the

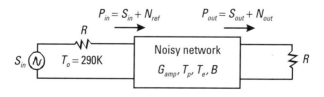

Figure 13.4 Noise figure.

total noise applied to the input of an ideal, noiseless amplifier is N_{ref} plus N_{amp}.

$$N_{in} = kT_oB + kT_{amp}B \qquad (13.9)$$

The output of the receiver consists of signal and noise components

$$G_{amp}(S_{in} + N_{in}) = S_{out} + N_{out} \qquad (13.10)$$

where $S_{out} = G_{amp}S_{in}$ and $N_{out} = G_{amp}N_{in} = G_{amp}k(T_o + T_{amp})B$. Substituting into (13.6) the noise figure for the receiver, F_{amp}, is determined in terms of its equivalent temperature, T_{amp}.

$$F_{amp} = \frac{S_{in}}{kT_oB} \frac{G_{amp}k(T_o + T_{amp})B}{G_{amp}S_{in}} = 1 + \frac{T_{amp}}{T_o} \qquad (13.11)$$

Although this is expressed specifically with respect to the amplifier in Figure 13.4, (13.11) holds for any component or subsystem. The general relationship is expressed as

$$F = 1 + \frac{T_e}{T_o} \qquad (13.12)$$

To reiterate, T_e is an equivalent temperature, not a physical one. F is commonly referred to as the noise factor in many texts. The noise figure in dB is obtained as follows.

$$F_{dB} = 10\log_{10}F \qquad (13.13)$$

For more compact notation, the dB subscript will not be used in the sections to follow; that is, if F is specified as a number, the form of (13.12) is implied; if the noise figure is specified in dB then the form of (13.13) is assumed. This shorthand notation is necessary because the noise figure of each and every component in a cascaded system will be required in Section 13.4.

Example 13.1: Relationship Between Noise Figure and Temperature

A preamplifier attached to the terminals of the receive antenna has an equivalent noise temperature $T_{amp} = 90K$. If the bandwidth is $B = 10$ MHz, what is the noise figure of the amplifier, F_{amp}, and associated noise power, N_{amp}? The reference temperature is T_0 = 290K.

Using the specified value for the noise temperature of the amplifier, the resulting noise figure is

$$F_{amp} = 1 + \frac{T_{amp}}{T_0} = 1 + \frac{90}{290} = 1.31$$

Converting to dB, the result is F_{amp} = 1.17 dB. The noise power generated in a B = 10 MHz bandwidth is

$$N_{amp} = kT_{amp}B = 1.38 \times 10^{-23}(90)(10^7) = 1.242 \times 10^{-14}\,\text{W}$$

Alternatively this can be expressed in dBm,

$$N_{amp} = 10\log_{10}\left(\frac{1.242 \times 10^{-11}\,\text{mW}}{1\,\text{mW}}\right) = -109.1\,\text{dBm}$$

13.3 Noise Figure of Passive Elements with Loss

The components and subsystems of the receiving system may be interconnected with devices such as transmission lines or filters having dissipative loss. These can have a significant impact upon the noise performance of a system and as such are important to characterize. For a component at a physical temperature, T_p, with a loss factor, L_L, the equivalent noise temperature is determined as [1]

$$T_L = (L_L - 1)T_p \qquad (13.14)$$

L_L is the loss factor defined in the same manner as the cable loss in (11.1). It is important to note that T_p is the physical temperature of the device in kelvins and T_L is the equivalent noise temperature. For components with loss, the equivalent temperature will be directly related to the physical temperature. The corresponding noise figure is obtained by using the general form of (13.12).

$$F_L = 1 + \frac{T_L}{T_0} = 1 + (L_L - 1)\frac{T_p}{T_0} \qquad (13.15)$$

It is apparent that the loss of components in the receiving system should be minimized to reduce the noise figure.

Example 13.2: Lossy Transmission Line

A transmission line has a physical temperature, T_p = 305K, and the vendor-supplied data indicates that its attenuation is 26 dB/100 ft at f = 2251.5 MHz. What is the corresponding noise figure in dB if its length is 10 ft?

To begin, the attenuation for a 10-ft-long section of transmission line is

$$A_{dB} = \frac{26}{100} 10 = 2.6 \text{ dB}$$

Using (11.1)

$$L_L = 10^{\frac{A_{dB}}{10}} = 10^{0.26} = 1.819$$

Using (13.15)

$$F_L = 1 + (1.819 - 1)\frac{305}{290} = 1.862$$

Expressed in decibels, F_L = 2.7 dB. It is interesting to note that the value for the noise figure is approximately equal to the loss factor when the component is near the reference temperature of 290K.

13.4 Noise Figure and Temperature of a Cascaded System

As stated in Section 13.2, the noise added by a component, which actually appears at its output, is related to an equivalent noise power at the input port to the device. The question remains: How can the noise at the input to each component in a cascade be referenced to a common point? A cascade of two components is depicted in Figure 13.5; the first is characterized by G_1 and T_1, the second by G_2 and T_2. Again, a reference input noise, N_{ref} = kT_oB, is used. For convenience, the two components are shown to have gain; however, lossy components can also be characterized in the same manner by setting the gain equal to the reciprocal of their loss factor.

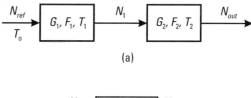

(a)

(b)

Figure 13.5 Noise of cascaded components: (a) cascade of two elements, and (b) equivalent element.

The equivalent noise added by the first component of Figure 13.5(a), referenced to its input, is characterized by $kT_1 B$. The noise at the output of an ideal, noiseless amplifier with gain G_1 is found to be

$$N_1 = G_1(kT_o B + kT_1 B) \qquad (13.16)$$

The equivalent noise added by the second component, referenced to its input, is characterized by T_2 and the total output noise power of the cascade is found

$$N_{out} = G_2(N_1 + kT_2 B) \qquad (13.17)$$

To represent the cascade of two components with an equivalent element, as shown in Figure 13.5(b), the total output noise power is the reference input plus that added by both components, represented by T_{cas}

$$N_{out} = G_{cas}(kT_o B + kT_{cas} B) \qquad (13.18)$$

Equating (13.17) and (13.18), substituting (13.16), and taking into account that $G_{cas} = G_1 G_2$ the equivalent temperature of the cascade is

$$T_{cas} = T_1 + \frac{T_2}{G_1} \qquad (13.19)$$

By extension of the same technique, the equivalent noise temperature can be determined for any number of components or subsystems. For n components the result is

$$T_{cas} = T_1 + \frac{T_2}{G_1} + \frac{T_3}{G_1 G_2} + \ldots + \frac{T_n}{G_1 G_2 \ldots G_{n-1}} \quad (13.20)$$

The corresponding noise figure using (13.12) is

$$F_{cas} = F_1 + \frac{(F_2 - 1)}{G_1} + \frac{(F_3 - 1)}{G_1 G_2} + \ldots + \frac{(F_n - 1)}{G_1 G_2 \ldots G_{n-1}}$$
$$(13.21)$$

13.4.1 Noise Figure and Temperature Summary

Although (13.20) and (13.21) were derived for gain blocks, they can also be applied to a cascade including components with loss. In this case, the lossy component has its equivalent noise temperature and noise figure determined by (13.14) and (13.15); the corresponding value of gain used is $G_L = 1/L_L$. When the first stage of the cascade has a gain much greater than unity, that is, $G_1 \gg 1$, F_1 is the dominant term in (13.21). Clearly, an element with loss, such as a transmission line, should be avoided. A lossy transmission line or filter will have a relatively high value for its noise figure and will have a deleterious effect on the overall system because gain values less than unity will have a multiplicative effect on the remaining terms. Based upon these observations, two conclusions are drawn: (1) the lowest noise components should be placed first in the receiver cascade, and (2) it is critical for the first component to have at least a moderate amount of gain (sufficient gain to minimize the impact of the remaining terms in (13.21)). Characterization of external noise entering the system is the topic of Section 13.5.

Example 13.3: Cascaded System I

Figure 13.6 depicts two components that comprise the receiver attached to the terminals of an antenna: a lossy transmission line and a commercial receiver. A low-noise preamplifier will be added in Examples 13.4 and 13.5. The 10-ft-long piece of cable has 26dB/100 feet attenuation at $f = 2{,}251.5$ MHz and is connected to a commercial receiver having a noise figure of $F_R = 12$ dB, a net gain of 35 dB and a selectable bandwidth. For the chosen application, the bandwidth is set to $B = 2$ MHz. All components are at a physical temperature of $T_p = 305$K. Find the overall receiver noise temperature and noise figure as well as the equivalent noise power at the reference plane, the terminals of the antenna.

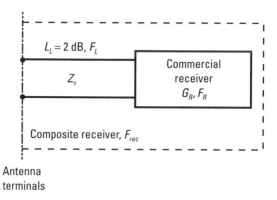

Figure 13.6 Receiver cascade.

Using the results of Example 13.1 in conjunction with (13.14) and (13.15), the noise figure and temperature of the transmission line are

$$F_L = 1.862$$

and

$$T_L = 250\text{K}$$

The loss factor used in these calculations was $L_L = 1.819$, the corresponding value of gain is $G_L = 0.5498$. To apply the cascade formula of (13.21), F_R must be converted from decibels.

$$F_R = 10^{12/10} = 15.85$$

For the cascade receiver system the noise figure, F_{rec}, is

$$F_{rec} = F_L + \frac{(F_R - 1)}{G_L} = 1.862 + \frac{(15.85 - 1)}{0.5498} = 28.87$$

The resulting value in decibels is $F_{rec} = 14.6$ dB. The corresponding equivalent temperature is

$$T_{rec} = (F_{rec} - 1)T_o = (28.87 - 1)\,290 = 8,082.3\text{K}$$

Alternatively, noise figures could have been converted to noise temperatures and this latter result achieved using (13.20). The equivalent noise power added at the input to a noiseless receiving system is then

$$N_{rec} = kT_{rec}B = 1.38 \times 10^{-23}(8,082.3)(2 \times 10^6) = 2.23 \times 10^{-13}\,\text{W}$$

The input noise power may also be expressed in dBm, N_{rec} = −96.5 dBm.

Example 13.4: Cascaded System II

To improve the noise figure of the configuration presented in Example 13.3, a low-noise amplifier is attached between the cable and the commercial unit. The receiver now consists of the components shown in Figure 13.7: a lossy transmission line, a low-noise preamplifier, and a commercial receiver. The transmission line and commercial receiver have the same specifications as in the previous example, the low-noise amplifier has an equivalent noise temperature of T_{LNA} = 90K and gain G_{LNA} = 23 dB. Again, the bandwidth is B = 2 MHz and all components are at a physical temperature of T_p = 305K. Find the overall receiver noise temperature and noise figure as well as the equivalent noise power at the reference plane, the terminals of the antenna of this modified receiver system.

Using the results of Example 13.1 and

$$G_{LNA} = 10^{23/10} = 199.5$$

$$F_{LNA} = 1 + \frac{T_{LNA}}{T_0} = 1.31$$

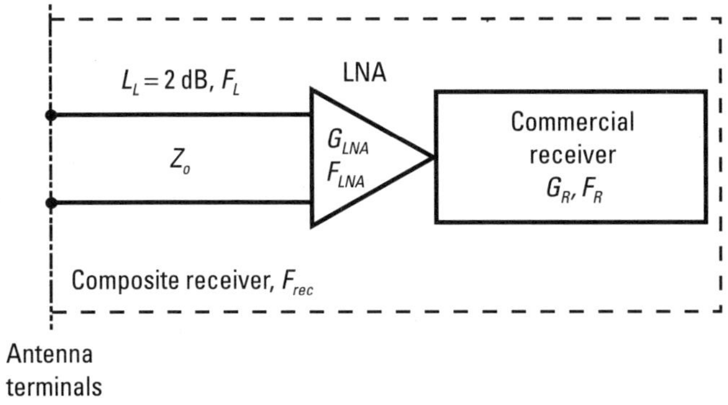

Figure 13.7 Modified receiver cascade.

Then

$$F_{rec} = F_L + \frac{(F_{LNA} - 1)}{G_L} + \frac{(F_R - 1)}{G_L G_{LNA}}$$

$$= 1.862 + \frac{(1.31 - 1)}{0.5498} + \frac{(15.85 - 1)}{0.5498(199.5)} = 2.561$$

The resulting value in dB is F_{rec} = 4.08 dB. The corresponding equivalent temperature is

$$T_{rec} = (F_{rec} - 1) T_o = (2.561 - 1) 290 = 452.7K$$

Alternatively, the noise figure of individual components could have been converted to noise temperature and the latter result achieved using (13.20). The equivalent noise power added at the input to a noiseless receiving system is then

$$N_{rec} = k T_{rec} B = 1.38 \times 10^{-23}(452.7)(2 \times 10^{6}) = 1.249 \times 10^{-14} W$$

The input noise power may also be expressed in dBm, N_{rec} = −109.3 dBm. This is a considerable improvement compared to the result of Example 13.3.

Example 13.5: Cascaded System III

One might ask how much improvement could be achieved if the position of the transmission line and low-noise amplifier are interchanged. Using the same components as the previous two examples but interconnected as shown in Figure 13.8, find the noise figure, equivalent temperature, and noise power.

Upon rearranging the components the noise figure expression becomes

$$F_{rec} = F_{LNA} + \frac{(F_L - 1)}{G_{LNA}} + \frac{(F_R - 1)}{G_{LNA} G_L}$$

$$= 1.31 + \frac{(1.861 - 1)}{199.5} + \frac{(15.85 - 1)}{199.5(0.5498)} = 1.45$$

The noise figure in decibels is F_{rec} = 1.61 dB and the equivalent temperature of this system is

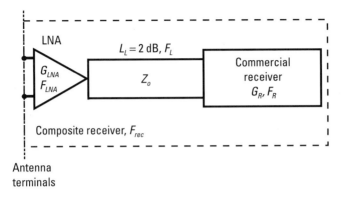

Figure 13.8 Improved receiver cascade.

$$T_{rec} = (F_{rec} - 1)T_o = (1.45 - 1)290 = 130.7\text{K}$$

The corresponding input noise power is then

$$N_{rec} = kT_{rec}B = 1.38 \times 10^{-23}(130.7)(2 \times 10^6) = 3.61 \times 10^{-15}\,\text{W}$$

Or, $N_{rec} = -114.4$ dBm, an improvement of 5.1 dB over the result of Example 13.4. This reinforces the need for the first element in the chain to have a good noise figure with adequate gain to minimize the impact of other components in the cascade.

13.5 Antenna Temperature

As introduced in Section 13.2, in addition to the noise generated by the receiver, a certain amount is received from external sources and a nonideal antenna attached to the receiver terminals. Noise produced by a structure with loss is due to the same physical mechanisms discussed in Section 13.3. External to the receiver there are noise sources such as the Earth and sky. Exclusive of the noise generated within the receiver referenced to the input terminals, that entering from the antenna itself is characterized in terms of an equivalent temperature, T_A. It is defined as

$$T_A = T_b e_{rad} + (1 - e_{rad})T_p \qquad (13.22)$$

Where e_{rad} is the radiation efficiency discussed in Section 11.4, T_p is the physical temperature of the antenna in kelvins and T_b is the integrated brightness in kelvins. One notes that a lossless antenna would have a radiation efficiency $e_{rad} = 1$ and the antenna structure itself would generate no noise; however, it could still receive noise from external sources. The concept of brightness temperature is depicted in Figure 13.9; T_b is a function of the apparent temperature of the background, weighted by the normalized antenna pattern.

$$T_b = \frac{\displaystyle\int_0^{2\pi}\int_0^{\pi} T_{ap}(\theta, \phi)\,|F(\theta, \phi)|^2 \sin\theta\,d\theta\,d\phi}{\displaystyle\int_0^{2\pi}\int_0^{\pi} |F(\theta, \phi)|^2 \sin\theta\,d\theta\,d\phi} \qquad (13.23)$$

T_{ap} is the apparent temperature of the background in kelvins. At a fixed instance in time, the apparent temperature of the background changes as a function of angle with respect to the antenna coordinate system; thus, T_b is the weighted average of the noise background in the field of view of the antenna. For example, T_{ap} for the Earth has a nominal value of about 330K, while the sky is substantially colder; however, the Sun is a hot source. Its apparent temperature is about 6,000K at S-band for a quiet Sun (it is substantially hotter when there is significant solar activity [3]). The apparent brightness of the sky is dependent upon many factors including frequency and elevation angle. In accordance with (13.23), the integrated brightness can be quite large when a high-gain, narrow-beam antenna is pointed at a hot source such as the Sun. It should be emphasized that the apparent temperature distribution expressed in the local antenna coordinate system,

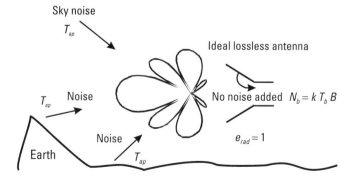

Figure 13.9 Integrated brightness temperature.

$T_{ap}(\theta, \phi)$, will be a function of time when tracking is employed. The impact of this is that T_b can vary significantly during the course of a mission, affecting the antenna temperature and system performance.

When characterizing a dynamic configuration, such as that associated with telemetry applications, it is important to use a representative value of T_b. For example, assuming a best-case situation (such as a narrow-beam antenna pointed at a cold-sky region) might produce an overly optimistic value of the integrated brightness temperature. On the other hand, use of a worst-case value might result in significant receiver overdesign. Due to the fact that the transmitter and receiver are in relative motion to each other, the integrated brightness will vary while tracking the target and (13.23) is often approximated in practice. For most systems, the approach is to estimate T_b knowing the radiation pattern characteristics of the receive antenna and apply (13.22) from there.

For omnidirectional antennas, or nearly isotropic radiators, the brightness temperature may be approximated by the average of the Earth and sky apparent temperatures.

$$T_b^{omni} = \frac{T_{ap}^{sky} + T_b^{Earth}}{2} \tag{13.24}$$

When a hot source of relatively small angular extent, such as the Sun, is within the field of view of the receive antenna, an approximation for the brightness temperature is

$$T_b^{hot} \approx T_{ap}^{hot} \frac{\Omega_{hot}}{\Omega_A} \tag{13.25}$$

where Ω_A is the beam solid angle of the receive antenna in steradians, and Ω_{hot} is that subtended by the source as viewed from the receiver location. The expression in (13.25) assumes the peak of the antenna beam is pointed toward the hot source.

Typical values for the apparent temperature of selected background objects are listed in Table 13.1.

Example 13.6: Antenna Temperature for Omnidirectional Antenna Case

A representative background apparent temperature distribution for an omnidirectional receive antenna is shown in Figure 13.10. An omnidirectional unit can be used as the ground station receive antenna in some applications. In this case, tracking is unnecessary—the pattern is similar to that depicted

Table 13.1
Apparent Temperatures

Source	Apparent Brightness, T_{ap} (Kelvins)	Extent, Ω (Steradians)
Sun	6,000	1.1×10^{-4}
Cassiopeia	3,700	$<3 \times 10^{-10}$
Cygnus	2,650	$<3 \times 10^{-10}$
Mercury	613	1.2×10^{-9}
Venus	235	1.1×10^{-8}
Earth	330	Depends upon antenna location
Mars	217	5.5×10^{-9}
Jupiter	138	5×10^{-10}
Nominal Sky*	5–95	Depends upon antenna location

Note: The sky temperature depends upon frequency, angle of incidence, and the portion of the sky observed; the values cited are for operation in S-band telemetry channels.

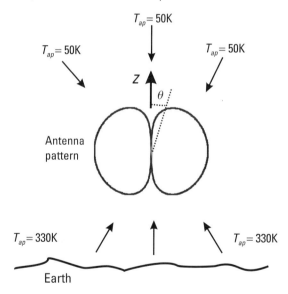

Figure 13.10 Apparent temperature distribution and radiation pattern for Example 13.6.

in Figure 11.8; that is, the azimuth pattern is uniform, the pattern varies in the elevation plane. The gain and radiation efficiency are specified by the antenna vendor: $G_r = 1$ dBi and $e_{rad} = 0.85$. The physical temperature is $T_p = 310$K.

For this simplified example, the apparent temperature, T_{ap}, can be approximated by the following expression.

$$T_{ap} = \begin{cases} 50\text{K} & 0 \le \theta \le \dfrac{\pi}{2} \quad \text{Sky} \\[2ex] 330\text{K} & \dfrac{\pi}{2} \le \theta \le \pi \quad \text{Earth} \end{cases}$$

The simplified antenna pattern can be written as in (11.30)

$$|F(\theta, \phi)|^2 = \sin^2(\theta)$$

Computing T_b using (13.23)

$$T_b = 190\text{K}$$

Note that the calculation of T_b was facilitated by an approximate apparent temperature distribution specified in the receive antenna coordinate system. Due to symmetry in the antenna pattern, the value obtained using the approximate expression, (13.24), is 190K as well. The corresponding antenna temperature is

$$T_A = e_{rad} T_b + (1 - e_{rad}) T_p = 0.85(190) + (1 - 0.85)310 = 208\text{K}$$

Note that the antenna temperature is not explicitly a function of the gain; however, since G depends upon the directivity, D, and, in turn, upon the pattern characteristics, T_A will vary indirectly with the gain.

Example 13.7: Antenna Temperature for Directive Antenna Case

The apparent temperature distribution is the same as in Example 13.6, but now a relatively narrow-beam antenna, which has the ability to track the device under test, is used. The gain, radiation efficiency, and half-power beamwidth are specified to be $G_r = 23.6$ dBi, $e_{rad} = 0.75$, and $HPBW = 10.2°$, respectively. The physical temperature is $T_p = 310$K. The apparent temperature distribution and radiation pattern are depicted in Figure 13.11. Assume the nominal Earth and sky temperatures are identical to Example 13.6.

When the device is directly overhead, a representative antenna pattern can be expressed as

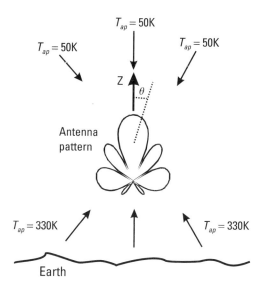

Figure 13.11 Apparent temperature distribution and radiation pattern for Example 13.7.

$$|F(\theta, \phi)|^2 = \frac{\sin(5\pi\sin(\theta)\cos(\phi))}{10\sin\left(\dfrac{\pi}{2}\sin(\theta)\cos(\phi)\right)} \frac{\sin(5\pi\sin(\theta)\sin(\phi))}{10\sin\left(\dfrac{\pi}{2}\sin(\theta)\sin(\phi)\right)}$$
$$\times \frac{1 + \cos(\theta)}{2}$$

Upon integration using (13.23), T_b is found to be

$$T_b = 51.7\text{K}$$

The corresponding antenna temperature is

$$T_A = e_{rad}T_b + (1 - e_{rad})T_p = 0.75(51.7) + (1 - 0.75)310 = 116.3\text{K}$$

It is important to note that the antenna temperature decreased due to the fact that the directive antenna was pointed skyward, in a generally cold region, and the Sun was not in the field of view.

Example 13.8: Antenna Temperature for Directive Antenna with Hot Source

In this case, the relatively narrow-beam antenna of Example 13.7 is considered, but a hot source, the Sun, is directly overhead. The remainder of the apparent temperature distribution remains as in the previous example. The apparent temperature distribution and radiation pattern are depicted in Figure 13.12. With the exception of the Sun, the Earth and sky temperatures are the same as in Example 13.6.

The antenna pattern is the same as in Example 13.7 and upon integration T_b is found to be

$$T_b = 81.2\text{K}$$

The corresponding antenna temperature is

$$T_A = e_{rad} T_b + (1 - e_{rad}) T_p = 0.75(81.2) + (1 - 0.75)310 = 138.4\text{K}$$

The integrated brightness increased by about 20% due to the presence of the Sun within the field of view. The increase was not greater because the angular extent of the Sun is about 0.6° while the *HPBW* of this antenna is approximately 10.2°.

Example 13.9: Antenna Temperature for Narrow-Beam Antenna with Hot Source

In this example a very narrow-beam antenna will be considered. The pattern when the device is directly overhead, is given by

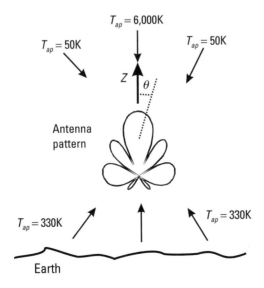

Figure 13.12 Apparent temperature distribution and radiation pattern for Example 13.8.

$$|F(\theta,\ \phi)|^2 = \frac{\sin(50\pi\sin(\theta)\cos(\phi))}{100\sin(\pi\sin(\theta)\cos(\phi))} \frac{\sin(50\pi\sin(\theta)\sin(\phi))}{100\sin(\pi\sin(\theta)\sin(\phi))}$$
$$\times \frac{1+\cos(\theta)}{2}$$

The gain and radiation efficiency are specified to be G_r = 44.6 dBi and e_{rad} = 0.63. The beamwidth of this antenna is about $HPBW \approx 1.02°$. The physical temperature of the receive system is T_p = 310K. The apparent temperature distribution is the same as shown in Figure 13.12; the Sun is directly overhead.

Upon integration using (13.23), T_b is found to be

$$T_b = 1,597\text{K}$$

The corresponding antenna temperature is

$$T_A = e_{rad}T_b + (1 - e_{rad})T_p = 0.63(1,597) + (1 - 0.63)310 = 1,089.8\text{K}$$

For the sake of comparison, the brightness temperature is approximated using (13.25). The beam solid angle for the normalized power pattern given above is

$$\Omega_A = 2.772 \times 10^{-4}\ \text{sr}$$

The angular extent of the Sun, as viewed from the receive antenna location, as specified in Table 13.1 is

$$\Omega_{Sun} = 1.1 \times 10^{-4}\ \text{sr}$$

Using (13.25) the approximate brightness temperature is

$$T_b^{Sun} = 6,000\,\frac{1.1 \times 10^{-4}}{2.772 \times 10^{-4}} = 2,381\text{K}$$

The approximate value overestimates the actual one by a factor of two; this discrepancy is resolved in practice by the correction factor discussed in Section 13.7. It is apparent that tracking targets in the vicinity of the Sun will have a deleterious affect upon the noise performance of the system. For tracking antennas with a moderate beamwidth, say $10°$ and *not* pointed directly at the Sun, a nominal value of T_b between 200 and 270K is probably most representative in the majority of telemetry applications.

13.6 System Noise Figure

The noise associated with the overall system is due to the equivalent temperature of the antenna in addition to that of the composite receiver. Once these are determined, the overall performance of the telemetry system can be characterized at the reference plane—the terminals of the receive antenna. The noise due to external sources as well as the nonideal receive antenna is

$$N_A = kT_A B \qquad (13.26)$$

The noise associated with the composite receiver, referenced to the terminals of the receive antenna, is given by a form similar to (13.8) as

$$N_{rec} = kT_{rec} B \qquad (13.27)$$

The equivalent total noise power at the input of the system is then

$$N_{sys} = N_A + N_{rec} = kT_{sys} B \qquad (13.28)$$

It must be emphasized that the bandwidth, B, is the predetection value; that is, the smallest bandwidth in the cascade prior to detection. This is usually the video bandwidth B_v discussed in Chapter 2. Based upon the results of (13.26) and (13.27) the equivalent temperature of the system can be written as

$$T_{sys} = T_A + T_{rec} \qquad (13.29)$$

The resulting system noise figure is then

$$F_{sys} = 1 + \frac{T_{sys}}{T_o} \qquad (13.30)$$

Any interconnecting transmission lines should be included as part of the receiver noise temperature, T_{rec}. This is illustrated in the following example.

Example 13.10: System Noise Figure I

The composite receiver of Example 13.3, consisting of the transmission line and commercial receiver, is used in conjunction with the zenith-looking directive antenna of Example 13.7. The effect of the Sun on system performance is ignored in this example. Find the system temperature and noise floor assuming a bandwidth of $B = 2$ MHz.

In Example 13.3, the receiver temperature was found

$$T_{rec} = 8{,}082.3 \text{K}$$

The antenna temperature given by Example 13.7 was $T_A = 116.3$K. Therefore, the system temperature is

$$T_{sys} = T_A + T_{rec} = 116.3 + 8{,}082.3 = 8{,}198.6 \text{K}$$

The corresponding noise floor of the system is

$$N_{sys} = kT_{sys}B = 2.263 \times 10^{-13} \text{W}$$

Or, $N_{sys} = -96.4$ dBm. This is virtually the same result as Example 13.3, and it is due to the fact that the equivalent temperature of the composite receiver dominates that of the antenna.

Example 13.11: System Noise Figure II

What is the system noise temperature if the same zenith-looking, directive antenna of Example 13.7 is connected to the improved composite receiver discussed in Example 13.5?

The receiver temperature was found in Example 13.5

$$T_{rec} = 130.7 \text{K}$$

Again, neglecting the Sun, the antenna temperature is the same as in Examples 13.7, $T_A = 116.3$K. The equivalent system temperature is then given by

$$T_{sys} = T_A + T_{rec} = 116.3 + 130.7 = 247 \text{K}$$

The corresponding noise floor of the system (assuming the same bandwidth, $B = 2$ MHz) is

$$N_{sys} = kT_{sys}B = 6.817 \times 10^{-15}\,\mathrm{W}$$

Or, $N_{sys} = -111.7$ dBm. Note the 15-dB decrease in noise level, which is due to the significantly reduced receiver noise temperature, compared to that in Example 13.10.

13.7 *C/N* and *G/T*

To quantify telemetry system performance, the C/N is computed; this ensures that there is adequate signal to achieve a specified level of data quality. For example, the C/N ratio required to achieve a certain BER depends upon the modulation format as discussed elsewhere in this as well as other texts [4]. The received signal, or carrier power, is determined as in Section 11.4. Using the results of (11.79) and (11.82), the carrier is P_{line}, the power delivered to the line attached to the terminals of the antenna.

$$C = P_{line} = \frac{P_s G_t G_r}{L_{tc} L_{path}} pq \tag{13.31}$$

The noise of the system, referred to the same point (the terminals of the receive antenna) is given by (13.25). Taking the ratio and rearranging yields

$$\frac{C}{N_{sys}} = \left(\frac{P_s}{L_{tc}} G_t \right) \left(\frac{1}{L_{path}} \right) \left(\frac{G_r}{T_{sys}} \right) \left(\frac{pq}{kB} \right) \tag{13.32}$$

The terms in the first set of parentheses are solely transmitter parameters and those in the second depend only upon the path. The quotient in the third set of parentheses is a function of the receive system alone; in fact, G_r/T_{sys} is taken to be a figure of merit for receiving systems. The last set is dependent upon the bandwidth and other system parameters such as impedance and polarization mismatch, which are related to the quality of the transmitter and receiver components and how well they are designed to operate together.

Since the antenna and receiver are often supplied by different vendors, it should be noted that the figure of merit for a receive antenna, G_r/T_A, is sometimes specified rather than that for the overall system, G_r/T_{sys}. Based upon the results of Examples 13.10 and 13.11, it is evident that care should be exercised when discussing G/T to ensure that a consistent definition is used; for example, the antenna and system temperatures in Sections 13.6.1 and 13.6.2 were more than an order of magnitude different. In some instances, a low-noise down converter is an integral part of the receive antenna and the noise associated with those electronic components is usually included in the G_r/T_A value. It is sometimes convenient to transform (13.32) to decibel form.

$$
\begin{aligned}
\frac{C}{N_{sys}} \text{ (dB)} = &\{P_s \text{ (dBm)} + G_t \text{ (dBi)} - A_{tc} \text{ (dB)}\} \\
&- A_{path} \text{ (dB)} + \frac{G_r}{T_{sys}} \text{ (dB/K)} \\
&- \{PML \text{ (dB)} + MML_{tot} \text{ (dB)} + kB \text{ (dBm/K)}\}
\end{aligned}
\tag{13.33}
$$

Using the methods of Section 13.5 in conjunction with analyses of Section 11.3.3, G_r/T_A or G_r/T_{sys} can be calculated. In addition, G/T is often measured, this is addressed in Section 13.7.1.

The bandwidth required to support a particular data requirement depends upon the modulation scheme employed. The following representation is generally used to characterize the receive system independent of the bandwidth.

$$
\frac{C}{N_o} = \frac{P_r}{kT_{sys}}
\tag{13.34}
$$

For digital modulation formats, the important parameter is the ratio of carrier energy per bit to noise spectral density ratio.

$$
\frac{E_b}{N_o} = \left(\frac{C}{N_o}\right) T_{bit}
\tag{13.35}
$$

This parameter is determined once the bit interval is specified.

13.7.1 *G/T* Measurement

In practice G/T is often determined via measurement. One method is the solar calibration technique that requires the system to point at a hot source, for example the Sun, to provide a reference. The technique outlined below is for relatively narrow-beam antennas; that is, those producing half-power beamwidths less than about 10°. In this case, the equivalent temperature of the Sun, T_{Sun}, is related to the gain of the antenna, G_r, the impinging solar flux, SF, the wavelength, λ_o, Boltzman's constant, k, and the correction factor, L_{cor}, as [5]

$$T_{Sun} = \frac{SF\lambda_o^2}{8\pi k L_{cor}} G_r \qquad (13.36)$$

The other parameters have been discussed previously, and L_{cor} is necessary to account for the fact that the Sun is not a point source. Dividing by the system noise temperature, T_{sys}, the system figure of merit may be written as

$$\frac{G_r}{T_{sys}} = \frac{8\pi k L_{cor}}{SF\lambda_o^2} \frac{T_{Sun}}{T_{sys}} \qquad (13.37)$$

The flux, SF, is determined from daily measurements at solar observatories [6]. L_{cor} is known given the beamwidth of the receive antenna and the angular extent of the source, in this case the Sun. It remains to make two measurements: (1) a power measurement at the output of the receiver with the mainbeam pointed at the Sun, and (2) a second one with the mainbeam pointed at the cold sky.

$$P_{cold} \approx k T_{sys} B \qquad (13.38)$$

$$P_{hot} = k(T_{Sun} + T_{sys})B \qquad (13.39)$$

Thus, the temperature ratio in (13.37) can be written as

$$\frac{T_{Sun}}{T_{sys}} = \frac{P_{hot} - P_{cold}}{P_{cold}} = \frac{P_{hot}}{P_{cold}} - 1 \qquad (13.40)$$

Then (13.37) can be written as

$$\frac{G_r}{T_{sys}} = \frac{8\pi k L_{cor}}{SF\lambda_o^2}\left(\frac{P_{hot}}{P_{cold}} - 1\right) \qquad (13.41)$$

Of course, the value in decibels per kelvin is obtained by taking $10\log_{10}$ of the quantity in (13.41).

The correction factor, L_{cor}, in terms of the solid angles Ω_{Sun} and Ω_A (both in steradians) is given by

$$L_{cor} \approx 1 + 0.38\left(\frac{\Omega_{Sun}}{\Omega_A}\right) \qquad \text{for } \frac{\Omega_{Sun}}{\Omega_A} < 1 \qquad (13.42)$$

The solar flux, SF, is determined by measurements at two different frequencies, 1,415 and 2,695 MHz. The value to use in (13.41) is calculated by the following equation for operation between 1.5 and 3 GHz, SF_{1415} and SF_{2695} are the solar flux values at $f = 1,415$ and 2,695 MHz, respectively. The carrier frequency of the telemetry link is f_c in MHz.

$$SF = SF_{2695}\left[\frac{SF_{1415}}{SF_{2695}}\right]^{N_S} \qquad (13.43)$$

The exponent is calculated as

$$N_S = \frac{\ln\left(\dfrac{f_c}{2,695}\right)}{\ln\left(\dfrac{1,415}{2,695}\right)} \qquad (13.44)$$

Example 13.12: Receive System G/T

A telemetry system uses a paraboloidal reflector antenna connected through a transmission line to a commercial receiver. The system operates at $f = 2,251.5$ MHz, the receive antenna is 3 ft in diameter and the remainder of the system is identical to the components considered in Example 13.3; that is, the antenna is connected to a receiver with noise figure, $F_R = 12$ dB, through a 10-ft-long piece of cable having a total attenuation, $A_{tc} = 2.6$ dB. The approach is

1. Compute the free space wavelength, λ_o.

2. Calculate the gain of the receive antenna.

3. Estimate the antenna temperature.

4. Determine the noise temperature of the composite receiver (transmission line and commercial receiver).

5. Compute the system noise temperature.

6. Calculate G/T.

The free space wavelength at $f = 2{,}251.5$ MHz is

$$\lambda_o = 13.324 \text{ cm}$$

Convert the diameter of the dish to centimeters and use (11.75) to estimate its gain.

$$G_r \approx \frac{1}{2}\left(\frac{\pi d_{dish}}{\lambda_o}\right)^2 = \frac{1}{2}\left(\frac{\pi(91.44)}{13.324}\right)^2 = 232.4 \ (23.7 \text{ dBi})$$

The antenna temperature can't be calculated using (13.23) because the patterns aren't known. Proceed by estimating the beamwidth and compare it to the results of Example 13.7. The approximate beamwidth of the 3-ft diameter dish is found using (11.74) by replacing W with d_{dish}.

$$HPBW = 65° \frac{\lambda_o}{d_{dish}} = 65° \frac{13.324}{91.44} = 9.5°$$

This value is close to the 10.3° half-power beamwidth of the pattern considered in Example 13.7; thus, if the radiation efficiency can be considered to be $e_{rad} = 0.75$, it would be realistic to use the value $T_A = 116.3$K determined in that exercise. For the purposes of illustration, this value will be used; however, a more conservative approach would have been to use a higher temperature such as 200 to 270K as suggested in Section 13.5. The equivalent temperature of the composite receiver was determined in Example 13.3.

$$T_{rec} = 8{,}082.3 \text{K}$$

The system noise temperature is the same as that calculated in Example 13.10.

$$T_{sys} = T_A + T_{rec} = 116.3 + 8{,}082.3 = 8{,}198.6K$$

The G/T for the system is computed

$$\frac{G_r}{T_{sys}} = \frac{232.4}{8{,}198.6} = 0.0283 \ (-15.47 \ dB/K)$$

For the sake of comparison, compute the result for the antenna temperature alone.

$$\frac{G_r}{T_A} = \frac{232.4}{116.3} = 2.0 \ (+3.0 \ dB/K)$$

The impact of the commercial receiver noise figure is evident in this example.

Example 13.13: Improved Receive System G/T

As suggested in Example 13.5, a low-noise amplifier could be attached directly to the terminals of the receive antenna to provide improved system performance. The equivalent noise temperature of this configuration, discussed in Section 13.4.4, was $T_{rec} = 130.7K$. The resulting system noise temperature is then

$$T_{sys} = T_A + T_{rec} = 116.3 + 130.7 = 247K$$

The G/T for the system is computed

$$\frac{G_r}{T_{sys}} = \frac{232.4}{247} = 0.941 \ (-0.26 \ dB/K)$$

This value is more than an order of magnitude better than the system value in the previous example. It is important to use a high-quality receiving system.

Example 13.14: Receive System with High-Gain Antenna

To extend the range of the telemetry system, the improved receiver system of Example 13.5 is attached to a dish with a diameter, $d_{dish} = 10m$. The apparent temperature distribution is the same as that depicted in Figure 13.11. Find G/T for this antenna/receiver combination. Convert the diameter of the dish to centimeters and use (11.75) to estimate its gain.

$$G_r \approx \frac{1}{2}\left(\frac{\pi d_{dish}}{\lambda_o}\right)^2 = \frac{1}{2}\left(\frac{\pi(1,000)}{13.324}\right)^2 = 27{,}797.2 \ (44.4 \ \text{dBi})$$

The antenna temperature can't be calculated using (13.23) because the patterns aren't known. Estimating the beamwidth of the 10-m dish using (11.74) by replacing W with d_{dish}

$$HPBW = 65° \frac{\lambda_o}{d_{dish}} = 65° \frac{13.324}{1,000} = 0.9°$$

This value is close to the $1.02°$ half-power beamwidth of the pattern considered in Example 13.9; however, that calculation assumed that the receive antenna was pointed directly at the Sun. If the same narrow-beam antenna of Example 13.9 is pointed toward a region of the sky away from the Sun and the calculation using (13.23) is repeated, the brightness temperature is $T_b = 72.6$K. If the same radiation efficiency is used as in Example 13.9, $e_{rad} = 0.63$, the antenna temperature is $T_A = 129.4$K, assuming a physical temperature of $T_p = 310$K. The resulting system noise temperature is then

$$T_{sys} = T_A + T_{rec} = 129.4 + 130.7 = 260.1\text{K}$$

The G/T for the system is

$$\frac{G_r}{T_{sys}} = \frac{27{,}797.2}{260.1} = 106.85 \ (+20.3 \ \text{dB/K})$$

Example 13.15: Measurement of G/T

Determine G/T of the same telemetry receiving system as in Example 13.14 using the measurement procedure outlined in Section 13.7.1. The beamwidth of the 10-m dish was found to be about $HPBW = 0.9°$. The receiver is set up for operation at $f = 2{,}251.5$ MHz and the first measurement made is with the antenna pointed at the Sun for maximum signal level. The gain of the telemetry receiver is adjusted to set this point at the high end of the linear range. The antenna is then pointed at least three $HPBW$s away from the Sun and the cold sky measurement is taken; the power level drops by 18 dB. That is,

$$10 \log_{10}\left(\frac{P_{hot}}{P_{cold}}\right) = 18 \ \text{dB}$$

The solar flux data is available from radio observatories at two frequencies, 1,415 and 2,695 MHz. The results are $SF_{1415} = 103 \times 10^{-22}$ and $SF_{2695} = 142 \times 10^{-22}$ W/m^2/Hz, respectively. Using (13.43) and (13.44)

$$N_S = \frac{\ln\left(\dfrac{2{,}251.5}{2{,}695}\right)}{\ln\left(\dfrac{1{,}415}{2{,}695}\right)} = 0.279$$

The solar flux is

$$SF = 142 \times 10^{-22}\left(\frac{103 \times 10^{-22}}{142 \times 10^{-22}}\right)^{0.279} = 142(0.9142) = 129.82 \; \frac{W}{m^2 Hz}$$

Next, the beamwidth correction factor must be computed. The beam solid angle associated with the 10-m dish may be approximated using its *HPBW*.

$$\Omega_A \approx \left(HPBW \frac{\pi}{180}\right)^2 = \left(0.9 \frac{\pi}{180}\right)^2 = 2.467 \times 10^{-4} \; sr$$

Utilizing the value for the solid angle of the Sun from Table 13.1,

$$L_{cor} = 1 + 0.38\left(\frac{1.1 \times 10^{-4} \; sr}{2.467 \times 10^{-4} \; sr}\right) = 1.169$$

Finally computing G/T using (13.41)

$$\frac{G_r}{T_{sys}} = \frac{8\pi k L_{cor}}{SF \lambda_o^2}\left(\frac{P_{hot}}{P_{cold}} - 1\right)$$

$$= \frac{8\pi(1.38 \times 10^{-23})(1.169)}{(129.82 \times 10^{-22})(0.13324)^2}(10^{18/10} - 1) = 109.24$$

Converting to dB the result is

$$\frac{G_r}{T_{sys}} = 20.4 \; dB/K$$

Example 13.16: Calculation of C/N

The calculation of C/N will be illustrated by combining the results of Example 11.7 with that of Example 13.13. The gain of the receive antenna in Example 13.13 was $G_r = 23.7$ dBi. The calculated carrier power at the terminals of the antenna was found to be $C = -79.5$ dBm. Calculate C/N to determine if there is sufficient signal strength. Using (13.33), C/N is calculated and the details are provided in Table 13.2.

Table 13.2
C/N for Example 13.16

Source power	P_s	20.0	(dBm)
Transmit antenna gain	G_t	+2.0	(dBi)
Transmit cable attenuation	$-A_{tc}$	−0.5	(dB)
Path attenuation	$-A_{path}$	−119.5	(dB)
Gain-to-temperature ratio	G_r/T_{sys}	0.3	(dB/K)
Polarization mismatch	$-PML$	−4.8	(dB)
Impedance mismatch	$-MML_{tot}$	−0.4	(dB)
kB	$-kB$	+198.6 − 63 = 135.6	(dBm/K)
Carrier-to-noise ratio	$C/N =$	+32.7	(dB)

Note that this is the same value obtained if one computes the system noise power in Example 13.13 as

$$N_{sys} = kT_{sys}B = 1.38 \times 10^{-23} (215.9) 2 \times 10^6 = 5.959 \times 10^{-15} \text{W}$$

Converting to dBm, $N_{sys} = -112.2$ dBm, This noise power value is 32.7 dB below the carrier.

13.8 Link Margin

Once the analyses of Sections 13.7 and 11.4 have been completed, the signal-to-noise ratio can be determined. It is common to cite the minimum C/N required for a system to achieve a certain level of performance; the difference between the minimum value and that achieved by the actual system is referred to as the link margin. This is usually expressed in decibels. For example, if the minimum carrier-to-noise ratio for proper system performance is $(C/N)_{min}$, the link margin can be defined as

$$\text{Margin (dB)} = \left(\frac{C}{N_{sys}}\right) \text{(dB)} - \left(\frac{C}{N}\right)_{min\,dB} \tag{13.45}$$

Problems

Problem 13.1

An amplifier has an equivalent noise temperature of T_{amp} = 70K. What is: (a) the associated noise power in dBm, N_{amp}, for a bandwidth B = 2 MHz, and (b) the noise figure in dB, F_{amp}?

Problem 13.2

A preselection filter has an attenuation of 2 dB across its pass-band where the reflection coefficient is extremely low; that is, the loss is due to dissipation, not reflection. The enclosure containing the filter, along with other receiver electronics, has a cooling fan but the nominal physical temperature is still T_p = 320K, what are its: (a) equivalent noise temperature, T_{fil}, and (b) noise figure, F_{fil}?

Problem 13.3

An amplifier has a gain of G_{amp} = 21 dB and a noise figure of F_{amp} = 1.5 dB. If a noise source at T_s = 300K is applied to the input, what is the output noise power, N_{out}, in dBm?

Problem 13.4

A composite system consists of a low-noise amplifier, an interconnecting transmission line, and a receiver. The LNA is intended to be attached directly to the terminals of the receive antenna. The receiver itself will be inside a mobile telemetry van. The LNA has a gain of G_{LNA} = 27 dB and a noise figure of F_{LNA} = 0.6 dB. The cable has attenuation of 15 dB across the frequency of operation. The receiver has a noise figure of F_R = 6 dB. The frequency of operation is f = 2,272.5 MHz and the bandwidth is B = 10 MHz. The antenna and LNA are exposed to relatively high temperatures; on hot days the surface temperature of the externally mounted components often exceeds 55°C. Calculate the overall noise figure for the composite system.

Problem 13.5

A somewhat omnidirectional antenna is used in an application requiring minimal receive antenna gain. The receive antenna is mounted 3m above the Earth. The normalized field pattern is given by:

$$F_{tot}(\theta, \phi) = \sin^2\theta$$

Estimate the antenna temperature, T_A, assuming the efficiency is $e_{rad} = 0.75$. State any necessary assumptions that you make.

Problem 13.6

The antenna of Example 13.7 is used in conjunction with the improved receiver cascade of Example 13.5 to form a telemetry ground station that will attempt to track targets to within $10°$ of the horizon. That is, the peak of the beam will be steered to an angle $10°$ above the plane of the Earth. The frequency and bandwidth will remain the same as in Example 13.5, $f = 2{,}251.5$ MHz and $B = 2$ MHz. Calculate the following: (a) the worst-case antenna temperature, T_A, (b) your estimate of the system noise temperature, T_{sys}, and (c) the corresponding value of the system noise floor in dBm, N_{sys}.

Problem 13.7

A simple receiving system is implemented using a horn antenna to acquire telemetry data in a weapons test program. The link is designed to operate at $f = 2{,}251.5$ MHz and the horn has dimensions $W = 40$ cm by $H = 27.2$ cm. To acquire the data from initiation through midflight, the horn is pointed toward the launch canister, which is mounted 5m above the ground. The receiver of Example 13.5 is connected to the antenna. Find G_r / T_{sys} in dB/K.

Problem 13.8

The solar calibration method is used to quantify the performance of a telemetry receiving system that operates at 2,295.5 MHz. The antenna is an 8m-diameter dish. The operator checks to make sure the receiver is operating in its linear region when the mainbeam is pointed at the Sun. When it is pointed toward a cold region of the sky, the level drops 14.5 dB. A solar flux reading is obtained from the radio observatory at 2,695 MHz, $SF_{2695} = 136 \times 10^{-22}$ W/m^2/Hz. Determine the system G_r / T_{sys} in dB/K.

Problem 13.9

Calculate the carrier to noise ratio, C/N, if the receiving system of Example 13.4 is used in conjunction with the link analysis presented in Example 11.5.

Problem 13.10

The modulation scheme employed in Example 13.16 requires a minimum carrier-to-noise ratio of $\left(\dfrac{C}{N}\right)_{min\,dB}$ to maintain good quality data. What is the link margin?

References

[1] Pozar, D. M., *Microwave and RF Design of Wireless Systems,* New York: Wiley, 2001.

[2] Ulaby, F. T., R. K. Moore, and A. K. Fung, *Microwave Remote Sensing—Volume I,* Reading, MA: Addison-Wesley, 1981.

[3] Kraus, J. D., *Radio Astronomy,* New York: McGraw-Hill, 1966.

[4] Sklar, B., *Digital Communications—Fundamentals and Applications,* Englewood Cliffs, NJ: Prentice-Hall, 1988.

[5] Telemetry Group, Inter-Range Instrumentation Group, "Test Methods for Telemetry Systems and Subsystems—Volume I: End-to-End Test Methods for Telemetry Systems," IRIG Document 118-98, 1998.

[6] Space Environment Center, National Oceanic and Atmospheric Administration, U.S. Dept. of Commerce, http://sec.noaa.gov or http://www.sel.noaa.gov/ftpdir/lists/radio/rad.txt.

14

Telemetry Link RF System Design

Chapter 14 addresses the selection of RF components in telemetry system design; that is, the choice of transmitters, transmit antennas, receive antennas, and receivers. Due to the availability of commercial off-the-shelf transmitters and receivers, previously developed for a wide variety of applications, the emphasis here is upon transmit antennas and radiating systems.

14.1 Learning Objectives

Upon completing this chapter the reader should understand the following:

- Telemetry link design process;
- Transmitter selection;
- Candidate transmit antennas and their associated characteristics;
- Receive antenna selection.

14.2 RF Telemetry System Design

Telemetry systems are used to support a variety of applications ranging from transmission of data for spacecraft science and housekeeping purposes, flight-testing of high-performance aircraft, and system testing of projectiles and missiles. The operational environment depends upon whether it is a space-to-ground, air-to-ground, or air-to-air link.

While a range of transmitters, receivers, and associated components are available to implement an end-to-end design, the selection of such devices and their integration into a telemetry system are naturally application-dependent since data requirements and physical environments vary. The keys to a successful system are the appropriate choices of frequency, modulation, transmit power, and antenna pattern coverage for use in conjunction with a high-performance receiver. The frequency, or frequencies, allocated on a particular test range are normally limited to reserved telemetry bands; there is usually little or no latitude for change. The bands generally reserved for telemetry applications are: L Band (1,435 to 1,535 MHz), S Band (2,200 to 2,300 MHz) and Upper S Band (2,310 to 2,390 MHz) [1].

Modulation issues are discussed elsewhere in the text; the major impact is upon transmitter selection—the unit chosen must have sufficient tuning range to be modulated over the band of interest. In this chapter it will be assumed that the required bandwidth to support the data rate, using the selected modulation format, is known. Thus, the issues to be addressed here are primarily with respect to transmit power and pattern coverage, as well as receive antenna characteristics. In some cases, transmit power is restricted by regulation; in others it is simply due to limited prime power on the article under test. Desirable transmit antenna pattern characteristics depend upon the projected trajectory and the data requirements of the mission.

In some instances, transmit/receive (Tx/Rx) subsystems are designed to operate as an overall telemetry system for a given project. In others, the receive subsystem antennas and electronics are in place on an established range and the task is to design the transmit subsystem to operate within the capabilities of the existing equipment. The latter application is the primary focus of this chapter; the emphasis will be on the design of the test article transmit subsystem. In terms of the receive subsystem, the primary task is to quantify link performance to ensure good quality data transmission.

One possible approach to the RF system design is:

- Select modulation format commensurate with data rate (must be compatible with ground station capability if existing infrastructure is to be used);

- Obtain frequency allocation and confirm that channel bandwidth is sufficient;

- Determine range of source transmit power levels (constrained by prime power consumption or radiated power limits);

- Outline transmit pattern coverage requirements;

- Document ground station parameters, G/T, RF and video bandwidth, et cetera;
- Perform preliminary link analysis using nominal transmit source power and anticipated transmit antenna gain along with receiver subsystem characteristics;
- Finalize design, then quantify C/N and link margin.

Along with the electrical parameters, the design of the telemetry system must be accomplished so that it can operate when exposed to the specified environmental conditions. Spacecraft and airborne and gun-launched projectiles are exposed to differing acceleration, vibration, and temperature environments. Temperature fluctuations due to solar heating are generally an important consideration for spacecraft systems, and aerodynamic heating can be significant in reentry applications. Gun-launched projectiles can experience high acceleration rates. Specialized environmental testing of telemetry system components is often required.

A sketch of a simplified end-to-end telemetry system is shown in Figure 14.1. Based upon the results summarized in Section 11.4, it is important to minimize transmit and receive cable loss. In some instances, an amplifier is required to provide sufficient transmit power to obtain satisfactory C/N. Also, as illustrated by Examples 13.3 to 13.5, the receive cable loss (between the antenna terminals and the first gain block) must be minimized to reduce the system noise. It was shown that a low-noise preamplifier could significantly improve the noise floor and increase the link margin.

14.3 Transmitter Selection

The selection of an appropriate transmitter is based upon several factors. Of course, it must be suitable for the modulation format selected. Additionally,

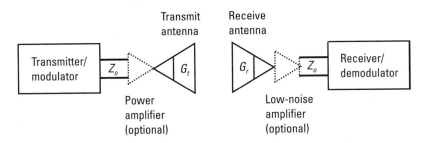

Figure 14.1 Telemetry system elements.

it must produce the required transmit power to provide an acceptable link margin, and, in many instances, flight qualification is required. Furthermore, the transmitters must operate in accordance with IRIG telemetry standards (IRIG 106-01) and within electromagnetic interference limits (MIL-STD-461E) [1, 2]. Currently, transmitters are available to support traditional FM/FM and PCM/FM systems as well as BPSK and QPSK formats. Typical source power levels range from about 150 mW to 10W and are designed for use with supply voltages between 12 and 28 VDC.

The transmitters must operate in the specified temperature and vibration environments. Specialized units are available to operate in high-shock environments. Flight-qualified power amplifiers are also available for use in conjunction with transmitters to increase the output power levels to 10 → 20W.

An important consideration is the prime power requirement; the supply voltage and current requirements must be compatible with the platform capabilities. The transmitter efficiency should be high enough to allow for continuous battery operation over the mission lifetime. The efficiencies of common telemetry transmitters range from 10% to 20%.

Generally, the units have an integrated isolator or circulator to provide *VSWR* protection so that large impedance mismatches can be tolerated without disabling the unit. Care must be exercised during integration and system test of the units at the high end of the power range cited above to ensure that a radiation hazard does not exist.

14.4 Antenna Selection

Antenna selection is a critical element of telemetry system design. The gain of the transmit/receive antennas in conjunction with the transmitter power, P_s, must be sufficient to maintain positive link margin for two general cases: (a) a stabilized vehicle with a known trajectory, and (b) an article under test that may be in an arbitrary and changing orientation as well as deviating from the projected flight path. In the former situation, case (a), the transmit antenna pattern can be shaped to provide the necessary coverage in the direction of the ground station. This generally produces a higher transmit gain, $G_t(\theta_t, \phi_t)$, with a concomitant increase in link margin, or decrease in the required transmitter power level, P_s. However, when a directive antenna is used, deviation from the predicted orientation and/or flight path can cause telemetry dropouts. In case (b), the radiation pattern on the test article is usually chosen to be as isotropic as possible; for example, the desired coverage is the uniform spherical pattern depicted in Figure 11.2(a). Often telemetry and telecommunication engineers refer to this as an omnidirectional pattern. In practice, many so-called omnidirectional patterns have at least one null (usually two) and are better represented by the pattern depicted in three different forms as Figures 11.6 to 11.8; that is, they are uniform along one direction but have nulls in orthogonal cuts. To improve performance, some telemetry systems use polarization diversity; in this case, nulls can be filled with a complementary polarization to get more uniform coverage; this will be discussed in further detail below. In summary, the transmit antenna problem results in a gain versus coverage tradeoff. While the stabilized platform case will be discussed, the emphasis for the remainder of this chapter will be on air-to-ground applications, which generally require omnidirectional coverage.

Receive ground stations are implemented in a variety of forms ranging from simple stationary antennas to sophisticated tracking pedestals. It is most common to have receive antennas with moderate to high gain to maintain positive link margin. In this case, the ground station requires a tracking capability to keep the mainbeam pointed toward a moving test article. This can be accomplished by: (1) mechanical pointing, or (2) electronic scanning. The former requires a tracking pedestal and can range from simple hand track units to program track and up to high-performance autotrack systems. Electronic scanning can be accomplished in two ways: (a) phased arrays to track the test article, and (b) beam switching implementations. The latter will be discussed in Section 14.6.

14.5 Transmit Antennas

The telemetry engineer has many problems to solve to achieve an operational system. This is complicated by the fact that there are often severe restrictions placed upon the interior volume available for the transmitter as well as the exterior area available for the placement of antennas. The commonality from system-to-system of components such as transmitters and their associated electronics increases their availability; however, due to dependence of the antenna patterns upon the platform, specialized design must often be done for the radiating systems. The desired goal of the transmit antenna system is to provide a signal propagating toward the ground station, regardless of the location or orientation of the article under test, during the course of its flight. It is particularly important to maintain link margin during mission-critical events.

14.5.1 Pattern and Gain

The two scenarios introduced in Section 14.4, related to the specification of transmit antennas, can be restated as: (a) telemetry data is only required while the platform is stabilized, and (b) telemetry data is required for any and all orientations of the object. In the former case, the telemetry system engineer can increase link margin and/or reduce transmitter power by use of a directive antenna such that $G_t > 0$ dBi. In extreme cases, the transmit antenna could be a high-gain unit pointed toward the ground station; however, the size and pointing requirement limit this solution to large platforms. Such an antenna would be designed to achieve a certain gain or beamwidth. Based upon the results of Section 11.3, it is apparent that the electrical size of the transmit antenna would have to be large to obtain a narrow beamwidth. The discussion of case (a) in this section is limited to individual radiating elements with moderate gain such as the antennas discussed in Section 12.2. The maximum transmit gain, G_t, that can be realized is limited by the area/volume available for the transmit antenna, and the stability of the platform. For example, if the orientation of the platform is only known within $\pm 30°$ it would be totally inappropriate to have an antenna with a half-power beamwidth of $5°$; when the platform pointing error is maximum, the directive gain toward the ground station might be reduced by 30 dB or more.

Case (b) applies in the majority of telemetry applications; that is, it is generally desirable for the antenna pattern to be as isotropic as possible. Again, based upon the concepts introduced in Section 11.3, this implies that the transmit antenna gain will necessarily be low. In fact, in many applications, the peak gain of the airborne telemetry antennas is less than 0 dBi. Consider the projectile shown in Figure 14.2; there are two mounting options: (1) have the transmit antenna(s) protrude, or (2) use a conformal configuration. In the former case, one can use antennas such as blades, quadraloops, or spikes. Figure 14.3 shows a photograph of antennas of this type. These are designed to present a relatively small cross section; however, in some applications they can affect the aerodynamic performance, and it is necessary to use conformal antennas. These are typically thin, printed-circuit antennas such as microstrip patches or stripline slots, although cavity-backed slots and helices are often used as well. Figure 14.4 is a photograph of some S-band conformal antennas. They can be attached to the outside surface of the vehicle or mounted in a thin depression machined such that the outer surfaces of the antenna and vehicle are coincident.

Individual radiating elements, such as those discussed in Sections 12.2, can be used when the attitude of the article under test is stable and well

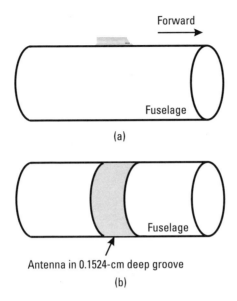

Figure 14.2 Transmit antenna area: (a) low-profile antenna, and (b) conformal antenna.

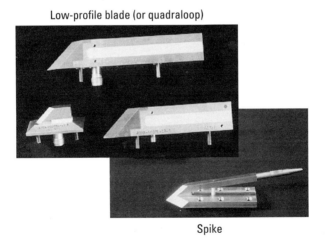

Figure 14.3 Low-profile antennas. (Courtesy of Physical Science Laboratory.)

known. Typical patterns were shown in Figures 12.3, 12.5, 12.7, and 12.14. For an LHCP microstrip patch, the gain at the peak of the beam shown in Figure 14.5 is about $G = 5.7$ dBic. Versions printed on higher dielectric constant substrates are smaller and tend to have less gain. The radiation distribution plot for the same LHCP patch is shown in Figure 14.6. Use of

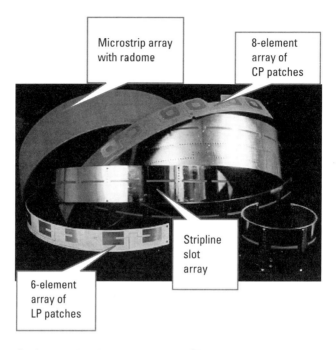

Microstrip array
with radome

8-element
array of
CP patches

Stripline
slot
array

6-element
array of
LP patches

Figure 14.4 Conformal printed-circuit antennas. (Courtesy of Physical Science Laboratory.)

single radiators with patterns such as these, mounted to the bottom (or one side) of the vehicle provide a larger peak gain than that achievable with the cylindrical array configuration discussed in Section 12.3; however, as stated above, it is restricted to applications where its beamwidth is wide enough to ensure sufficient gain toward the ground station as the location/orientation of the test article varies.

When the conformal antenna completely surrounds the vehicle, an array is used and the analysis of Section 12.3 is applied. A conformal array is simply a collection of antennas connected through a set of transmission lines; a planar view of a typical printed-circuit configuration is sketched in Figure 14.7. This can also be implemented with coaxial transmission lines and power dividers, but the expense is usually greater than fabricating it in the printed-circuit form.

The polar radiation patterns of a generic linearly polarized microstrip array were shown in Figures 12.20 and 12.21. That array consisted of 16 antennas equispaced around the circumference of a metallic cylindrical. The individual radiating elements were fed with equal amplitudes. All elements had the same phase; that is, the excitation was in-phase, or $\Phi^i = 0$,

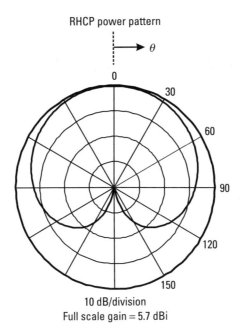

RHCP power pattern

10 dB/division
Full scale gain = 5.7 dBi

Figure 14.5 Typical LHCP patch pattern.

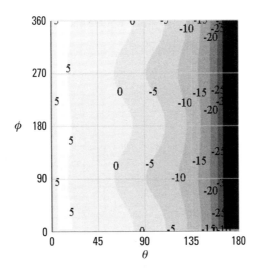

Figure 14.6 LHCP patch radiation distribution plot.

CP microstrip patch

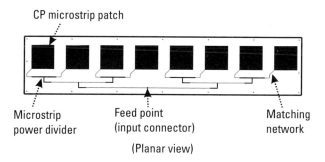

Microstrip   Feed point   Matching
power divider   (input connector)   network

(Planar view)

Figure 14.7 Conformal microstrip array configuration.

i = 1, 2, . . . 16. In practice, a set of polar patterns is often used to give a rudimentary estimate of the coverage. A more complete representation of the coverage for Example 12.3 is the contour plot shown in Figure 14.8. The three-dimensional pattern shown in Figure 14.9 presents the same information but only in a qualitative form. Other than the null evident along the z-axis, the pattern looks relatively uniform. (One also exists along the negative z-axis; this is consistent with the null shown at $\theta = 180°$ in Figure 12.20.) The patterns presented in Figures 12.20, 12.21, 14.7, and 14.8 are representative of a conformal array mounted on a cylinder with small fins. If wings and/or a tail are present, the results can be somewhat modified; however, it has been demonstrated that overall impact on pattern

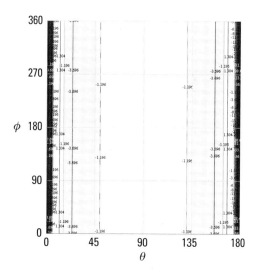

Figure 14.8 Radiation distribution plot for conformal array of Example 12.3.

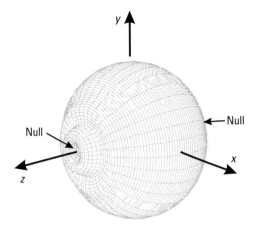

Figure 14.9 Three-dimensional pattern for conformal array of Example 12.3.

coverage is minimal if the antennas are fuselage-mounted as far as possible from the blocking structures [3].

To eliminate the deep nulls, the phasing of the array elements can be modified as was done in Example 12.4. That is, the relative phase between the elements was adjusted to eliminate the nulls fore/aft of the vehicle. For the 16-element array, the elements were placed around the circumference at locations $\phi_i' = (i - 1)\ 22.5°$, for $i = 1, 2, \ldots 16$. The respective phases were chosen to be $\Phi^i = -(i - 1)\ 22.5°$, for $i = 1, 2, \ldots 16$. Two polar patterns were included as Figures 12.22 and 12.23. Figure 14.10 depicts the corresponding radiation distribution contour plot for this configuration. The three-dimensional version is shown in Figure 14.11. Again, the patterns presented in Figures 12.22, 12.23, 14.10, and 14.11 are representative of those of a wrap-around antenna mounted on a cylinder with only small fins. It is apparent that the nulls have filled, and the coverage appears to be more uniform. A quantifiable comparison is required to make engineering tradeoffs; a method to do so is presented in the next section.

14.5.2 Percent Spherical Coverage

Antenna gain itself may not adequately quantify the performance of the antenna system, particularly in telemetry applications where spherical coverage is desired. In this case, the telemetry engineer often specifies the antenna performance in terms of percent spherical coverage. That is, regardless of vehicle orientation, the system designer can have a quantifiable measure of the transmit antenna's performance. What does this mean? Basically, the

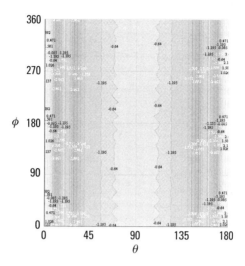

Figure 14.10 Radiation distribution contour plot for conformal array of Example 12.4.

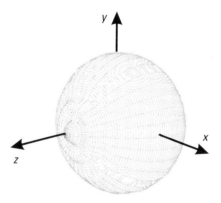

Figure 14.11 Three-dimensional pattern for conformal array of Example 12.4.

interpretation is that the gain is greater than or equal to a specified value, G_{level}, for a certain percentage of radiation sphere, or 4π steradians. For example, the specification might be that the transmit gain has to be greater than -8 dBi over 95% of the radiation sphere; explicitly stated as $G(\theta, \phi) \geq -8$ dBi over a solid angle of 3.8π sr. In other applications, -6 dBi might be required over 98%, or for some, -20 dBi over 99%. It is important to note that percent spherical coverage does not indicate the direction(s) for which $G(\theta, \phi) \geq G_{level}$.

Percent spherical coverage is computed based upon the radiation patterns of the transmit antenna. In order to obtain reasonable accuracy, a full radiation distribution plot, rather than a limited set of polar patterns, must be calculated or measured. The percent spherical coverage, *PC*, is often determined to different polarizations; that is, the percent coverage can be different at a specified gain level to linear as compared to circular polarization. Thus, polarization must be considered when this type of diversity is used. Percent spherical coverage is computed as

$$PC_{pol} = \int_0^{2\pi} \int_0^{\pi} \{\text{If}(G_{pol}(\theta, \phi) \geq G_{Level}; 1, 0)\} \sin\theta\, d\theta\, d\phi \frac{100}{4\pi} \quad (14.1)$$

The term inside the brackets of (14.1) is unity if $G_{pol}(\theta, \phi) \geq G_{level}$ and zero otherwise. G_{level} is a specified value to a particular polarization. That is, $G_{pol}(\theta, \phi)$ is computed as

$$G_{pol}(\theta, \phi) = G_t |F(\theta, \phi)|^2 p_{pol}(\theta, \phi) \quad (14.2)$$

The mismatch factor, p_{pol}, is the same as the one presented in Section 11.3. The subscript *pol* relates to the polarization mismatch between the transmit antenna (whatever its polarization state) and a specific receive antenna polarization; for example, linear E_θ, slant linear, or RHCP. The mismatch, p_{pol}, is computed using (11.59) accounting for the polarization of the radiated wave and the receive ground station antenna. The polarization state of the former generally varies throughout space as the article under test moves. The latter can normally be considered to be constant.

The percent spherical coverage is calculated to quantify the performance of the three antenna configurations addressed in the previous section. The results, for the individual patch, the in-phase excitation array of Example 12.3, and the progressive-phase array of Example 12.4, are computed using (14.1) and plotted in Figures 14.12 through 14.14.

One way of interpreting the information in Figure 14.12 is that the total power gain is greater than −7 dBi over 70% of the radiation sphere (regardless of the orientation of the vehicle); however, this requires the ground station to receive the total power (thick solid line on the plot) in the signal regardless of its polarization state. The plot also includes three other curves: the thin solid line for the percentage coverage to LHCP, the dotted line for the coverage to LHCP *or* RHCP, and the dot-dash line for coverage to RHCP only. For LHCP, the gain is greater than −11 dBic 80% of the time (dotted line). These gain values may be unacceptable to maintain

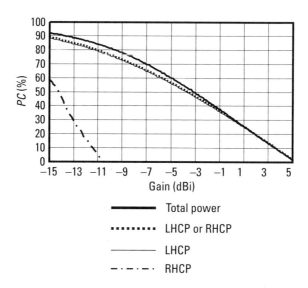

Figure 14.12 Percent spherical coverage for individual CP patch antenna.

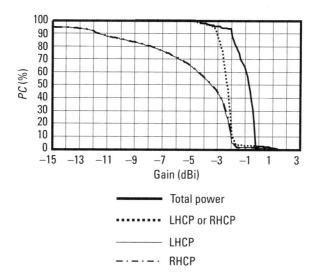

Figure 14.13 Percent spherical coverage for conformal array of Example 12.3.

satisfactory link margin; however, a higher level, for example −3 dBic, is only achieved for about 50% of the radiation sphere for any of the receive polarizations. To obtain the coverage to LHCP *or* RHCP requires the ground station to be able to switch between the two signals and determine the higher-level one in a dynamic fashion. The coverage to RHCP is no better

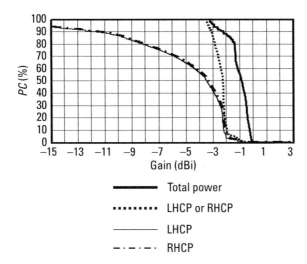

Figure 14.14 Percent spherical coverage for conformal array of Example 12.4.

than 60% at the −15 dBic level (dot-dash line) so the LHCP or RHCP options provide no real improvement over LHCP for this antenna.

Figures 14.13 and 14.14 show that the use of a conformal array, such as those presented in Examples 12.3 through 12.4, can increase the coverage. For example, at the −3-dBic level it improves to about 90% for the coverage to LHCP *or* RHCP. This requires the ground station to detect the power level in the two orthogonal channels and select the greater one. The coverage can be further increased if the two components can be combined coherently, that is, in amplitude and phase to obtain the total power. These are two forms of polarization diversity techniques to be discussed in Section 14.6.

In summary, quantification of transmit antenna performance in terms of percent spherical coverage allows the system designer to predict the link margin in terms of the probability of the data being received.

14.5.3 Transmit Antenna *VSWR*

Analogous to the definition of total mismatch loss, MML_{tot}, in Section 11.3.4, the performance of the transmit antenna alone can be considered as well. The transmit mismatch loss in decibels, MML_t, is given by (14.3).

$$MML_t = -10\log_{10}(q_t) = -10\log_{10}(1 - |\Gamma_t|^2) \qquad (14.3)$$

It may also be determined in terms of the antenna *VSWR* inverting (11.44).

$$MML_t = -10 \log_{10}\left(1 - \left(\frac{VSWR - 1}{VSWR + 1}\right)^2\right) \qquad (14.4)$$

The mismatch loss factor in dB is plotted versus $VSWR$ in Figure 14.15. In many applications the acceptable $VSWR$ limit is a maximum of 2; the corresponding mismatch loss is about 0.5 dB. When the maximum allowable $VSWR$ drops below 1.5, the associated value of mismatch loss is less than 0.2 dB. As discussed in Section 11.3.4, the primary impact of a $VSWR$ greater than 1 is a reduction in radiated power due to the reflection; however, the $VSWR$ can also affect the operation of the transmitter—the load $VSWR$ can pull the carrier away from the intended center frequency if load isolation is not provided.

14.6 Ground Station Antennas

As discussed above, the emphasis in this chapter is on air-to-ground applications. Thus, the receive end of the telemetry system is termed the ground station in this section. Similar implementations are possible on an air-to-air link (such as a high-altitude balloon-to-aircraft telemetry link); however, one limiting factor is the maximum size of the antenna possible in airborne receive applications.

There are a variety of sophisticated tracking pedestals that use directive antennas to achieve high data rates with large link margins. These units can be operated via program track or autotrack. In the former, the predicted trajectory is used to point the receive antenna system for the duration of the flight test. In the latter, as the name implies, the system automatically tracks the article under test. Typically, a conical scan, monopulse, or pseudo-

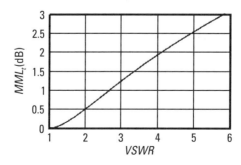

Figure 14.15 Mismatch loss factor, MML_t, versus $VSWR$.

monopulse tracking system processes the incoming data stream through sum and difference channels to keep the ground station antenna pointed at the vehicle.

A simplified form of program track can be implemented as shown in Figure 14.16. It uses a switch to connect the receiver to a series of antennas pointed in different directions to support the telemetry link during various portions of flight. The switch is generally preprogrammed to switch at prescribed times according to the predicted trajectory of the vehicle. This can be a low-cost solution but obviously has the liability that the data quality can be poor if the vehicle deviates from the predetermined trajectory. The configuration depicted in Figure 14.16 can be implemented with discrete antennas or as an array with a Rotman lens to form multiple beams.

A photograph of two tracking pedestals is shown in Figure 14.17. The unit in the foreground is a 10-ft diameter dish, the one in the background is a 30-ft diameter unit. The nominal gains are 33 and 43 dBic at the middle of the 2,200 to 2,300 MHz band, respectively. The larger unit has a five-antenna monopulse tracking system. The smaller one has a three-element tracking array mounted to the edge of the dish; one element is at the center, one is displaced in the azimuth direction, the other in elevation to provide error signals to the servo-controller that points the pedestal toward the article under test.

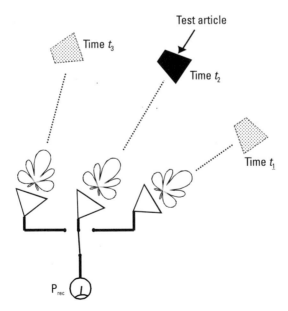

Figure 14.16 Switched-beam program track.

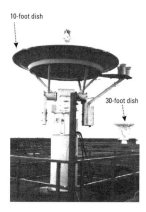

Figure 14.17 Tracking pedestals. (Courtesy of Physical Science Laboratory.)

14.6.1 Pattern and Gain

The ground segment is normally where directive antennas are used to increase the passive gain of the telemetry system. The use of a high-gain receive antenna can result in the use of a lower-power transmitter on the article under test. However, the use of a directive antenna necessitates a tracking system. The system must be capable of tracking at specified rates and in the presence of interference and clutter. That is, tracking a slow moving vehicle well above the horizon is a much simpler task than that of a high-velocity one near the ground.

The selection of a receive antenna and tracking pedestal is dependent upon the frequency, data rate, and overall link geometry. The nominal mid-band gain of some typical S-band receive antennas are tabulated in Table 14.1.

In addition to the increased gain values, a primary advantage of dish antennas, relative to a horn or helix, is that they have low sidelobe levels.

Table 14.1
Receive Antennas

Antenna	Gain (dBi)
30-ft diameter dish	43
10-ft diameter dish	34
4-ft diameter dish	26
Pyramidal horn	20
10-turn helix	15

For example, a dish antenna will have sidelobes 20 to 30 dB below the beam peak, whereas for a horn or helix this degrades to about 8 to 12 dB down.

14.6.2 Use of Diversity

As discussed in Sections 11.4 and 11.5, the signal available at the terminals of the receive antenna can be somewhat degraded from the free space value by multipath fading and other factors. The degradations are usually short-term dropouts, that is, a temporary loss of data; however, they can also cause the receiver to lose lock. To overcome this problem, the system can receive multiple signals and combine them to reconstruct the original data stream [4]. The same information can be received using different frequencies, at varying locations or with orthogonal polarizations. These are termed frequency, spatial, and polarization diversity, respectively. There are two methods to add them—prior to detection and afterward; these are termed predetection and postdetection combination, respectively.

Frequency diversity requires both transmit and receive ends of the telemetry system to operate at more than one frequency; this consumes more of the available spectrum, and, as discussed in Section 11.7, its use generally requires an additional frequency allocation. Spatial diversity can be used to overcome the effects of multipath fading and is a good solution in many applications. In Section 14.5 it was found that use of more than one polarization could provide wider transmit pattern coverage. In this case, polarization diversity must be implemented in the ground segment.

Frequency diversity utilizes the same transmit and receive equipment but operates with two carriers to achieve fading decorrelation [4]. The fading characteristics of the particular configuration should be well understood to specify the frequency separation of the carriers. Figure 14.18 demonstrates the concept of frequency diversity.

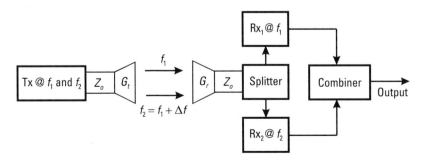

Figure 14.18 Frequency diversity.

Spatial diversity relies upon separation between two independent receive antennas to mitigate the effects of multipath. In the most usual case, the antennas are separated vertically to reduce the fading due to the primary specular reflection. A spatial diversity system is shown in Figure 14.19. A separation distance of many wavelengths, generally 100 to $200\lambda_o$, is required [4].

As stated above, there are two basic methods to add the diversity signals: (a) predetection, or coherent combination, and (b) postdetection. In the former, the vector signals are combined in amplitude and phase (see Figure 14.20). In the latter, the signals are detected then combined. One possible postdetection scenario is depicted in Figure 14.21; it switches between the two signals, depending upon which one is larger. The possible improvement in signal level for a variety of combination techniques is quantified in [4, 5].

As discussed in Section 14.5, multiple polarizations are often used to achieve a near-isotropic, total-power, transmit antenna pattern. Figure 14.22

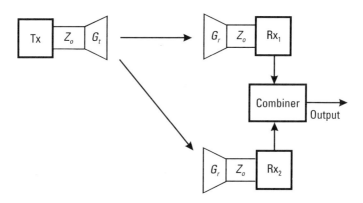

Figure 14.19 Spatial diversity.

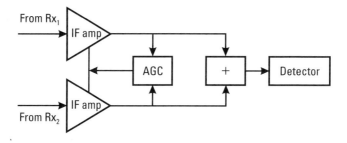

Figure 14.20 Predetection combination.

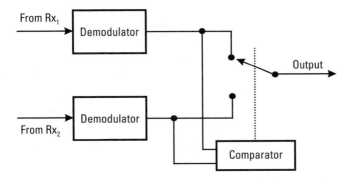

Figure 14.21 Postdetection selection.

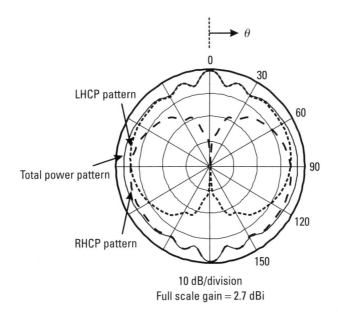

10 dB/division
Full scale gain = 2.7 dBi

Figure 14.22 LHCP, RHCP, and total-power patterns of Example 12.4.

replots the total power pattern of the conformal array of Example 12.4; for comparison, the patterns to LHCP and RHCP are graphed on the same plot. The figure clearly shows the spatial variation due to the polarization characteristics. In the region where the polarization is LHCP or RHCP the total-power pattern is well matched to the LHCP or RHCP pattern, respectively. At $\theta = 90°$ and 270°, the overall pattern is linearly polarized and the 3-dB mismatch loss in the LHCP and RHCP plots is evident. Of course, in the regions where the sense of the radiating antenna is opposite there are deep nulls in both the LHCP and RHCP patterns. It is apparent that the use of polarization diversity would greatly enhance the link margin; if the ground station had the capability of postdetection combination, the received signal would never be more than 3 dB below the total pattern envelope in Figure 14.22. If predetection diversity combination was implemented, it would allow the coverage of the total-power pattern (solid line) to be fully utilized.

A receive antenna system used to implement polarization diversity would be configured to produce LHCP and RHCP outputs simultaneously. One possible configuration (shown as a horn for simplicity) is a dual linearly polarized antenna interconnected as shown in Figure 14.23. Such an antenna system could be used in conjunction with either of the combiners shown in Figures 14.20 or 14.21.

14.7 Receiver Selection

Although the transmit portion is usually the most unique section from system to system in telemetry applications, the key to successful transmission of data is a high-performance receiver. On many ranges a ground station can often be utilized by different test programs; this requires it to operate over

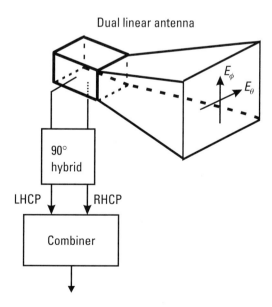

Figure 14.23 Polarization diversity implementation.

all allocated telemetry bands. It must have the capability to adjust its bandwidth on a mission-by-mission basis to achieve the best noise floor and support the required data rate. For these reasons, versatile telemetry receivers have been designed for a variety of applications. These units can typically be reconfigured to change carrier frequency, RF, and IF bandwidth.

Due to the varying nature of the missions to be supported, telemetry ground stations are often implemented as mobile units so that their location can be optimized for individual applications. A typical mobile telemetry unit is depicted in Figure 14.24. The associated instrumentation is ruggedized to ensure operability when the equipment is moved from location to location. Some units have roof-mounted pedestals; others have antennas with positioners that are set on the ground outside the van.

14.8 System Example

A telemetry system is required for an atmospheric sounding rocket; it will be used in conjunction with an established ground station on an existing test range. The current frequency allocation is for a link at $f = 2,251.5$ MHz and the bandwidth selected must be large enough to support standard telemetry as well as a video link. The bandwidth required, using the selected modulation format, is $B = 20$ MHz. The maximum altitude of the flight at apogee is 300 miles; the predicted trajectory will result in a maximum range of $R = 350$ miles.

Figure 14.24 Mobile telemetry unit. (Courtesy of Physical Science Laboratory.)

The primary ground station uses a 10-m diameter tracking dish with integrated low-noise amplifiers whose equivalent noise temperature is T_{LNA} = 90K. The receiver noise temperature of T_{rec} = 130.7K was obtained in Example 13.5; the resulting system noise temperature calculated in Example 13.14 was T_{sys} = 260.1K (when the peak of the beam was not pointed within 3 *HPBWs* of the Sun). The gain of this large dish antenna is estimated to be G = 44.4 dBi using (11.75). Using the solar calibration method, the G/T of this system was determined to be 20.4 dB/K in Example 13.15. (The result from Example 13.14 was 20.3 dB/K.) For the specified bandwidth of B = 20 MHz, the system noise floor is N_{sys} = −101.4 dBm. A 10-ft diameter dish with a gain of G = 34.1 dBi and a radiation efficiency of e_{rad} = 0.43 is also used to support the mission. To approximate the pattern of the 10-ft diameter dish, the pattern expression in Example 13.7 is modified

$$|F(\theta, \phi)|^2 = \frac{\sin(25\pi\sin(\theta)\cos(\phi))}{50\sin\left(\frac{\pi}{2}\sin(\theta)\cos(\phi)\right)} \frac{\sin(25\pi\sin(\theta)\sin(\phi))}{50\sin\left(\frac{\pi}{2}\sin(\theta)\sin(\phi)\right)}$$

$$\times \frac{1 + \cos(\theta)}{2}$$

Using the same distribution as that shown in Figure 13.11, the integrated brightness is T_b = 39.2K. The resulting antenna temperature is T_A = 193.6K. If an identical receiver is used, the system noise temperature is T_{sys} = 324.3K. The G/T is then about 9 dB.

The data requirements stipulate telemetry coverage throughout the flight. To accommodate this requirement for vehicle of radius r_v = 0.219m, the antenna system of Example 12.4 is selected; the conformal array is fed in phase progression eliminating the nulls fore/aft of the sounding rocket. As shown in Figure 14.14, this configuration approaches full spherical coverage at the gain level, $G_t \approx$ −3.5 dBic.

At the frequency of 2,251.5 MHz, the path attenuation is essentially the free space loss, the atmospheric losses are assumed to be negligible. The free space loss at the maximum range is *FSL* = 154.5 dB. The free space value reduces to about 110 dB when the sounding rocket is on the launcher; however, in the latter case, it is relatively close to the ground and multipath fading can reduce the signal level by about 40 dB. To somewhat ameliorate the impact of multipath, polarization diversity is used in conjunction with the two separate ground stations.

The ground station antennas are well matched; their *VSWR* is less than 1.2, resulting in an impedance mismatch loss of less than 0.05 dB. The low-noise amplifiers are attached to the monopulse feed antennas through a cable with negligible loss. The transmit antenna has a maximum *VSWR* of 2 across the ±10 MHz band centered at f = 2,251.5 MHz. The impedance mismatch loss on the transmit end is found to be 0.5 dB, resulting in a total impedance mismatch loss of MML_{tot} = 0.55 dB. The coaxial cable interconnecting the transmitter and conformal array has an attenuation of A_{tc} = 0.2 dB.

Both the primary and secondary ground stations have dual, circularly polarized antennas (LHCP and RHCP ports are available simultaneously), while the polarization state of the transmit antenna varies throughout space as discussed in Sections 14.5.2 and 14.6. The 16 elements are phased to produce LHCP forward, RHCP aft, and in the vicinity of the roll plane linear polarization. If postdetection polarization diversity combination is used, the worst-case polarization mismatch of the system will be *PML* = 3 dB.

An off-the-shelf 5W transmitter is available and has been flight qualified. It draws 1.2A at 28 VDC. The corresponding source power is P_s = 37 dBm. The resulting carrier-to-noise ratio at maximum range, using the primary 10-m diameter ground station is given in Table 14.2.

The carrier-to-noise ratio decreases when the secondary ground station is used; the result for the 10-ft diameter dish is given in Table 14.3.

The path attenuation used to calculate the carrier-to-noise ratio presented above does not include multipath effects. When the sounding rocket is on the launcher, the free space loss decreases to 110 dB and the multipath loss factor is about *MPL* = 40 dB; it improves the carrier-to-noise ratio to about *C/N* = 25.8 dB for the 10-m dish. In the presence of multipath, it changes to *C/N* = 14.4 dB for the 10-ft diameter antenna.

Table 14.2
C/N for 10-m Antenna in the System Example

Source power	P_s	37.0	(dBm)
Transmit antenna gain	G_t	−3.5	(dBi)
Transmit cable attenuation	$-A_{tc}$	−0.2	(dB)
Path attenuation	$-A_{path}$	−154.5	(dB)
Gain-to-temperature ratio	G_r/T_{sys}	20.4	(dB/K)
Polarization mismatch	$-PML$	−3.0	(dB)
Impedance mismatch	$-MML_{tot}$	−0.55	(dB)
kB	$-kB$	+198.6 − 73 = 125.6	(dBm/K)
Carrier-to-noise ratio	C/N	+21.3	(dB)

Table 14.3
C/N for 10-ft Antenna in the System Example

Source power	P_s	37.0	(dBm)
Transmit antenna gain	G_t	−3.5	(dBi)
Transmit cable attenuation	$-A_{tc}$	−0.2	(dB)
Path attenuation	$-A_{path}$	−154.5	(dB)
Gain-to-temperature ratio	G_r/T_{sys}	9	(dB/K)
Polarization mismatch	$-PML$	−3.0	(dB)
Impedance mismatch	$-MML_{tot}$	−0.55	(dB)
kB	$-kB$	+198.6 − 73 = 125.6	(dBm/K)
Carrier-to-noise ratio	C/N	+9.9	(dB)

It is interesting to plot the carrier-to-noise ratio, C/N, as a function of range. Once the rocket is more than a couple of half-power beamwidths above the ground, multipath effects can largely be ignored and the variation in signal power is primarily due to the varying free space loss. A plot of C/N for the 10-m dish receiving system is shown in Figure 14.25.

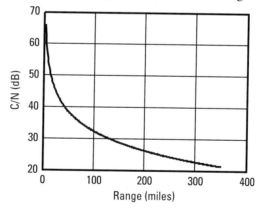

Figure 14.25 *C/N* versus range.

Problems

Problem 14.1

Perform a survey of commercially available flight-qualified S-band transmitters. Tabulate the results including the following information for each unit: (a) output power level, P_s, (b) characteristic impedance, Z_o, (c) maximum load *VSWR,* and (d) supply voltage and current. From the prime power requirement also calculate and tabulate the efficiency of the unit as well.

Problem 14.2

A cylindrical array of slanted dipoles produces RHCP everywhere in space. The normalized power pattern is given by:

$$P(\theta, \phi) = \sin^2 \theta$$

Part (a): If the radiation efficiency is $e_{rad} = 0.8$, determine the gain in dBic. Part (b): Compute the percent spherical coverage for the following cases: (i) total power, (ii) LHCP, (iii) RHCP, (iv) LHCP or RHCP, (v) linear E_θ, and (vi) linear E_ϕ.

Problem 14.3

The normalized total power pattern of an antenna is given by:

$$P(\theta, \phi) = \left(\frac{\sin\left(\frac{\pi}{2} \cos\theta \right)}{\frac{\pi}{2} \cos\theta} \right)^4$$

The polarization of the antenna is linear E_θ. Part (a): If the radiation efficiency of the antenna is $e_{rad} = 0.65$, determine the gain, G. Part (b): Compute the percent spherical coverage for the following cases: (i) total power, (ii) LHCP, (iii) RHCP, and (iv) LHCP or RHCP.

Problem 14.4

Perform a survey of diversity combiners. Tabulate the results and discuss their use in a telemetry application.

Problem 14.5

Identify an off-the-shelf receiver for use at S-band appropriate for use in the example of Section 14.8. Discuss the important parameters of the unit and how they impact the performance of the telemetry system.

Problem 14.6

A telemetry system is required for testing of gun-launched projectiles. The maximum range to impact is 35 km. The maximum altitude is 2 km. A frequency allocation has been obtained for operation at 2,295.5 MHz. Data is required from the instant the article under test leaves the barrel to impact. The prime power available for the entire transmit subsystem is 3.5W (28 VDC at 125 mA). Design the system including the number and location of ground stations. State any and all assumptions you make. Quantify the performance of the system.

References

[1]   Telemetry Group, Range Commanders Council, *Part I—Telemetry Standards,* IRIG 106-01, White Sands Missile Range, NM, Jan. 2001.

[2]   Department of Defense, *Interface Standard, Requirement for the Control of Electromagnetic Interference Characteristics of Subsystems and Equipment,* MIL-STD-461E, Aug. 20, 1999.

[3]   Jedlicka, R. P., et al., "X-34 Vehicle Mockup Configuration for Percent Coverage Determination—Simulations and Measurements," *Proceedings of the 22nd Annual Antenna Measurement Techniques Association Meeting,* Philadelphia, PA, Oct. 2000.

[4]   Freeman, R. L., *Radio System Design for Telecommunications (1–100 GHz),* New York: Wiley, 1987.

[5]   Telemetry Group, Range Commanders Council, *Telemetry Applications Handbook,* White Sands Missile Range, NM, Feb. 1988.

15

Synchronization

The overall system, from data collection and transmission to receiving and separation of data, was shown in Figure 1.1. At the ground station, it is necessary to separate the data from the N sensors, whether using FDM or TDM. Synchronization is of paramount importance in the TDM-PCM/FM system. It is necessary to establish bit synchronization, word synchronization, and frame synchronization.

This chapter will discuss the demultiplex system of the ground station for TDM and FDM of Figure 15.1(a). Specifically, this chapter will discuss this block for time divisional demultiplexing with the subsystems, shown in Figure 15.1(b).

The input to the bit synchronizer is the noisy output from the carrier FM demodulator. The bit synchronizer must evaluate this noisy and distorted waveform representing a one or a zero and make a decision on whether a logic one or zero was transmitted and establish the bit synchronization clock. Once bit synchronization is established, the output of the bit synchronizer is fed into the frame synchronizer. Once frame synchronization is established, the demultiplexer routes valid data to the N data channels.

15.1 Learning Objectives

This chapter is concerned with the concepts of synchronization and the implementation of synchronizing modules. Upon completing this chapter, the reader should understand the following topics:

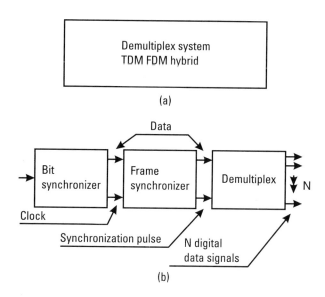

Figure 15.1 Demultiplex system: (a) overview, and (b) TDM-PCM/FM.

- Bit, word, and frame synchronization processes;
- Generation of synchronizing waveforms;
- Open-loop and closed-loop hardware synchronizers;
- Digital correlators;
- Code words for frame synchronizers;
- Frame synchronization logic;
- Overall operation of the frame synchronizer and demultiplexer.

15.2 Functions of the Bit Synchronizer

For optimal bit detection, all digital communication system receivers must establish bit synchronization. A functional block diagram of a bit synchronizer is illustrated in Figure 15.2. The function of the blocks in the bit synchronizer will be discussed next.

15.2.1 Bit Detector Function

One important function of the bit synchronizer is to evaluate the incoming noisy and distorted waveform from the carrier demodulator during each bit

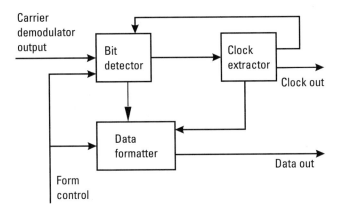

Figure 15.2 Bit synchronizer.

period and determine if it is a zero or a one. The block that performs this operation is referred to as the bit detector. The bit detector takes in a noisy distorted wave and makes the decision as to whether it is a one or a zero and outputs crisp and sharp waveforms representing the binary sequence. This regenerated sequence is applied to the clock extractor. If the bit detector is using an integrate-and-dump filter, it must know when to start and stop the integration and reset or dump. The clock signal is fed back into the bit detector from the clock extractor and provides this timing.

15.2.2 Clock Extractor Function

The purpose of the clock extractor is to establish bit synchronization by extracting a clock sequence from the regenerated incoming data sequence. The output of the clock extractor goes to the bit detector in order that the bit detector may operate on the incoming waveform during exactly one bit period. The clock also goes to the frame synchronizer and other points.

15.2.3 Data Formatter Function

The data formatter converts the data into the format desired, say, NRZ-L from BiΦ.

15.2.4 Control Signals

The control signals to the bit detector to set the type of response to the incoming bit sequence needed for optimum detection whenever this option is designed into the equipment. If a matched filter is used, the filter must change to match the bit waveform. That is, if the incoming data changed from a NRZ-L to BiΦ during the next experiment, then the matched filter must change. The control signal must also set the data formatter such that it outputs the desired format.

15.3 Hardware Block Implementation

15.3.1 Bit Detector Implementation

The implementation of the bit detector is usually achieved with a matched filter or, equivalently, an integrate-and-dump filter. A clock from the clock extractor is absolutely necessary.

15.3.2 Clock Extractor Implementation

Clock extractors are classified as open loop or closed loop. In practice, almost all of the clock extractors operate closed loop.

15.3.2.1 Open-Loop Clock Extractors

An open-loop clock extractor, using transition or edge detection, is shown in Figure 15.3. The idealized bit sequence input, $b(t)$, and waveforms at other points in the clock extractor are shown in Figure 15.4. The first block is a differentiator and whenever the waveform changes level, the differentiator produces a pulse as shown. If $b(t)$ represents NRZ-L sequence, then a pulse out of the differentiator occurs only when a one or zero is followed by a zero or a one, respectively. The pulses from the differentiator are rectified and applied to a bandpass filter with a center frequency of Rb. These pulses are represented with a Fourier series with a fundamental frequency of Rb, which is passed by the filter. The hard limiter is a threshold device that gives a constant voltage, positive or negative, whenever the input voltage exceeds 0V or falls below 0V, respectively. The last waveform represents the clock. A sequence of all ones or all zeros would eliminate the clock.

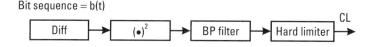

Figure 15.3 Transition detection with open-loop clock extractor.

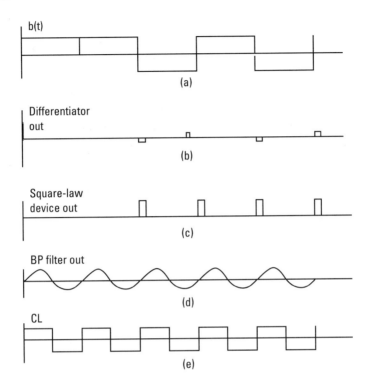

Figure 15.4 Open-loop clock extractor waveforms.

An open-loop digital data transition tracking clock extractor is shown in Figure 15.5. The associated waveforms are shown in Figure 15.6. The regenerated bit sequence, $b(t)$, from the bit detector is applied to the *digital data transition tracking loop* (DDTTL). This waveform is differentiated, squared, and multiplied by the clock from the *voltage-controlled clock* (VCC). The output from the multiplier, shown in Figure 15.6(d), is integrated, applied to a lowpass filter, and applied to the VCC as a control voltage. Inspection of Figure 15.6 shows that the output of the multiplier has equal area whenever the clock transition is located exactly in the center of the pulse resulting from the differentiation. Otherwise, a positive or negative voltage will result for control of the VCC.

15.3.2.2 Closed-Loop Clock Extraction

One type of closed-loop clock extractor used extensively is referred to as the *early-late* (E-L) gate type and is shown in Figure 15.7. The waveforms that indicate the operation of the E-L gate are shown in Figure 15.8. The operation of the E-L gate on a per-bit basis is as follows. The VCC produces a clock

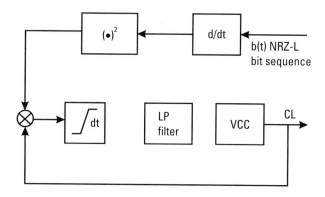

Figure 15.5 Digital data transition tracking loop.

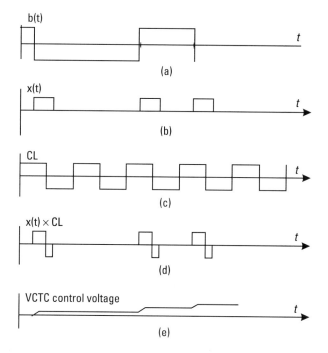

Figure 15.6 DDTL waveforms.

pulse whose leading edge is an estimate of the starting time, $0e$, of the bit
pulse. This clock starts the integration of the early integrator and integrates
to $Te - d$, where Te is an estimate of the end of the bit pulse. The clock
is delayed by d seconds and starts the late integrator at d seconds and
integrates to Te seconds. All timing assumes that the particular bit starts at
time equal to zero and ends at T. For this example NRZ-L is assumed.

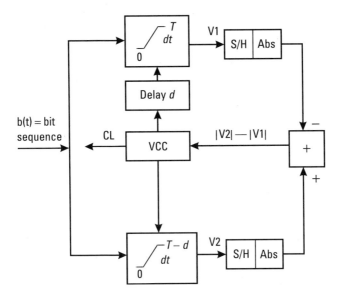

Figure 15.7 Early-late gate clock extractor.

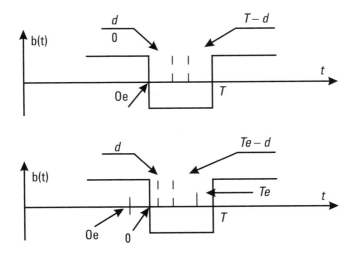

Figure 15.8 Early-late gate integration waveforms.

Whenever the estimate of $0e$ and Te are correct and align with 0 and T, the outputs of the two integrators are exactly the same. The two output voltages are sampled and held. The absolute values of the voltage, $|V1|$ and $|V2|$, are summed plus and minus, respectively, and applied to the VCC. When the gates start at zero and end at T, the control voltage is zero.

The lower waveform of Figure 15.8 indicates what happens when the clock is running too fast. The explanation also assumes alternating zeros and ones. The early gate integrates over a positive voltage initially, and then over a negative voltage since Oe falls in the previous bit period. The late gate integrates over the interval from $Oe - d$ to Te, which does not align with the interval defined by the actual O and T, but all the integration is performed on a negative waveform. This results in $V1$ being larger than $V2$, and a negative control voltage results and slows down the VCC. It is important to note that the control voltage is dependent upon an alternating logic sequence whenever NRZ-L is used.

15.4 Frame Synchronizer

The purpose of the frame synchronizer is to establish both major and minor frame synchronization. The functional operation of the frame synchronizer is to search for the synchronization word that has been inserted periodically into the data bit sequence indicating the start of a minor frame. (Suggested synchronizing words are listed in Appendix B.) The search is implemented by a digital correlator whose output is a function of the number of bit agreements between the stored sync word and the group of bits under observation. For a detailed probabilistic analysis of false lock and the impact of bit errors on the probability of lock, see [1].

15.4.1 Synchronization Steps

The synchronization process is divided into three steps, as indicated in Figure 15.9—search, check, and lock. The output of the digital correlator, C, is equal to the number of bit agreements minus the number of disagreements. Ts is the minimum value of C that will allow the synchronization process to proceed to the check mode from the search mode. Ts is the threshold value of C in order to move to the check stage. Once the threshold Ts is reached or exceeded, the process moves into the check mode. With the proper logic and control supporting the digital correlator in the check mode, the correlator can look at the interval in the bit sequence where the synchronization word is expected to be based on the search routine and determine if the synchronizing word is indeed there. If at this time, C equals or exceeds Tc, then the process may move into the lock mode. The lock mode usually has a Tl less than the first two thresholds so that once the system is locked, it can experience additional errors during a short time period without loss

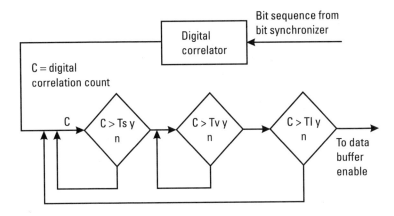

Figure 15.9 Frame synchronization logic.

of lock. Once the lock mode is established, an enable signal goes to a data buffer and the data stream is fed into the output data lines.

15.4.2 Digital Correlator

The digital correlator plays a major role in the frame synchronization process. A simplified digital correlator is shown in Figure 15.10.

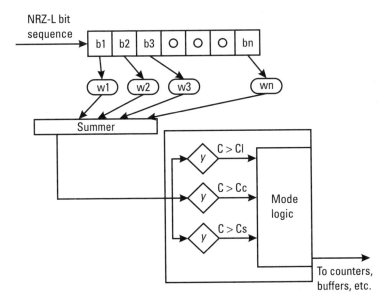

Figure 15.10 Digital correlator.

The operation of the digital correlator is as follows. The bit stream is shifted into the serial n-cell register. Each bit in the register is then multiplied by the corresponding bit, w_i, of the stored synchronizing word. This is a parallel operation. The output of this operation is fed into a summer. The output of the summer is given by

$$C = \sum_{i=1}^{n} b_i w_i$$

This is the number of agreements minus the number of disagreements, since NRZ-L is the format. That is, when b_i and w_i agree, both are +1 or −1, and their product is a +1. If they disagree, then their product is a −1.

15.5 Demultiplexer

A frame synchronizer, including a digital correlator and the mode check module, is shown with the demultiplexer and their functional interaction in Figure 15.11. The operation of the two is as follows: once major and minor frame sync is established, the mode-check module signals the counters to start. The counter will start counting and establish a virtual pointer, such that at any time it will be known for a particular count where the single bit of the pointer is in a word, where the word is in a minor frame, and where the minor frame is in a major frame. The counter will indicate this count to the control module, which will give control signals and a clock to the decommutator.

The decommutator will connect the incoming data word to the correct output data line. The data word comes from the indicated shift register. The output of the shift register is enabled by the mode check whenever frame lock is obtained. The purpose of the shift register module is to take the serial data and convert it into a parallel word. The length of the parallel word is set by the controller knowing where in the major frame and minor frame the particular incoming word is located. The parallel word is shifted into the decommutator block and connected to the correct data line. The word may be converted into a serial word if it is required by the application.

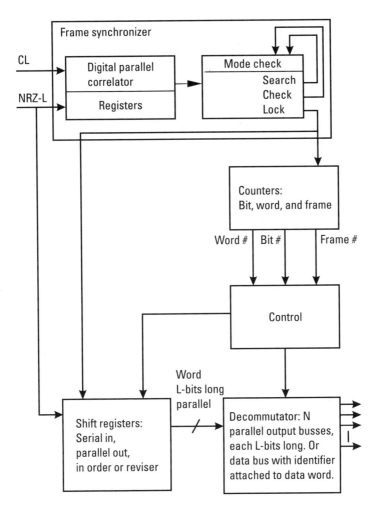

Figure 15.11 Frame synchronizer and demultiplexer.

Problems

Problem 15.1

Draw a block diagram of an end-to-end PCM/FM TDM system.

Problem 15.2

List the functions of a bit synchronizer.

Problem 15.3

Draw an open-loop bit synchronizer and label the functions of each block.

Problem 15.4

Draw a close-loop bit synchronizer and label the functions of each block.

Problem 15.5

Sketch the waveforms of an E-L gate and show the integration period in both gates when the generated clock and the bit sequence are together.

Problem 15.6

Sketch the waveforms of an E-L gate and show the integration period in both gates when the generated clock is ahead of the bit sequence.

Problem 15.7

Sketch the waveforms of an E-L gate and show the integration period in both gates when the generated clock is behind the bit sequence.

Problem 15.8

Draw a block diagram of an E-L gate. For Problems 15.5, 15.6, and 15.7, determine the polarity of the voltage going to the VCC.

Problem 15.9

For frame synchronization, select a 16-bit frame synchronizing word from Appendix B. (a) If there is one bit error, what would be the output of the digital correlator of Figure 15.10 if $n = 16$? (b) two bit errors? (c) three bit errors?

Problem 15.10

What is the output of the digital correlator in Figure 15.10 if $n = 16$ and the frame sync word of Problem 15.9 slips (a) one bit? (b) two bits?

Problem 15.11

What is the output of the digital correlator in Figure 15.10 if $n = 16$ and the frame sync word is created by alternate ones and zeros and the word slips (a) one-bit? (b) two bits?

Problem 15.12

Set up the logic, that is, search, check, and lock, for a frame synchronizer based upon one, two, and three errors and discuss.

Reference

[1] Roden, M. S., *Digital Communication System Design*, Englewood Cliffs, NJ: Prentice-Hall, 1988, pp. 262–270.

16

Hybrid Systems: PCM/FM + FM/FM, PCM/FM/FM

Systems that combine PCM/FM and FM/FM, referred to as hybrid systems, will be discussed in this chapter, and the design process for combining the two processes will be developed. PCM/FM + FM/FM combine PCM at baseband with the subcarrier package, and the combined system is modulated onto the carrier. In PCM/FM/FM, the bit sequence is frequency modulated onto one or more subcarriers, analog signals from the sensors are modulated onto subcarriers, and all subcarriers are summed and used to modulate a carrier.

16.1 Learning Objectives

Upon completing this chapter, the reader should understand the following topics and processes:

- Concept of PCM/FM + FM/FM systems;
- Spectrum of PCM/FM + FM/FM;
- Interference of the bit sequence in the subcarrier channels;
- Degradation of the bit error rate due to subcarrier interference;
- Design of PCM/FM + FM/FM systems;
- Spectrum of PCM/FM + FM/FM;

- Concept of PCM/FM/FM system;
- Spectrum of PCM/FM/FM;
- PCM/FM/FM system design;
- Setting receiver parameters based upon modulation design parameters.

16.2 PCM/FM + FM/FM System Design

A PCM/FM + FM/FM multiplex system is depicted in Figure 16.1. This modulation format is the result of summing a filtered bit sequence and the FM subcarriers at baseband and then modulating the carrier. The design of the PCM/FM + FM/FM system involves six design considerations:

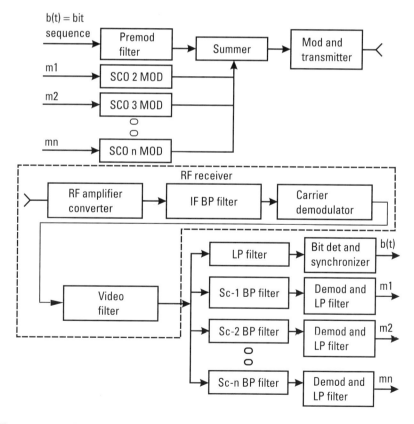

Figure 16.1 A PCM/FM + FM/FM multiplex system.

1. Filter specifications;

2. Interference of the baseband bit sequence spectrum on the FM/FM subcarrier channels;

3. Impact of the interference term on the signal-to-noise ratio in the channel output;

4. Signal-to-noise ratios for composite modulation;

5. Relative magnitude of the PCM sequence and of the highest frequency subcarrier carrier deviation;

6. The impact of the subcarriers on the BER of the bit sequence.

16.2.1   Filter Specification

The premodulation filter for the PCM of the PCM/FM + FM/FM system is usually a six-pole Bessel filter in contrast to PCM/FM, where the filter is usually a three-pole to six-pole filter. The reason for the higher-order filter in the composite system is not only to limit the RF spectrum but to limit the interference in the subcarrier channels. The corner frequency of the six-pole filter should be between $0.5R_b$ and R_b. The 3-dB point of the lowpass filter, $f_{3dB}(lp)$, used after the carrier demodulator to separate the bit sequence should be set equal to the R_b. Both filters are shown in Figure 16.1. The video filter must be set to pass the highest-frequency subcarrier and half of the modulation spectrum of the highest-frequency subcarrier with minimum attenuation, just as in FM/FM.

16.2.2   Interference

Not only must the signal-to-noise ratios be maintained in each channel by considering the noise component, but the effect of the baseband sinc f spectrum of the PCM bit sequence on the subcarriers must be analyzed. The interference in the lowest-frequency subcarrier channel bandpass filter should be approximately 35 dB below the power of the subcarrier. This will create less than 0.2 dB degradation in the lowest frequency subcarrier channel output signal-to-noise ratio. Typically, the lowest frequency subcarrier is chosen such that the interference due to the sinc f spectrum is acceptably small. The spectrum of this modulation format after the summer of Figure 16.1 is shown in Figure 16.2.

The spectrum of Figure 16.2 is the result of combining a 25-Kbps bit stream and four subcarriers from the CBW channels, starting with the 56-kHz subcarrier. Inspection of Figure 16.2 shows that at the receiver a fraction of

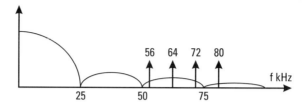

Figure 16.2 Spectrum of PCM/FM + FM/FM.

the spectrum of the bit sequence will pass through the bandpass filter and interfere with the message demodulated from the subcarrier. The spectrum of the bit sequence (assumed to be unfiltered by the premodulation filter) is given by the sinc fT_b function, where T_b is the bit period.

The following approach may be used to determine the power of the interference. After carrier demodulation, the unfiltered spectrum of the bit sequence is given by

$$S_b(f) = (\Delta f)^2 T_b \operatorname{sinc}^2 fT_b \qquad (16.1)$$

where,

T_b = bit period;

Δf = deviation of the carrier by the bit sequence.

If the bit sequence is filtered by a premodulation filter to limit the RF spectrum, as it normally is, then the interfering spectrum is given by

$$S_I(f) = |H_{pm}(f)|^2 (\Delta f)^2 T_b \operatorname{sinc}^2 fT_b \qquad (16.2)$$

where $H_{pm}(f)$ is the transfer function of the premodulation filter.

The premodulation filter for PCM/FM + FM/FM is usually a six-pole Bessel filter in contrast to PCM/FM, where the filter is usually a single-pole filter. The actual interference power, P_I, will be the power of $S_I(f)$ passed by the bandpass filter separating the lowest frequency subcarrier and is given by

$$P_I = 2 \int_{f_1}^{f_2} |H_{bp}(f)|^2 |H_{pm}(f)|^2 (\Delta f)^2 T_b \operatorname{sinc}^2 fT_b \, df \qquad (16.3)$$

where $H_{bp}(f)$ is the transfer function of the bandpass filter for the lowest-frequency subcarrier and f_1 and f_2 are the lower and upper 3-dB points. Note that a two precedes the integral since the spectrums are two-sided and symmetrical about zero.

The power of the lowest frequency subcarrier is given by

$$P_L = \frac{f_{dL}^2}{2}|H_{bp}(f)|^2 \tag{16.4}$$

where f_{dL} = deviation of the carrier by lowest frequency subcarrier.

The ratio of P_I to P_L should be less than -35 dB. The expression of (16.3) is tedious to evaluate in its integral form; however, an approximate expression for the integral may be obtained by assuming the two filter transfer functions and the power spectrum of the bit stream are constant at the frequency of the lowest frequency subcarrier. This assumption is very good except at the null points of the spectrum [1, 2, 3]. Making this assumption, (16.3) becomes

$$P_I = 2|H_{bp}(f_{sc})|^2|H_{pm}(f_{sc})|^2(\Delta f)^2 T_b \,\text{sinc}^2 f_{sc} T_b \int_{f_1}^{f_2} df \tag{16.5}$$

$$= 2|H_{bp}(f_{sc})|^2|H_{pm}(f_{sc})|^2(\Delta f)^2 T_b \,\text{sinc}^2(f_{sc}T_b)(f_2 - f_1)$$

Let R be the interference ratio in dB, then

$$R = 10 \log P_I/P_L \tag{16.6}$$

Substituting (16.4) and (16.5) into (16.6) gives,

$$R = 10 \log\left(\frac{2\Delta f}{f_{dL}}\right)^2 + 10 \log(T_b \,\text{sinc}^2(f_{sc}T_b)) + 10 \log|H_{pm}(f_{sc})|^2$$

$$+ 10 \log(f_2 - f_1) \tag{16.7}$$

$$= 20 \log\frac{2\Delta f}{f_{dL}} + 20 \log(\sqrt{T_b}\,\text{sinc}\, f_{sc}T_b) + 20 \log|H_{pm}(f_{sc})|$$

$$+ 10 \log(f_2 - f_1)$$

Noting that by Carson's rule, the bandwidth of the subcarrier bandpass filter is given by

$$B_{bpsc} = (f_2 - f_1) = 2f_{ds}$$

Using this, (16.7) becomes

$$R = 20 \log \frac{2\Delta f}{f_{dL}} + 20 \log(\sqrt{T_b} \operatorname{Sinc} f_{sc} T_b) + 20 \log|H_{pm}(f_{sc})| \quad (16.8)$$
$$+ 10 \log 2f_{ds}$$

The attenuation due to the premodulation filter may be found from a plot (Figure 16.3) of the transfer function normalized to the 3-dB point of the filter. That is, the attenuation of the normalized filter should be evaluated at

$$f = \frac{f_{sc}}{f_{3dB}}$$

The actual value, I_3(dB), of attenuation for this value of normalized interfering frequency is read from the ordinate in Figure 16.3.

16.2.3 Impact of the Interference on the FM/FM Channel's Output $[S/N]_o$

This case may be considered to be the bandpass-lowpass FM discriminator case and the equations developed accordingly. The value of the spectrum of

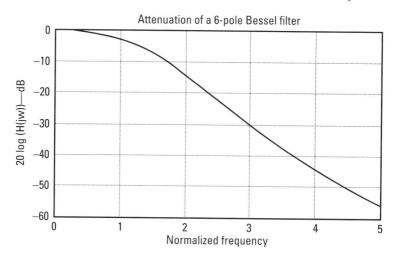

Figure 16.3 Attenuation of a lowpass 6-pole Bessel filter versus frequency normalized to 3-dB frequency.

the interference at the center of the subcarrier bandpass filter can be thought of as white noise with the PSD constant of the white noise given by

$$N_o = \eta = S_I(f_{scL}) \tag{16.9}$$

where

$$S_I(f) = 2|H_{bp}(f_{sc})|^2 |H_{pm}(f_{sc})|^2 (\Delta f)^2 T_b \, \text{sinc}^2 f_{sc} T_b$$

Note that in f_{scL}, the subscript L is for the lowest-frequency subcarrier. Figure 16.4 depicts this spectrum and concept.

For thermal noise and for the bandpass to lowpass FM discriminator case, the output signal-to-noise is given by [see (2.42)].

$$[S/N]_o = 3D^2 \overline{m(t)^2} \frac{B_{IF}}{f_m} [S/N]_i \tag{16.10}$$

Equation (16.10) is developed by finding the output noise power for the case of carrier plus noise; then the output signal power is found for the case of a carrier frequency modulated by a message with no noise present. Using the explicit parameters to expand this equation gives

$$[S/N]_o = \frac{3f_{dsi}^2}{f_{mi}^2} \frac{B_{IF}}{f_m} \frac{A^2}{2B_{IF}N_o} \overline{m(t)^2} \tag{16.11}$$

Letting $\overline{m(t)^2}$ be the signal and all else in the expression be the thermal noise, N_r, (16.11) can be written as

$$[S/N]_o = \frac{\overline{m(t)^2}}{N_r} \tag{16.12}$$

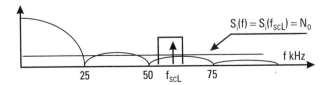

Figure 16.4 Approximate white noise spectrum derived from the interference spectrum.

The equation and the concept for output signal-to-noise for the thermal noise case may be used to predict the signal-to-interference ratio for the interference case by treating the interference as white noise and the lowest frequency subcarrier as the carrier, and noting that B_{IF} becomes B_i, the bandwidth of the subcarrier bandpass filter.

In this case, the index is changed from the ith to L, for the lowest-frequency subcarrier. (This model holds unless the subcarrier is within 10% of a zero crossing of the sinc PSD of the random bit sequence; then this method will predict more noise than is present.)

Using (16.9) and (16.11) and multiplying numerator and denominator by 2, the output signal-to-noise (interference) becomes

$$[S/N]_{oI} = \frac{3}{2} \frac{f_{dsL}^2}{f_{mL}^2} \frac{B_L}{f_{mL}} \left[\frac{f_{dsL}^2}{2B_L S_I(f_{scL})} \right] \frac{\overline{2m(t)^2}}{1} \qquad (16.13)$$

The bracket term in the expression is the interference ratio after the integral approximation is made. Thus, (16.13) becomes

$$[S/N]_{oI} = \frac{3}{2} \frac{f_{dsL}^2}{f_{mL}^2} \frac{B_L}{f_{mL}} \frac{1}{R} \frac{\overline{2m(t)^2}}{1} \qquad (16.14)$$

For single-tone modulation, this expression becomes

$$[S/N]_{oI} = \frac{3}{2} \frac{f_{dsL}^2}{f_{mL}^2} \frac{B_L}{f_{mL}} \frac{1}{R} \qquad (16.15)$$

Equation (16.15) determines the signal-to-noise out for the subcarrier channel in this model since the subcarrier is a single tone. The following derivation uses $m(t)^2$ in order to show how two independent signal-to-noise ratios combine in general, but if sinusoids were used, then $m(t)^2$ would be replaced by a 1/2 in the following without any loss in generality. Letting $m(t)^2$ be the signal and all else in the expression be the interference noise N_I, (16.14) can be written as

$$[S/N]_{oI} = \frac{\overline{m(t)^2}}{N_I} \qquad (16.16)$$

Denote the output signal-to-noise ratios for the thermal and the interference cases as $[S/N]_{or}$ and $[S/N]_{oI}$ respectively. Since the final signal-to-noise is the result of degradation due to both the thermal noise and the interference, which are independent processes, and the message is the same for the same subcarrier channel, the final signal-to-noise ratio, $[S/N]_f$, can be expressed as follows;

$$[S/N]_f = \frac{\overline{m(t)^2}}{N_r + N_I} = \frac{1}{\dfrac{N_r + N_I}{\overline{m(t)^2}}}$$

$$= \frac{1}{\dfrac{N_r}{\overline{m(t)^2}} + \dfrac{N_I}{\overline{m(t)^2}}} \tag{16.17}$$

$$[S/N]_f = \frac{1}{\dfrac{1}{[S/N]_{or}} + \dfrac{1}{[S/N]_{oI}}}$$

Equation (16.17) is true for any final signal-to-noise that is the result of two independent noise processes in a single channel.

16.2.4 Signal-to-Noise Ratios for Composite Modulation

In the PCM/FM + FM/FM composite modulation format, the signal-to-noise ratios in the FM/FM subcarrier channels are computed first based upon the design procedure for FM/FM. Adding PCM/FM at baseband usually does not significantly increase the required IF bandwidth for the computation of the signal-to-noise ratios in the subcarrier channels. However, after the interference term is determined, (16.17) must be used to determine the reduced and final output signal-to-noise ratio in the lowest-frequency subcarrier channel.

Further, in the equation predicting the signal-to-noise ratio in the PCM channel, the IF bandwidth term will be significantly increased. This will increase the output signal-to-noise ratio in the PCM channel. Increasing the IF bandwidth will increase the noise power; hence, the carrier power must be increased to maintain the minimum required $[S/N]_i$ of 12 dB. That is, $[S/N]_i$ will then remain the same, but B_{IF} will increase. In going to the PCM/FM + FM/FM format, then, the $[S/N]_{Opcm}$ will usually be greater than the minimum 15 dB required. Therefore, Δf can have a value between $0.35R_b$ and $0.41R_b$, but Section 16.2.5 must also be considered with respect to Δf.

16.2.5   Carrier Deviation by the PCM Sequence and the Highest-Frequency Subcarrier

It is necessary to compare the relative magnitude of the carrier deviation by the PCM sequence and the carrier deviation by the highest frequency subcarrier. The deviation of the carrier by the bit sequence should be at least one 1/10 the deviation of f_{dn} or one 1/10 the deviation of the carrier by the highest frequency subcarrier, whichever is larger. This is necessary to take care of small parameter variations, intermodulation products, and the dynamic range of the receiver.

16.2.6   Impact of the Subcarriers on the BER of the Bit Sequence

In considering the impact of the subcarriers on the BER of the bit sequence, it has been found that if $R < -35$ dB, the effect will be minimal. For a more detailed analysis of this problem, see [1–3].

To minimize the BER problem, two receivers can be used to receive PCM/FM + FM/FM with the bandwidth of one set just wide enough to receive the PCM/FM only. For more information on the two-receiver case, see [4]. The two-receiver case will reduce the impact of the subcarriers on the bit sequence and is used frequently. However, the receiver set to receive the FM/FM package must also pass the bit sequence and the interference of the bit sequence on the FM subcarriers must be considered.

16.3   PCM/FM + FM/FM Design Example

A design example will be worked using the above concepts and the resulting mathematical models. Although the example will use CBW channels, the procedure is equally applicable to proportional channels.

16.3.1   Constant Bandwidth Channels

The design of the PCM/FM + FM/FM package will start with the 14 CBW subcarriers used in Chapter 3. It will be assumed that the 14 subcarriers meet the sensor requirements and that the preemphasis schedule designed in Chapter 3 is satisfactory with respect to bandwidth utilization and output signal-to-noise ratios.

Example 1: CBW Subcarriers

The subcarrier frequencies start with 56 kHz and go to 160 kHz. Table 3.6 shows derived parameters. A 25-Kbps bit sequence will be added to the subcarriers at baseband. The spectrum of the four lowest subcarriers and the bit sequence is shown in Figure 16.2.

In the design procedure of a composite PCM/FM + FM/FM modulation format, the first step is the design of the FM/FM portion, as was done in Chapter 3. The specifications for the FM/FM portion consisting of the 14 subcarriers will be repeated here for convenience.

Specified parameters for the FM/FM modulation:

$$[S/N]_c = 12 \text{ dB (3.98 numerical);}$$
$$D_{si} = 5 \text{ nominal;}$$
$$[S/N]_{oi} = 46 \text{ dB (199 numerical) for subcarrier channels.}$$

Specified parameters for the PCM/FM modulation:

$$[S/N]_c = [S/N]_i = 12 \text{ dB (3.98 numerical);}$$
$$[S/N]_{opcm} = 15 \text{ dB in the bandwidth of the LP filter;}$$
$$R_b = 25 \text{ Kbps separating the bit sequence;}$$
$$\Delta f = 0.41 R_b \text{ (anticipating the need for } \Delta f \geq f_{dn}).$$

It will be necessary to compute the bit sequence interference in the lowest-frequency subcarrier channel using (16.8). The deviation, f_{dL}, of the carrier by the lowest-frequency subcarrier, say f_{s56}, is taken from Table 3.5, and is 17.065 kHz. The premodulation filter for the PCM channel will have a 3-dB frequency of 0.7 $R_b = (0.7)(25 \text{ kHz}) = 17.5$ kHz. The lowpass filter following the carrier discriminator will have $f_{3dB}(lp) = R_b = 25$ kHz.

Step 1

Compute the interference ratio R using (16.8).

$$R = 20 \log \frac{2\Delta f}{f_{dL}} + 20 \log(\sqrt{T_b} \text{ sinc } f_{sc} T_b) + 20 \log|H_{pm}(f_{sc})| + 10 \log 2f_{ds}$$

The first term is given by $I_1 = 20 \log \dfrac{2\Delta f}{f_{dL}}$, where

$\Delta f = 0.41 R_b = (0.41)(25,000) = 10.25 \text{ kHz}$

$I_1 = 20 \log \dfrac{2\Delta f}{f_{dL}} = +1.6 \text{ dB}$

$I_2 = 20 \log \sqrt{T_b} \operatorname{sinc} f_{sc} T_b$

$\quad = 20 \log(0.006 \operatorname{sinc}(56/25)) = 20 \log((0.006)(0.085))$

$\quad = -65.8 \text{ dB}$

For the normalized six-pole filter with the 3-dB corner frequency at 0.7 R_b, I_3 is given by

$I_3 = 20 \log |H_{pm}((f_{sc})/.7R_b)| = 20 \log H_{pm}(56/17.5)$

$\quad \text{(from Figure 16.3)} = -33 \text{ dB}$

$I_4 = 10 \log 2 f_{ds} = +36 \text{ dB}$

$R = I_1 + I_2 + I_3 + I_4 = -61 \text{ dB}$

This low value of interference indicates that the design is satisfactory with a large margin to work with. If another subcarrier was needed, Step 1 would be repeated for the next lower subcarrier. Alternately, the bit rate might also be increased.

Step 2

Compute the new required bandwidth, f_{dn}.

From Chapter 2, the estimate of required bandwidth is given by

$$B_c = 2[f_{dn} + f_{s1}] \qquad (16.18)$$

where

f_{dn} = rms of the deviations of the carrier by each subcarrier;

f_{s1} = highest frequency subcarrier.

That is,

$$f_{dn} = \sqrt{f_{dc1}^2 + f_{dc2}^2 + + + f_{dcn}^2} \qquad (16.19)$$

For the composite modulation format, f_{dnc} becomes

$$f_{dnc} = \sqrt{(\Delta f)^2 + f_{dc1}^2 + f_{dc2}^2 + + + f_{dcn}^2} \tag{16.20}$$

$$f_{dnc}^2 = (\Delta f)^2 + [f_{dc1}^2 + f_{dc2}^2 + + + f_{dcn}^2] \tag{16.21}$$

The bracketed term in (16.21) is f_{dn}. From Design Example 3 in Chapter 3, the value of f_{dn} is 128.7 kHz and gives

$$f_{dnc}^2 = (\Delta f)^2 + (128.7)^2 = (10.25)^2 + (128.7)^2 \tag{16.22}$$

$$= 16,668.8$$

$$f_{dnc} = 129.1 \text{ kHz} \tag{16.23}$$

The new estimate of bandwidth becomes

$$B_c = 2(f_{dnc} + f_{s1}) = 2(129.1 + 160) = 578 \text{ kHz} \tag{16.24}$$

The 578 kHz is a negligible increase over the 577 kHz of the FM/FM system for the transmission or IF bandwidth.

Since the IF bandwidth for the FM/FM system did not increase significantly, it is not necessary to recalculate the signal-to-noise ratios in the subcarrier channels. It will be necessary to calculate the signal-to-noise ratio for the PCM channel

Step 3

Calculate the $[S/N]_o$ for the PCM/FM channel.

Repeating (6.18) in voltage form gives

$$[S/N]_{opcm} = \frac{\Delta f}{f_m} \sqrt{\frac{3B_{IF}}{2f_m}} [S/N]_i \tag{16.25}$$

The parameters appropriate for use of (16.25) in the PCM/FM + FM/FM case are as follows:

$$\Delta f = 0.41 R_b = 0.41(25 \text{ kb/sec}) = 10.25 \text{ kHz} \tag{16.26}$$

$$\sqrt{B_{IF}} = \sqrt{578} \tag{16.27}$$

$$f_m = f_{3dB}(LP) = R_b = 25 \text{ kHz} \tag{16.28}$$

Substituting (16.26), (16.27) and (16.28) into (16.25) gives

$$[S/N]_{opcm} = 5.54 \text{ (numerical)}$$
$$= +14.88 \text{ dB}$$

The $[S/N]_{opcm}$ of the PCM channel in the composite modulation system is usually above or close to the minimum value of 15 dB, since B_{IF} must be wider to allow the FM/FM portion through, but this also allows more noise. Hence, the S_i, which is the input carrier power, must be increased to maintain the minimum $[S/N]_i$ required.

Step 4

Determine the final output signal-to-noise ratio for the lowest-frequency subcarrier channel using (16.17). Equation (16.8) must first be used to determine the interference ratio, R, and then (16.15) used to determine the signal-to-interference ratio in the lowest-frequency subcarrier channel.

$$[S/N]_{oI} \text{ (dB)} = 10 \log \frac{3}{2} \frac{f_{dsL}^2}{f_{mL}^2} \frac{B_L}{f_{mL}} \frac{1}{R}$$

The numerical version of this equation is

$$[S/N]_{oI} = 1.22 D_{sL} \sqrt{\frac{B_L}{f_{mL}}} \frac{1}{R}$$

The parameters for this channel are

$D_{sL} = 5;$

$B_L = 2f_{dsL} = 2(2 \text{ kHz}) = 4 \text{ kHz};$

$f_{mL} = 400 \text{ Hz};$

$\dfrac{1}{R} = 1,258.$

Then

$$[S/N]_{oI} = (1.22)(5)\sqrt{10}\,(1,258) = 24,267 \text{ (voltage)}$$
$$= (24,267)^2 = 5.8 \times 10^8 \text{ power}$$

The system was designed such that the thermal signal-to-noise was 46 dB or 199 in terms of voltage and 3.96×10^4 power ratio. Using (16.16), and noting that this equation requires power additions, gives

$$[S/N]_f = \frac{1}{\dfrac{1}{[S/N]_{or}} + \dfrac{1}{[S/N]_{oI}}}$$

where

$$1/[S/N]_{or} = 1/39{,}600 = 0.253 \times 10^{-4}$$

and

$$1/[S/N]_{oI} = 1/5.8 \times 10^8 = 0.171 \times 10^{-8}$$

Then

$$[S/N]_f = \frac{1}{0.253 \times 10^{-4} + 0.171 \times 10^{-8}} \approx \frac{1}{0.253 \times 10^{-4}}$$

$$= 3.96 \times 10^4$$

or

$$[S/N]_f = 10 \log[3.96 \times 10^4] = 46 \text{ dB}$$

Had the final design of the system been such that the interference term was −35 dB, the decrease in the desired $[S/N]_o$ of 46 dB would be as follows:

$$1/[S/N]_{or} = 1/39{,}600 = 0.253 \times 10^{-4}$$

and

$$1/[S/N]_{oI} = 1/(1{,}084)^2 = 8.5 \times 10^{-7}$$

$$[S/N]_f = \frac{1}{\dfrac{1}{[S/N]_{or}} + \dfrac{1}{[S/N]_{oI}}} = \frac{1}{0.253 \times 10^{-7} + 8.5 \times 10^{-7}}$$

$$= 3.85 \times 10^4$$

$$[S/N]_f = 10 \log 3.85 \times 10^4 = 45.9 \text{ dB}$$

Requiring a −35-dB interference term will decrease the specified signal-to-noise or calculated signal-to-noise ratio by about 0.1 dB.

Step 5

Specify the premodulation filter and lowpass filter following the carrier discriminator bandwidths in the PCM channel. The premodulation filter should be a six-pole Bessel filter with a 3-dB down point of $0.7R_b = 17.5$. The lowpass filter should have $f_{3dB}(lp) = R_b = 25$ kHz. In the analysis this was the assumed value of the 3-dB frequency.

Step 6

Check relative magnitude of the carrier deviation by the PCM sequence Δf, and f_{dn} and the carrier deviation by the highest-frequency subcarrier. The deviation of the carrier by the PCM sequence should not be less than 1/10 the norm of the deviation, f_{dn}, since it is larger than the deviation by the highest-frequency subcarrier. For the composite modulation format, $f_{dnc} = 125$ kHz and 1/10 is 12.5 kHz. Since $\Delta f = 10.25$ kHz, it is probably acceptable.

Summary

The parameters of the FM/FM portion are listed in Table 3.6. The PCM/FM parameters are as follows:

$R_b = 25$ kb/s;

$\Delta f = 0.41R_b = 0.41$ (25 kb/s) $= 10.25$ kHz;

$f_{3dB}pm$ (6-pole Bessel) $= 0.7R_b = 17.5$ kHz;

$f_{3dB}lp = R_b = 25$ kHz (LP filter to separate the bit sequence).

16.4 PCM/FM/FM

The design of a system with a bit sequence modulated directly onto a subcarrier, summed with other frequency modulated subcarriers and the resultant frequency modulated onto a carrier, will be considered next. This modulation format is referred to as PCM/FM/FM. The design procedure for this type of composite modulation involves four steps.

1. Select the subcarriers that will meet the FM/FM requirements.

2. Select an additional subcarrier that will satisfy the PCM requirement and determine the compatibility of the selected subcarriers. In this type of system, the channel with the PCM modulated subcarrier will have a bit synchronizer and frame synchronizer following the subcarrier discriminator (SCD) and lowpass filter.

3. Follow the FM/FM design procedure to set the carrier deviation by the subcarriers, including the one to be modulated by the data sequence, to meet the signal-to-noise requirements in the channels.

4. Design the PCM channel to meet or exceed the minimum 15-dB requirement of the bit synchronizer.

16.4.1 Selecting the Subcarriers for FM/FM

The selection procedure for the subcarriers to handle the analog signals is based upon the frequency requirements of sensors. This may mean selecting constant or proportional bandwidth subcarriers or a combination of the two.

16.4.2 Selecting the PCM Subcarrier Channel

In selecting the subcarrier channel to carry the bit sequence, there are two parameters to consider:

1. The peak deviation of the subcarrier by the bit sequence should not exceed f_d, the value specified by the IRIG specifications. That is, $0.35R_b = \Delta f$ should be equal to or less than f_d, the IRIG specified maximum specified deviation.

2. The subcarrier channel should be able to handle f_{3dB}, the 3-dB frequency of the premodulation filter for the bit stream; f_{3dB} should be between $0.5R_b$ and R_b. In the IRIG specifications, the maximum frequency a subcarrier channel can handle is determined assuming the modulation index is 1. However, for PCM/FM/FM, the modulation index is 0.7. Although this allows the channel to handle a larger frequency, it may require a nonstandard lowpass filter in the output of the subcarrier channel.

Since the subcarrier discriminator lowpass filter should be set somewhere between R_b and $2R_b$, it is desirable to assume a mod index of 1 and to select the channel to carry the data such that

$$R_b \leq f_m \qquad (16.29)$$

where f_m is the maximum frequency the channel will handle with a mod index of 1. This is also the 3-dB frequency of the subcarrier discriminator lowpass filter.

The compatibility of the PCM subcarrier and the FM/FM subcarriers is not a problem if the two requirements above are met.

16.4.3 FM/FM Preemphasis Design

The third step involves designing the preemphasis schedule in accordance with the design procedures developed in Chapter 3. This design includes the subcarrier or subcarriers selected to handle the bit sequence.

16.4.4 Designing for the Output Signal-to-Noise in the PCM Channel

In achieving the design objective of Step 4, it is necessary to determine the signal-to-noise ratio in the PCM channel. Since this is an FM/FM case and the bit sequence is filtered, producing in effect an analog signal modulating the subcarrier, VCO, (2.30) will be used to determine the signal-to-noise ratio for the PCM channel with the relevant parameters. Repeating (2.30) as (16.30), the equation with the FM/FM parameters is

$$[S/N]_{oi} = \sqrt{3/4} \, \frac{f_{dsi}}{f_{3p}} \, \frac{f_{dci}}{f_{si}} \, \frac{\sqrt{B_c}}{\sqrt{f_{3p}}} \, [S/N]_c \qquad (16.30)$$

where

> f_{dsi} = deviation of the ith subcarrier by the ith message and in this case the bit sequence;

> f_{dci} = deviation of the carrier by the ith subcarrier;

> f_{si} = center frequency of the ith subcarrier;

> B_c = center IF bandwidth;

> D_{ci} = modulation index of the carrier and the ith subcarrier;

$[S/N]_c$ = carrier-to-noise in the carrier IF;

> f_{3p} = 3-dB point of the lowpass filter following the PCM channel SC;

> D_{si} = modulation index of subcarrier and message in the ith channel. Note that this term is missing because the SCD lowpass output filter 3-dB point will be set at a higher frequency than the maximum frequency of the predetection filter in the PCM channel.

In the design procedure, the only parameter available to satisfy the specified $[S/N]_{oi}$ in the ith channel is f_{dci}. In the FM/FM subcarrier preemphasis schedule design procedure, f_{dci} was determined in order to satisfy the specified $[S/N]_{oi}$, say, 46 dB. A value of 5 was assumed for D_{si}, which will certainly decrease in the PCM channel. In fact, several things will change in the channel where the subcarrier is modulated by PCM. The modulation index of the bit sequence on the subcarrier, D_{si}, is replaced by the ratio f_{dsi}/f_{3p} which will be less than 5, probably around 1. Certainly, the lower ratio will lead to a lower $[S/N]_{oi}$, but a lower $[S/N]_{oi}$ is adequate in the PCM channel since the required signal-to-noise output ratio in this channel is only 15 dB going into the bit synchronizer. In general, the preemphasis schedule will specify a D_{ci} that will give a signal-to-noise ratio in the PCM channel that will meet or exceed this requirement.

The new parameter definitions or specifications are as follows:

$f_{dsi} = \Delta f = 0.35R_b$ = deviation of the ith subcarrier by the bit sequence;

$f_{3p} = R_b$ = 3 dB point of the lowpass filter following the PCM channel SCD;

f_{dci} = (determined by the FM/FM design) deviation of the carrier by the ith subcarrier;

f_{si} = center frequency of the ith subcarrier;

B_c = (determined by the FM/FM design) center IF bandwidth;

$f_{dsi}/f_{3p} = \Delta f/f_{3p} = 0.35R_b/R_b = 0.35$;

D_{ci} = (determined by the FM/FM design) modulation index of the carrier and the ith subcarrier;

$[S/N]_c$ = carrier-to-noise in the carrier IF;

16.4.5 PCM/FM/FM Design Example

An example of PCM/FM/FM design will be worked using 15 PBW subcarriers. It will be assumed that the first 14 channels for the FM/FM package were selected based on sensor frequency requirements.

Example 2: PBW Channels

Requirements:

Fourteen FM/FM subcarrier channels to carry analog data from 14 sensors;
One PCM/FM channel to handle a bit rate of 5 Kbps;
$[S/N]_{oi}$ in the 14 subcarrier channels = 46 dB;
$[S/N]_{oPCM} > 15$ dB.

Step 1

Determine the set of subcarriers needed based upon the data transmission requirement. It will be assumed that in the analysis of sensor data to be transmitted it was determined that fourteen 7.5% PBW subcarriers would be needed for the FM/FM portion and one subcarrier for the PCM channel. The 14 subcarriers will be the 1.3 kHz to the 70 kHz and the PCM channel will use the 93-kHz subcarrier.

Step 2

The 93-kHz subcarrier was selected for the PCM channel based upon the following process.

1. One requirement is that the bit rate be 5 Kbps; therefore, for a minimum BER the peak frequency deviation should be

$$\Delta f = 0.35 R_b = 0.35(5 \text{ kb/s}) = 1.75 \text{ kHz}$$

Inspection of the IRIG Tables shows that the 93-kHz subcarrier with a peak deviation of 6.975 kHz will certainly handle this.

2. From the IRIG tables, for an assumed mod index of 1, the maximum frequency the channel can handle is

$$f_{max} = 6.975 \text{ kHz}$$

This will be the 3-dB point for a standard SCD lowpass filter. By (16.29) this point should be between R_b and $2R_b$. This limits the maximum bit rate to 6,975 Kbps. This channel will handle the 5-Kbps data rate.

The 3-dB point of the SCD postdetection lowpass filter should be set to

$$f_{3p} = R_b = 5 \text{ kHz}$$

Step 3

Design the PCM/FM/FM system such that all 15 channels meet the minimum requirement of 46 dB. (It should be noted that only the lowest 14 channels will ultimately meet the 46-dB specification, since the channel with the bit sequence will end up with a lower mod index.) This process was completed in Examples 1 and 2, in Chapter 2. In Step 4 of selecting the subcarrier channel, a single-pole RC filter, with $f_{3dB} = R_b$, was used; if the bit rate is increased to the limit, a higher-order linear phase filter should be used.

Step 4

Determine the $[S/N]_{oPCM}$ for the PCM channel.

The parameters relevant to (16.30) are as follows.
For the PCM/FM channel:

f_{3dB} = (0.5 to 1) R_b (Premodulation filter, single-pole RC);

f_{3-93} = R_b = 3 dB frequency of the SCD LP filter;

$$\frac{f_{dsi}}{f_{3p}} = \frac{f_{ds93}}{f_{3-93}} = \Delta f / f_{3-93} = 0.35 R_b / R_b = 0.35.$$

For the FM/FM design:

$$D_{c-93} = 0.718;$$
$$B_c = 362 \text{ kHz};$$
$$[S/N]_c = 12 \text{ dB (3.98 numerical)};$$

$$[S/N]_{oPCM} = \sqrt{3/4} \, \frac{f_{ds93}}{f_{3-93}} \, D_{ci} \, \frac{\sqrt{B_c}}{\sqrt{f_{3-93}}} \, [S/N]_c;$$

$$= (0.866)(0.35)(0.718)(8.5)(3.89) = 7.19.$$

Then, in decibels,

$$[S/N]_{oPCM} \text{ (dB)} = 20 \log 10.16 = 17.14 \text{ dB}$$

The PCM channel signal-to-noise meets the minimum requirement of 15 dB.

Step 5

Review the composite modulation spectrum.

From (6.7), an estimate of the bandwidth (BW) of the bandpass filter of the 93-kHz subcarrier modulated by the PCM sequence is given by

$$\text{BW of 93 kHz subcarrier BP filter} = R_b.$$

The 3-dB points of the filter will be $93 \pm R_b/2 = 90.5$ kHz and 95.5 kHz. An estimate of the bandwidth of the bandpass filter of the 70-kHz subcarrier modulated by an analog signal is

$$BW_{70} = 2(f_{ds70} + f_m) = 2(5.25 + 1.05) = 12.6 \text{ kHz}$$

The 3-dB points of this BP filter will be $70 \pm 12.6/2 = 63.7$ kHz and 76.3 kHz.

There is not an appreciable overlap in the spectrums of the 93-kHz PCM modulated subcarrier and the next subcarrier, the 70-kHz subcarrier.

Summary

All channels meet the signal-to-noise specified ratios and the composite spectrum is compatible.

Problems

Problem 16.1

Convert the constant bandwidth subcarrier FM/FM package of Problem 3.11 into an FM/FM + PCM/FM by adding a 25-Kbps bit sequence. For the bit sequence, let $\Delta f = 0.41 R_b$, f_{3dB} (pm) $= 0.7 R_b$, and f_m (3-dB point of the lowpass filter separating the bit sequence) $= R_b$.

    Determine the following: (a) the interference ratio R, (b) f_{dn}, (c) B_c, (d) the composite $[S/N]_o$ for the lowest-frequency subcarrier channel, and (e) $[S/N]_{opcm}$. Compare Δf to f_{dnc}.

Problem 16.2

Repeat Problem 16.1, but continue to increase the bit rate until the interference ratio R is close to -35 dB.

Problem 16.3

Convert the constant bandwidth subcarrier FM/FM package of Problem 3.12 into a FM/FM + PCM/FM by adding a 25-Kbps bit sequence. For the bit sequence, let $\Delta f = 0.41 R_b$, f_{3dB} (pm) $= 0.7 R_b$ and six order filter, and f_m (3-dB point of the lowpass filter separating the bit sequence) $= R_b$. Determine: (a) the interference ratio R, (b) f_{dn}, (c) B_c, (d) the composite $[S/N]_o$ for the lowest frequency subcarrier channel, and (e) $[S/N]_{opcm}$. Compare Δf to f_{dnc}.

Problem 16.4

Repeat Problem 16.3, but continue to increase the bit rate until the interference ratio R is close to −35 dB.

Problem 16.5

Convert the constant bandwidth subcarrier FM/FM package of Problem 3.13 into a FM/FM + PCM/FM by adding a 25-Kbps bit sequence. For the bit sequence, let $\Delta f = 0.41 R_b$, f_{3dB} (pm) = $0.7 R_b$, and f_m (3-dB point of the lowpass filter separating the bit sequence) = R_b. Determine the following: (a) the interference ratio R, (b) f_{dn}, (c) B_c, (d) the composite $[S/N]_o$ for the lowest frequency subcarrier channel, and (e) $[S/N]_{opcm}$. Compare Δf to f_{dnc}.

Problem 16.6

Repeat Problem 16.5, but (a), use a first-order premodulation filter and determine the interference ratio R, and (b) continue to increase the bit rate until the interference ratio, R, is close to −35 dB.

Problem 16.7

Convert the constant bandwidth subcarrier FM/FM package of Problem 3.14 into a FM/FM + PCM/FM by adding a 25-Kbps bit sequence. For the bit sequence, let $\Delta f = 0.41 R_b$, f_{3dB} (pm) = $0.7 R_b$, and f_m (3-dB point of the lowpass filter separating the bit sequence) = R_b. Determine: (a) the interference ratio R, (b) f_{dn}, (c) B_c, (d) The composite $[S/N]_o$ for the lowest frequency subcarrier channel, and (e) $[S/N]_{opcm}$. Compare Δf to f_{dnc}.

Problem 16.8

Repeat Problem 16.7, but use a first-order premodulation filter and determine the interference ratio, R.

PCM/FM/FM Design Problems

Design Problem 16.9

Convert the proportional bandwidth subcarrier FM/FM package of Problem 3.1 into a PCM/FM/FM by adding a 6-Kbps bit sequence to the

93-kHz subcarrier. Set the parameters, $\Delta f = 0.35R_b, f_{3dB}$ (pm) $= 0.5R_b$, and $f_m = f_{3p} = R_b$. Determine the following: (a) $[S/N]_{opcm}$, and (b) modulated spectrums of the 93- and the 70-kHz subcarriers compatibility.

Design Problem 16.10

Convert the proportional bandwidth subcarrier FM/FM package of Problem 3.2 into a PCM/FM/FM by adding a 6-Kbps bit sequence to the 93-kHz subcarrier. Set the parameters, $\Delta f = 0.35R_b, f_{3dB}$ (pm) $= 0.5R_b$, and $f_m = f_{3p} = R_b$. Determine the following: (a) $[S/N]_{opcm}$, and (b) modulated spectrums of the 93- and the 70-kHz subcarriers compatibility.

Design Problem 16.11

Convert the proportional bandwidth subcarrier FM/FM package of Problem 3.3 into a PCM/FM/FM by adding a 6-Kbps bit sequence to the 93-kHz subcarrier. Set the parameters, $\Delta f = 0.35R_b, f_{3dB}$ (pm) $= 0.5R_b$, and $f_m = f_{3p} = R_b$. Determine the following: (a) $[S/N]_{opcm}$, and (b) modulated spectrums of the 93- and the 70-kHz subcarriers compatibility.

Design Problem 16.12

Convert the proportional bandwidth subcarrier FM/FM package of Problem 3.4 into a PCM/FM/FM by adding a 5-Kbps bit sequence to the 93-kHz subcarrier. Set the parameters, $\Delta f = 0.35R_b, f_{3dB}$ (pm) $= 0.5R_b$, and $f_m = f_{3p} = R_b$. Determine the following: (a) $[S/N]_{opcm}$, and (b) modulated spectrums of the 93- and the 70-kHz subcarriers compatibility.

Design Problem 16.13

Convert the proportional bandwidth subcarrier FM/FM package of Problem 3.5 into a PCM/FM/FM by adding a 6-Kbps bit sequence to the 93-kHz subcarrier. Set the parameters, $\Delta f = 0.35R_b, f_{3dB}$ (pm) $= 0.5R_b$, and $f_m = f_{3p} = R_b$. Determine the following: (a) $[S/N]_{opcm}$, and (b) modulated spectrums of the 93- and the 70-kHz subcarriers compatibility.

Design Problem 16.14

Repeat Problem 16.9, but also add a bit sequence to the 70-kHz subcarrier. Add the maximum bit rate sequence possible. Specify and determine all parameters.

Design Problem 16.15

Repeat Problem 16.10, but also add a bit sequence to the 70-kHz subcarrier. Add the maximum bit rate sequence possible. Specify and determine all parameters.

Design Problem 16.16

Repeat Problem 16.11, but also add a bit sequence to the 70-kHz subcarrier. Add the maximum bit rate sequence possible. Specify and determine all parameters.

Design Problem 16.17

Repeat Problem 16.12, but also add a bit sequence to the 70-kHz subcarrier. Add the maximum bit rate sequence possible. Specify and determine all parameters.

References

[1] Carden, F., and S. Ara, "PCM/FM + FM/FM Bit Error Rate Determination by Modeling and Simulation," *ITC Proc.* Vol. XXIX, 1993, pp. 605–613.

[2] Carden, F., and S. Ara, "Design Equations for a Specified BER for PCM/FM + FM/FM and Optimum Bandwidths," *ITC Proc.,* Vol. XXX, 1994, pp. 339–346.

[3] Carden, F., and S. Ara, *BER Determination of PCM/FM + FM/FM Systems and Coded PCM/FM Systems,* ECE Tech Report Series, No. 93-005, Las Cruces, NM: New Mexico State University, Dec. 1993.

[4] Osborne, W., and D. Whiteman, *Optimizing PCM/FM + FM/FM Systems Using IRIG Constant Bandwidth Channels,* ECE Tech Report Series, No. 92-014, Las Cruces, NM, Nov. 1992.

17

Convolutional Coding for Forward Error Correction

The designer of a digital communication system is charged with the task of providing a cost-effective system of sending data from a source to a receiver with acceptable reliability. In digital systems, data reliability is often defined in terms of BER, which is a function of E_b/N_o. The bit energy, E_b, is a function of many factors, including transmitted power, antenna gains, path length, and losses. The PSD, N_o, of the noise is determined by the receiving system and transmission channel. After link analysis (see Chapter 13), it may be determined that the system E_b/N_o is too small to give the desired BER. The telemetry system designer may not have the freedom to increase the transmitted power or the size of the antennas or to change the receiver; in other words, the system is power limited. Coding for forward error correction is an acceptable alternative. In fact, for a fixed E_b/N_o, coding is the only method of lowering the BER. Coding adds redundant bits at the transmitter in order for the receiver to perform the two tasks of error detection and error correction. A code that performs both of these tasks is referred to as a forward error correcting code. The bits added to the transmitted bit stream will increase the required bandwidth for a given information bit rate.

The two methods most often used for forward error correction are block and convolutional coding. As the name implies, a block code operates on blocks of data. Convolutional code structure is such that during the encoding process a sliding window operates continually on the bit stream, while the decoder operates on the received bit stream serially. Convolutional coding is particularly suited to space and telemetry systems that require simple and small encoders and that obtain excellent coding gain by using sophisticated decoders at the ground station. Hence, this discussion will be limited to convolutional codes and to selected examples that highlight the most relevant features. Most of the text will also be concerned with the encoding of BPSK, although Section 17.13 will give some recent results concerning PCM/FM.

Two types of errors can occur on a channel, burst errors and random errors. Convolutional codes are adept at handling random errors but not burst errors. If burst errors are anticipated, scrambling the bit sequence prior to transmission and descrambling before decoding will minimize this problem.

17.1 Learning Objectives

Upon completion of this chapter, the reader should understand the following concepts:

- Hamming distance;
- Code rates;
- Elemental convolutional encoding;
- States of an encoder;
- Trellis for a simple code;
- Algorithm for maximum likelihood decoding;
- Hard and soft decision;
- Coding gains for BPSK and PCM/FM;
- Selecting an off-the-shelf decoder.

17.2 Hamming Distance

Hamming distance and the concept of Hamming distance plays a central role in coding, in the decoding operational decisions, and in understanding the process. The Hamming distance between two binary code words, two binary code vectors, or two binary words of the same length is defined as the number of positions in which the respective elements differ. For example, consider the two code words [00011] and [00001].

The Hamming distance between these words is one. Consider the code words

[00110] and [11001]
[00000] and [11111]

The Hamming distance between the two words of each set is five. The Hamming weight of a binary code vector is the number of nonzero elements or the Hamming distance between the code word and the all-zero word.

17.3 Convolutional Encoding

Since convolutional coding may be used for forward error correction and the encoder is physically small, it is of interest to the telemetry engineer. A convolutional encoder is a finite-state machine consisting of an L-stage shift register, N modulo-2 adders, and a multiplexer to convert the output into a serial bit stream. Each modulo-2 adder is connected to particular cells of the register. The connections, in fact, determine the code. For an input bit sequence much longer than L, the code rate, r, is

$$r = 1/n$$

The constraint length, K, of a convolutional code is the number of shifts that an input bit will impact the output. For an L-stage encoder, $K = L + 1$.

17.3.1 Two-Stage Encoder

An example of a two-stage encoder, L-2, with two modulo-2 adders, and hence a rate of 1/2, is shown in Figure 17.1. For a rate of 1/2, one input

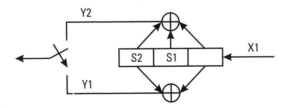

Figure 17.1 Convolutional encoder.

or information bit will produce two output or code bits. Only the last two cells of the register are considered as the encoder and determine the number of states of the encoder. The first cell of the register is shown to indicate the incoming bit.

Note that the upper modulo-2 adder is connected to each cell of the register, whereas the lower summer is connected to the first cell and the cell labeled S_2. The four states of this finite state machine are as follows: 00, 01, 10, and 11.

As the input bits enter the register, they are shifted through. Although the register goes through the four states according to the random bit input sequence, at any one time the register is in one of the four states.

17.3.2 Convolutional Code Trellis

A code trellis is a graphical representation designed to help in the study of convolutional encoders. The trellis is composed of nodes representing the states of the decoder, while connecting the nodes are branches that are the paths the decoder will follow from state to state. A one-step trellis for the convolutional encoder of Figure 17.1 is shown in Figure 17.2. The trellis shows the state the encoder is in, the next state after an input, and the output that is attached to a branch. The dashed line, a branch, leaving a node indicates the input is a 1, whereas the continuous line indicates an input 0. Note that two paths leave each node or state, since there are two possible inputs. Depending on the state of the decoder, each input bit will produce a decoder output of two bits. The particular output is attached to a branch emanating from the node representing the state.

The characteristics of the trellis in Figure 17.2 describe the encoder in Figure 17.1. The sequence 1110 will be applied to the register. The output will be determined by the state of the cells S_2, S_1, and F, where F is the first cell in the register. The bits will be listed in that order. Assume that the register is in state 00 and that the first 1 is shifted into the first cell of the encoder. The output, Y_1, is given by the modulo sum of $0 \oplus 0 \oplus 1 = 1$. The output, Y_2, is given by the modulo-2 sum of $0 \oplus 1 = 1$. Regardless of whether or not the next bit is a 0 or a 1, the 1 in the first cell will be shifted into S_1 and the register will be in state 01. This is depicted on the trellis: If the register is in the 00 state on the left side of the trellis and a 1 is received, the output is shown on the trellis to be a 11 along the dotted line leaving the 00 state and going to the 01 state on the right side of the trellis noted by "next state." The next state is achieved only whenever the next bit is inserted into the first cell of the register and the bit of the first cell is

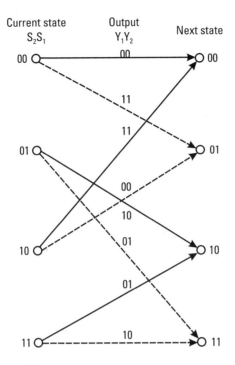

Figure 17.2 One-step trellis.

shifted into the S_1 cell. Specifically, after receiving the second logic 1, the current state of the register is then 01, and the index marker of the trellis is back to state 01 on the left side of the trellis. The output Y_1, is given by the sum $0 \oplus 1 \oplus 1 = 0$ and output Y_2, is given by the sum $0 \oplus 1 = 1$. This is shown on the trellis by the dashed line going from the 01 state on the left to the 11 state on the right with the output, $Y_1 Y_2$, being 01. The next state of the register, 11, becomes the current state, also 11, and the index marker moves back to the left side of the trellis. The state of the register, 11, is actually achieved when the third 1 is shifted into the initial cell and the 1s in the first cell and the S_1 cell are shifted. The third one going into the first cell produces the output Y_1, given by the sum of $1 \oplus 1 \oplus 1 = 1$. Y_2 is given by the sum $1 \oplus 1 = 0$. This event is illustrated by the dashed line going from the current state of 11 to the next state of 11 on the right side of the trellis. A 0, the last bit to enter the register will produce an output, Y_1, of $1 \oplus 1 \oplus 0 = 0$ and an output Y_2 given by $1 \oplus 0 = 1$. This is shown on the trellis by the solid line moving from the 11 current state on the left to the next state 10 on the right. This next state will actually only occur whenever the bit following the 0 is shifted into the first cell of the register. At that time the state of the register, $S_2 S_1$ will be 10. The input sequence, 1110 will produce the output sequence 11 01 10 01.

The one-step trellis of Figure 17.2 may expand to show the path generated by any bit sequence. Figure 17.3 is an expanded trellis, with the path for the input sequence 1110 shown by bold lines. The expanded trellis is obtained by repeating the single-step trellis. The states from 00 to 11 have been labeled a, b, c, and d, respectively. In moving through the trellis, the number of steps into the trellis is referred to as the level or depth of the decoder.

17.4 Free Distance of a Convolutional Code

The performance of a code is determined by evaluating the Hamming weight of all possible sequences, referred to as code vectors, of a given length and finding the sequence with the minimum weight. Since the weight, w, of a code vector is defined to be the number of nonzero elements, the Hamming distance between the code vector of interest and the all-zero vector is the weight of the vector. Because of the linearity and closure properties of the code, the minimum distance between two code vectors of a set may be found by comparing the code vectors to the all-zero vector. The weight of the minimum weight code vector is defined to be the free distance, d_f, that is, $d_f = w_{min}$.

17.5 The Code Vector

The code vectors are all possible output code sequences. The code vectors will be examined here by utilizing the expanded trellis to determine the free distance for the convolutional encoder of Figure 17.1.

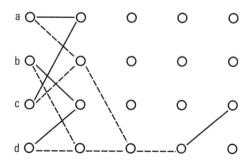

Figure 17.3 Expanded four-step trellis.

Consider the sequence abcaa of Figure 17.4. The weight of this output sequence is the Hamming distance between the all-zero vector [00 00 00 00 00] and the code vector [00 11 10 11 00]. That is, it is the number of non-zero elements in the code vector. The weight of this vector, w, is 5. Regardless of the length of the sequences allowed in this trellis, the path just described can be expanded to any length by adding an aaa . . . aa tail, an all-zero tail, which does not increase the weight of this code vector. Further, since this sequence with the aaa . . . a tail is an allowed sequence and no sequence can be found with less weight, the free distance of this code, d_f, is 5. The importance of the free distance in error correction will be demonstrated shortly.

17.6 Decoding

In forward error correction, a function of the decoder is to detect errors between the transmitted sequence and the received sequence and correct these errors. At the encoder, each input bit and the state of the encoder determine the output of the encoder. For a specific encoder and for a given input sequence, a unique path along the branches is traced through the trellis. The coded sequence generated is a code vector composed of the output elements associated with the branches. This sequence of coded bits, or code vector Y, is transmitted. Another function of the decoder is to reverse the process of coding and reproduce the original data sequence. Further, the received code vector, Z, may differ from Y due to bit errors. In the decoder, the trellis of the encoder is used for error correcting. That is, since each branch of the trellis indicates the correct elements of the transmitted code vector and a given path through the trellis gives a unique code vector, it is the purpose of the decoder to determine the correct path.

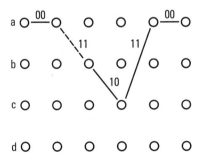

Figure 17.4 Free-distance path.

For maximum likelihood decoding, all possible sequences through the trellis and their associated code vectors must be compared to the received code vector. Then the code vector whose Hamming distance from the received code vector is minimum is declared the winner. If no errors were present, the Hamming distance between the received vector and one allowed trellis path would be zero. The path segment where the two differ allows the error to be detected, and choosing the correct path allows the error to be corrected.

The decoding procedure to be discussed next utilizes the Viterbi decoding algorithm and is, in fact, maximum likelihood decoding.

17.7 Viterbi Decoding

The Viterbi decoding algorithm is a procedure for comparing all possible allowed paths in the trellis, essentially but not actually, to the received sequence. The sequence whose Hamming distance from the received sequence is the least is declared the winner. To actually compare every possible path through the trellis with the received sequence would be an enormous task, since the number of paths doubles each time another step is taken into the trellis. The algorithm makes it possible to essentially compare all paths by discarding paths that could never have a smaller Hamming distance than the paths retained, referred to as survivors.

In the algorithm, the Hamming distances between the elements of the received code vector and the elements of the trellis branches are determined and attached to the branches. These are referred to as the branch metrics. The sum of the branch metrics for each path is the path metric.

Since there are two branches representing the end points of paths entering each node and each has an associated path metric at that point, surviving paths are determined by comparing the two path metrics and keeping the path with the smallest path metric. Since the two paths merged at the node and will be the same path from that point on, regardless of what occurs subsequently, the discarded path can never have a smaller metric than the survivor. Further, since one-half of the paths are discarded at each level, the doubling of paths does not occur; in fact, the number of paths remains constant. The sum of the branch metrics, or the path metric, for each survivor path is compared, and the path with the smallest path metric is chosen as the winner. In theory, this decision should be made after the entire received sequence is compared to the paths in the trellis. In practice, the decision is made by a sliding window that is N nodes or levels long. That is, a new node and the associated branch metric are added to the path metric, and the oldest node and branch metric is dropped off. The path metrics are compared, and the path with the minimum metric is determined. The binary elements of this branch from this path N nodes or levels back is selected as correct. The n-node window is again moved forward, picking up the next and newest node and dropping off the oldest node.

To aid the reader in better understanding the algorithm, an example will be worked and illustrated in Figure 17.5. The transmitted vector is $Y = 1110110000$ and the received vector is $Z = 1110111000$ with one bit error in the seventh bit position. It will be assumed that a preamble of zeros was transmitted, so that the decoder starts in the 00, or a, state. The decoder determines the Hamming distance between the received two-bit element and the two-bit branch element. This distance is the branch metric and is shown beneath the branch on the trellis. It requires three steps into the trellis before steady state is reached, that is, until there are two branches entering each node. Consider the path aaa. The received element is 11 and the branch element is 00. The Hamming distance between the two is 2. The 2 is added to the trellis just below the aa branch. The next two-bit element received is 10, and the Hamming distance is 1. This 1 is inserted below the second aa branch. The third received element is 11, and the distance between the received element and the branch aa is 2, which is inserted below the third aa branch. The path composed of three branches—

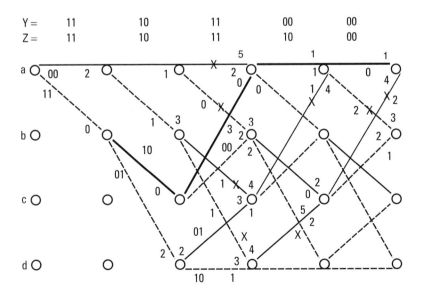

Figure 17.5 Decoder trellis and error correction.

aa aa aa—now has a path metric of 5. This is added to the trellis just above the third aa branch. At this point, there are two branches entering each node, and the path metrics must be compared and the path with the largest path metric discarded. The discarded path is marked with an x.

The path metric for the path abca will be determined next. The distance between the received element 11 and the branch element 11 of ab is 0, which is added just below the branch ab. The distance between the next received element 10 and the branch bc is also 0, which is added beneath the branch. Similarly, a 0 is added beneath the branch ca. The two paths enter the fourth node of the first row of nodes. The path metrics of the two are compared, and the aa aa aa path with a metric of 5 is discarded and the ab bc ca path is retained, becoming a survivor. The next element received is 10, which contains a bit error. The distance between this received element and the aa branch leaving the fourth node is 1, and this metric is added below the branch. The path metric on the path abcaa is 1. The distance between the next received element, 00, and the aa branch element, 00, is 0, which is added below the branch. The path metric for the path abcaaa is now $0 + 0 + 0 + 1 + 0 = 1$, which is displayed on the top of the branch. The branch entering from the c node has a path metric of $0 + 0 + 2 + 0 + 2 = 4$; therefore, when the two path metrics are compared, the path from c is discarded. At this point, the path metrics of the four surviving paths at the four nodes are compared, and the path with the minimum metric is declared the correct path. This is the bold path in the figure, abcaaa and the binary elements produced by this path 11 10 11 00 00. Thus, the error in the seventh position is corrected.

In this simple example, the correct elements were given at one time when the correct path was determined. In actual practice, after steady state is reached and path metrics have been obtained by summing branch metrics from N branches, the path metrics are compared for the fixed number of paths and the path with the smallest metric is selected as the winner. The branch element N nodes back is kicked out as correct, and the branches from all the paths at that distance back are deleted. At that point, a new branch is added; the branch metrics are calculated, added to the path metrics, and compared; and one of the two paths entering each node retained. The new path metrics of the fixed number of paths are compared, and a path winner is selected with the branch element N branches back kicked out as a winner. Actually, the branch element is not the output of the decoder, but rather the 0 or 1 corresponding to the selected path branch is the output. The process is then repeated. N, the number of nodes in the sliding window path metric adder, is referred to as the decoding depth. It has been determined empirically that the decoding depth should be about five times the constraint length. The operation is in real time, and after an initial delay, the bits are exiting the decoder at the same rate the coded elements are entering. That is, if the code rate is 1/2, then for every two coded input bits, there will be one information bit output.

Obviously, the encoder must store the path metrics and the code symbols or branch elements corresponding to a surviving path. For the example given here, there would be four paths. The process of adding the branch metric to the path metric, comparing the path metrics of the two paths entering a node, discarding the path with the largest, and storing the surviving path is referred to as, "add, compare, and store."

Decoders that employ the Viterbi decoding algorithm are available off the shelf, operating at an information bit rate of 25 Mbps, constraint length 7, and rate 1/2. Coding gains of 5 to 6 dB are achievable at BERs around 1×10^{-5}; however, as the BER increases due to a decreasing E_b/N_o, coding gains decrease until in a very noisy environment the uncoded system outperforms the coded system.

17.8 Detection Decisions

The impact of soft decisions on the decoding procedure will be discussed next. Up to this point, the discussion of the operation of the decoder employing the Viterbi algorithm has been limited to hard decisions. To understand soft decisions, it is necessary to review the output of the correlator and the integrate-and-dump operation in the bit synchronizer.

For BPSK, the output waveform of the correlator recovering the bit sequence from the modulated carrier is a constant voltage with fluctuating noise added to it. The constant voltage is a positive value for a logic one, say, and negative for a logic zero. This noisy waveform is applied to the integrator. The output of the integrator is a ramp that reaches a maximum value at the end of the bit period. For hard decisions, the output of the integrator is sampled at $t = T$, and this voltage is applied to a threshold device such as a comparator, which decides if the sampled voltage is less than zero or greater than zero.

The graphical representation of the sampled output of the integrator is shown in Figure 17.6. The sampled output of the integrate-and-dump is one dimension and falls somewhere on the horizontal axis of Figure 17.6. If no noise was present, and a logic one was transmitted, the ramp output of the integrator would reach a maximum of $+A$, say, at time $t = T$, when

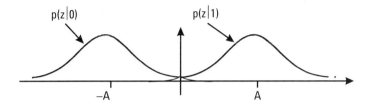

Figure 17.6 Probability density functions for the sampled integrator output.

it was sampled or a −A for a transmitted zero. However, when noise is present and is Gaussian, the output of the integrator is a Gaussian random variable, and the location of the sampled value is described by the probability density functions shown in Figure 17.6. The probability density functions are Gaussian with means of +A or −A. The variance of the distributions is determined by the noise power.

17.8.1 Hard Decisions

All things being equal, whenever the sampled output of the integrator falls above zero, the threshold device outputs a one. On the other hand, a value below zero indicates a zero was transmitted. The ones and zeros out of the threshold device go into the decoder and are used to calculate the branch metric by finding the Hamming distance between the received binary code elements and the branch elements of the trellis.

17.8.2 Soft Decisions

For binary soft decisions, the metrics are determined in the following way. Usually, the sampled output of the integrator is quantized to eight levels as shown in Figure 17.7. For a one, the levels are numbered from right to left, from zero to seven, starting with the output amplitude that would occur for the no-noise case. For a zero, the reverse procedure is followed. Say two

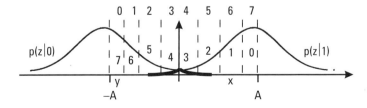

Figure 17.7 Eight-level quantization for soft decisions.

elements of a code vector are received, $z_1 = x$ and $z_2 = y$, as shown in Figure 17.7. For hard decisions, z_1 would be called a one and z_2 would be called a zero. However, it is desired to indicate the confidence we have in these decisions and incorporate that confidence in the metric. The closer a received point is to $+A$ or greater, the more confidence there is that a one was transmitted. Therefore, the metric selected is the distance from the received point to the amplitude A for a one and the distance from the received point to the $-A$ for a zero on the trellis branch.

For example, assume the first element $z_1 = x$, as shown in Figure 17.7, was received. For x, the distance from an exact one is the distance 1. Its distance from an exact zero is 6. These two distances will be the branch metrics, depending upon the two-bit code elements of the particular branch. Consider the 00 of the aa branch of the trellis of Figure 17.8. The distance between x and the first element 0 is 6, and the distance between y and the second element 0 is 0. The branch metric for this branch is $6 + 0 = 6$. For the ab branch with the allowed symbol, 11, the metric, m, equals the distance between x and $+A$ plus the distance between y and $+A$. Then m equals $1 + 7$, or 8. For the bc branch with the 10 element, the branch metric is given by $m = 7 + 6 = 13$. For the bd branch with element 01, the metric

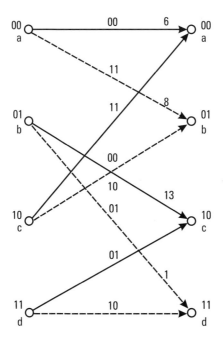

Figure 17.8 One-step trellis showing soft-decision branch metrics.

$m = 1 + 0 = 1$. The branch with the smallest distance from the (z_1, z_2) point and smallest metric is 01, whereas the branch with the element 10 has the largest distance and largest metric. Clearly there is more confidence that 01 was transmitted than that 10 was transmitted.

The branch metrics are added just as before to obtain the path metrics. Path metrics of two paths entering a node are compared, and the one with the largest metric is discarded. After N steps, the path metrics are compared, the path with the smallest path metric is selected as the winner, and the code symbol of this path N branches back is ejected as correct.

The best results are obtained with infinite quantization from the sampled output of the integrator. However, it has been determined that eight-level quantization only loses 0.25 dB.

17.9 Coding Gains and BERs

For BPSK with no coding, the BER is given by

$$BER_{nc} = Q\left(\sqrt{\frac{2E_b}{N_o}}\right) = 0.5\,erfc\left(\sqrt{\frac{E_b}{N_o}}\right) \tag{17.1}$$

The asymptotic BER occurs when E_b/N_o is large. For this condition, a good approximation to (17.1) is

$$BER_{nc} = 0.28\,\exp(-E_b/N_o)$$

For convolutional codes using hard decisions, a bound [1, pp. 412–413] on the BER is given by

$$BER_{chd} = K_{hd}\,\exp(-rd_f E_b/2N_o) \tag{17.2}$$

where

r = code rate;

d_f = free distance of the code;

K_{hd} is a constant composed of a number of factors determined by the specific code used.

Using soft decisions and infinite quantization, the BER is given [1, p. 413] by

$$BER_{sd} = K_{sd}\,\exp(-rd_f E_b/N_o) \tag{17.3}$$

17.9.1 Asymptotic Coding Gain

At high signal-to-noise ratios, which is the normal region of operation, the exponential terms of (17.1), (17.2), and (17.3) predominate; therefore, an asymptotic coding gain for hard decisions over an uncoded system is defined to be

$$G_{hd} = 10 \log(rd_f/2) \text{ dB} \tag{17.4}$$

The coding gain using soft decisions and infinite quantization over an uncoded system is

$$G_{sd} = 10 \log(rd_f) \text{ dB} \tag{17.5}$$

Inspection of (17.4) and (17.5) shows the coding gain for soft decisions and infinite quantization is 3 dB better than hard-decision decoding. In actual systems eight-level quantization is almost always used and it has been determined empirically that only about 0.25 dB is lost compared to infinite quantization.

17.10 Hardware Implementation

For convolutional encoding and decoding, the industrial standard is a rate 1/2 encoder, constraint length 7, and a decoder using the Viterbi decoding algorithm. A number of decoder chips are on the market with an information bit rate up to 25 Mbps and a channel rate of 50 Mbps. This is the upper end for the bit rate of off-the-shelf convolutional decoders using the Viterbi algorithm for decoding.

17.11 Comparison of Coded and Uncoded Systems

A coded system is an attractive alternative to enlarging the antenna size or increasing the transmitted power whenever it has been determined by link analysis that the E_b/N_o ratio is too small. Figure 17.9 illustrates an uncoded BPSK system and the resulting parameters. It should be noted that raised cosine filtering is assumed prior to modulation such that the transmission bandwidth required is approximately equal to the channel bit rate. In the uncoded system the channel bit rate is equal to the data bit rate.

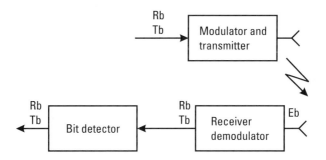

Figure 17.9 An uncoded BPSK system.

17.11.1 Relationship Between the Parameters of Coded and Uncoded Systems

The parameters of interest for the uncoded system are as follows. Information bits:

$$E_b = \text{bit energy of the uncoded or information bit;} \qquad (17.6)$$

$$T_b = \text{bit period of the information bit;} \qquad (17.7)$$

$$R_b = \text{bit rate of the information bits.} \qquad (17.8)$$

Also

$$E_b = CT_b \qquad (17.9)$$

where C = received carrier power
and
N_o = power spectral density of the white noise of the receiver and is composed of all the antenna and receiver noise components.

After link analysis, it may be determined that the ratio E_b/N_o is too small and produces an unacceptably high BER. A coded system can be an acceptable alternative. A block diagram of a coded PSK system is shown in Figure 17.10. The parameters of interest in the coded system are as follows.

Parameters of encoded system:

$$E_c = \text{bit energy of the channel bits or encoded bits;} \quad (17.10)$$

$$R_c = \text{channel bit rate after encoding.} \qquad (17.11)$$

r = encoding rate = ratio of information bits per coded bits For Figure 17.10, one information bit produces two coded bits, since $r = 1/2$.

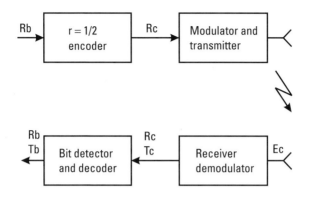

Figure 17.10 A BPSK coded system.

$$R_c = R_p / r.$$

Encoding increases the channel bit rate by adding redundant bits for error correction.

$$T_c = 1/R_c = \text{Bit period of the encoded or channel bits} \qquad (17.12)$$

$$T_c = rT_b \qquad (17.13)$$

$$E_c = rE_b \qquad (17.14)$$

$$E_c = CT_c \qquad (17.15)$$

Since in the coded system there are more bits in a time interval by a factor of $1/r$ than in the uncoded system, the channel bit energy, E_c is decreased by r as indicated in (17.14).

For comparison purposes, the energy, E_b, of the information bit is used as the base. Figure 17.11 shows the BER curves of an uncoded BPSK system and that of a coded BPSK system for $r = 1/2$, constraint length of $K = 7$, and soft decisions with eight-level quantization. Figure 17.11 is a composite of simulations and experimental results combined with curve fitting for the coded system. The curve for the uncoded system is a plot of (17.1). The curve of the coded system compares favorable with theoretical curves in the literature [2], and experimental curves from industry [3]. The theoretical curves for the coded system are for infinite quantization, but there is only a small 0.25-dB degradation for eight-level quantization, which is normally employed in actual systems. This is not discernible in the comparison. The $K = 7$, $r = 1/2$, is normally the off-the-shelf chip used in practice. There is a real coding gain of approximately 5.5 dB realized. For example,

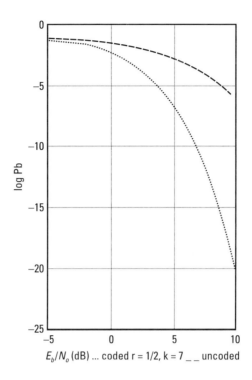

E_b/N_o (dB) ... coded r = 1/2, k = 7 _ _ uncoded

Figure 17.11 BER for a convolutional encoded system, r = 1/2, K = 7, and soft decisions.

in order to obtain a BER of 1×10^{-5} with an uncoded BPSK system, an E_b/N_o of 9.6 dB is required. From Figure 17.11 it is seen that for K = 7, an E_b/N_o of only 4.5 dB is required, which gives a coding gain of approximately 5 dB.

One note of concern is that the loops, used to lock onto the carrier and demodulate, must work on an E_c that is 1/2 of the original E_b. Further, for the rate 1/2 encoder, twice as much bandwidth is used. In effect, an exchange of bandwidth for a system that will produce a smaller BER for a given E_b/N_o has occurred. Going to a smaller code rate, say 1/3, will require about 0.4 dB less E_b/N_o but three times the bandwidth. The lock capability of the loop must be considered.

Example: Design a BPSK System
Required:

1. 10-Mbps information data rate;
2. BER $< 1 \times 10^{-5}$.

Available are 20 mHz of bandwidth. From link analysis: E_b/N_o = 4dB.

From Figure 17.11 it is seen that for E_b/N_o = 4 dB, a BER close to 1×10^{-6} can be achieved for K = 7 and r = 1/2. Select an off-the-shelf decoder with these parameters. After encoding at r = 1/2, the 10-Mbps information bit rate will create an encoded or channel bit rate of 20 Mbps, which will require the use of the entire 20 mHz.

17.12 Coded PCM/FM

It is of interest to determine the coding gain for convolutional encoding and maximum likelihood decoding in PCM/FM, a noncoherent communication system. In PCM/FM, the BER is closely related to the impulse rate in the IF, and since the height of each impulse is nominally the same for a fixed IF bandwidth, the effect of soft decisions on coding gain needs to be determined.

Using simulation techniques, bit error plots as a function of E_b/N_o or (C/N) were generated and are shown in Figure 17.12 for a rate 1/2 encoder, constraint length 7, and soft decisions. The off-the-shelf decoder for these parameters would be available from a vendor. In Figure 17.12, the two plots shown are BERs for the uncoded system and coded system versus C/N in the IF. Since the IF bandwidth is set equal to R_b, the C/N in the IF is also equal numerically to E_b/N_o in the uncoded system and E_c/N_o in the coded system. E_c is the energy of the coded bit or the energy of the channel bit. Since a rate 1/2 encoder is used, each channel or coded bit has one-half the energy of E_b, the information bit. In comparing coded systems to uncoded systems, the base for comparisons is always the information bit. Therefore, in Figure 17.12, 3 dB was added to the C/N in order to obtain E_b/N_o. For a bit error rate of 1×10^{-5}, it can be seen that a coding gain of about 3 dB is realized. These are preliminary results that show significant potential [4].

17.13 Hardware Decoder Implementation for PCM/FM

Convolutional encoding and Viterbi decoding may be achieved in PCM/FM with off-the-shelf hardware. Practically all the commercially available decoders are built and tested to work with BPSK or QPSK. However, as Section 17.11 indicated, the decoders will also work with PCM/FM if a quantized input is obtained from the bit synchronizer to use as an input to the decoder.

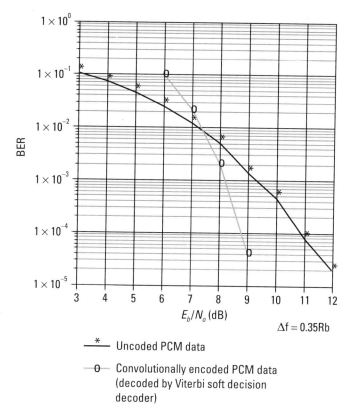

Figure 17.12 BER versus E_b/N_o for coded and uncoded PCM/FM systems using a rate of 1/2, constraint length 7, soft decisions, and convolutional encoding.

Once a decoder is chosen based upon the bit rate requirement, soft decisions should be used since there is usually significant coding gain using soft decisions rather than hard decisions. In order to employ a decoder using soft decisions, it is necessary for the output of the S/H device following the integrator or matched filter of the bit synchronizer to be quantized and used as an output of the bit synchronizer. (Normally this output would be used internally in the bit synchronizer as an input to a threshold device that makes a hard decision concerning whether a zero or one was received. The synchronizer then outputs a one or zero.) A number of commercial bit synchronizers offer a quantized 3- or 4-bit output for use in a decoder employing soft decisions.

Off-the-shelf soft-decision decoders are available that handle data rates up to 25 Mbps. Typically, a decoder may require two external clocks, one at the data rate and one at the channel bit rate. The decoder usually provides synchronization status on an output pin as well as an estimate of the channel BER on another output pin. For a more complete discussion of specifications, requirements and capabilities of decoders, see [3].

Problems

Problem 17.1

Discuss reasons for using coding.

Problem 17.2

Discuss how E_b/N_o is determined in terms of C/N_p.

Problem 17.3

Why is convolutional coding used in aerospace applications?

Problem 17.4

Discuss forward error correction.

Problem 17.5

For an encoder with six states, what would be the code rates for (a) two modulo-2 adders? (b) three modulo-2 adders?

Problem 17.6

Find the Hamming distance between the code vectors, (a) [001] and [000]; (b) [101010] and [010101]; and (c)[000111] and [111000].

Problem 17.7

Draw and label a one-step trellis for the encoder of Figure 17.1.

Problem 17.8

Create a four-step trellis by repeating the trellis of Problem 17.6 for four times. Trace the input [1010] through this trellis. Assume the coder starts in the zero state and the input vector is read from left to right.

Problem 17.9

For the input of Problem 17.8, what would be the coder output?

Problem 17.10

Discuss hard decisions and hardware implementation.

Problem 17.11

Discuss soft decisions and hardware implementation.

Problem 17.12

For BPSK and an E_b/N_o = 5 dB, what would be the BER for (a) an uncoded system? (b) for a coded system defined by the plot of Figure 17.11?

Problem 17.13

For a data rate of R_b = 5 Mbps and a rate 1/2 encoder, what would be the channel bit rate?

Problem 17.14

What transmission bandwidth would be required for Problem 17.13, assuming one bit requires 1 Hz of bandwidth?

Problem 17.15

For PCM/FM and an E_b/N_o = 9 dB, what would be the BER for (a) an uncoded system? (b) for a coded system defined by the plot of Figure 17.12?

References

[1] Haykin, S., *Digital Communications,* New York: Wiley, 1988.

[2] Clark, G. C., and J. B. Cane, *Error-Correction Coding for Digital Communication,* New York: Plenum, 1988.

[3] Qualcomm, Inc., *k=7 Multi-Code Rate Viterbi Decoder,* Technical Data Report Q1650, San Diego, CA: Qualcomm, Inc., June 1990.

[4] Carden, F., and S. Ara, *BER Determination of PCM/FM + FM/FM Systems and Coded PCM/FM Systems,* ECE Tech Report Series, No. 93-005, Las Cruces, NM: New Mexico State University, Dec. 1993.

18

Industrial Telemetry
by Dr. Brian Kopp[1]

Telemetry is used in industrial environments to both monitor and control. When these two functions coexist in an industrial telemetry system, as is usually the case in modern applications, the term *supervisory control and data acquisition* (SCADA) is often used to describe the system.

The history of industrial telemetry extends back 200 years, and events including the industrial revolution, World War II, and the development of the computer have had direct impacts on the technology utilized in industrial telemetry. Today, the array of field and office equipment available to the industrial telemetry design engineer is nearly as varied as the applications they support.

18.1 Learning Objectives

Upon completion of this chapter, the student should be familiar with the following aspects of industrial telemetry:

- How industrial telemetry began;
- Modern applications in industrial telemetry;
- Equipment used in industrial telemetry.

---

1. Dr. Kopp is the vice president of network design for Clifton, Weiss & Associates, Inc. of Philadelphia. He can be reached via e-mail at bkopp@cliftonweiss.com.

18.2 History of Industrial Telemetry

Steam-driven applications are generally considered to be the first industrial processes that were monitored and controlled deterministically and reliably, and therefore they represent the roots of industrial telemetry. In particular, the improvements made by James Watt on Thomas Newcomen's steam engine included several monitoring and control devices such as the mercury pressure gauge and the famous fly-ball governor.

The invention of commercial high-pressure steam gauges like the Bourdon tube pressure gauge in the mid-nineteenth century [1] enhanced performance of large steam engines and thus contributed to the industrial revolution. Similarly, hydraulic servos invented by Westinghouse came to replace the fly-ball governor on large-scale engines [2]. These early devices monitored and controlled the inhospitable internal process of a steam engine from a distance, albeit a short one. This was telemetry in its crude beginning.

While steam engines were used to run factories, they were also used to move resources to those factories and to take products from them. In the United States, the Baldwin Locomotive Works of Philadelphia began building practical railroad locomotives in 1845 [3] with the production of their Eight Wheeler. It would later become known as the American Standard locomotive. In 1807, Fulton's Folly demonstrated that a steam engine could be used to propel a ship reliably enough to offer steamship service between New York City and Albany.

By the mid-nineteenth century, the steamship began to replace the sailing ship for river, coastal, and even transatlantic shipping. The various pressures, water levels, and temperatures associated with these steam driven engines were monitored by their operators, and adjustments in the fuel and water consumption rate and the amount of generated steam used to do work were adjusted to optimize performance. The "feedback" communications path in these industrial "processes" was the operator, but early monitoring and control devices were present in the form of gauges and valves. Further, in those cases when a fly-ball governor was employed on a steam engine, the control of the amount of steam used to do work was somewhat automated.

The industrial revolution was ending at the close of the nineteenth century; however advances in industrial process monitoring and control were just starting. While simple mechanical and hydraulic devices had supported the industrial revolution, new sensors that used air to indirectly measure flow, level, or pressure were in development at the end of the century, and the invention of the pneumatic diaphragm-actuated valve in 1890 meant that the monitoring and control process could be automated with air as the linking medium [2].

These sensors and actuators were linked together directly and were collocated with the processing equipment. Prior to that time, as mentioned above, the link between monitoring device and control device had been the plant operator who toured the facility and made adjustments by hand after reading the appropriate mechanical or hydraulic gauge. These pneumatic systems monitored parameters and then performed rudimentary control loop functions using on-off or proportional control in an automated process.

Derivative control followed in 1935, paving the way for proportional-integral-derivative pneumatic control, which arrived in the 1940s [2] and is still commonplace today. Early applications of pneumatic control and monitoring included batch processing and railroad signal and switch machine operation.

The invention of the pneumatic transmitter in 1938 by the Foxboro Corporation meant that the process engineer could be located in a centralized control room [2]. The 3- to 15-pound-per-square-inch (psi) transmitter signal became the standard in the United States and created a robust means of relaying a transducer's state, while also demonstrating whether a fault condition existed. In the event of a leak, a value of 0 psi is detected as a fault condition since it is out of range. The use of pneumatic control and monitoring increased slowly through the first few decades of the twentieth century, peaking in the 50s and 60s when the 3- to 15-psi standard gained wide acceptance [4]. Because of its relative simplicity, inherent safety in flammable environments, and passive electrical classification, this type of mechanical instrumentation is still in use and available in the marketplace. A block diagram of a pneumatic control system is shown in Figure 18.1.

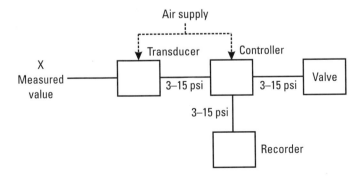

Figure 18.1 Pneumatic control system block diagram.

It should be noted that pneumatic devices operate below the speed of sound; therefore, long pneumatic runs translate into time delay in the control process. When fast response times are necessary in the control loop, pneumatics is seldom used. The 3- to 15-psi American standard is maintained by agencies such as the Instrumentation, Systems, and Automation Society, and the metric equivalent that is prevalent in Europe uses the range of 0.2 to 1.0 bars.

Several manufacturers continue to serve the pneumatic industry. Currently, the Fisher-Rosemount company offers power valve petitioners and air damper actuators that are based on proven 70-year-old designs by the Hagan company. Other current pneumatic controller and transmitter manufacturers include ABB Automation, Siemens-Moore Process Automation, Foxboro-Eckardt, and Honeywell.

One of the earliest applications of monitoring and control by electrical means was to improve railroad safety. In 1872, William Robinson patented the electric track circuit [5], which used the shunting action of the axles of a railroad train to close a circuit between the rails and indicate that a section of track was occupied. This indication was "telemetered" several miles down the rails themselves to a signal lamp, which told the engineer of an approaching train to wait until the section of track was clear.

Train control and monitoring has continued to improve at a steady pace in the twentieth century. Track circuits are still the predominant track occupancy indication technique, but their status is now sometimes communicated hundreds and even thousands of miles over copper and fiber-optic cables and sometimes via radio signal to the computers that control the "process" of train movement.

The legacy of Samuel Morse's 1836 invention of the telegraph communication system is also still present in track circuit technology. Certain track occupancy circuits are still referred to as "OS" circuits. This name is derived from the telegraph signal of "O-S" that was transmitted by a railroad station telegraph operator to other stations to indicate that a train was "on station" [6].

Currently, one of the most sophisticated forms of train control and monitoring ever conceived is being developed for use on the New York City Subway System by Siemens-Matra. Communications-based train control (CBTC) is being implemented on the Canarsie line of the New York Subway using wireless transponders mounted between the rails to tell subway train sets where they are. The train sets then communicate with wayside zone controllers, using low-power, license-free, spread spectrum radio transceivers, to monitor and control train movement. Since train location can be determined within inches, train sets can be spaced closer together ahead and behind each other, while preserving safe braking distances.

Electrical telemetry sensors that would eventually be used in industrial processes were in development just before and after the turn of the century [2]. In the 1930s, a practical strain gauge was developed, 80 years after Lord Kelvin demonstrated that metallic conductors subjected to mechanical strain alter their electrical resistance [7]. Experimental rocket and aviation research in the 1930s used the strain gauge extensively.

Similarly, effective weather monitoring of upper-atmosphere air temperature, humidity, and pressure began in the 1930s with the invention of the radiosonde [8]. This radio meteorograph replaced early aircraft weather soundings because it could reach altitudes of 50,000 feet. Other inventions during World War II advanced radio monitoring, and to some degree radio control, in military aviation and rocket research. It is the evolution of this effort that is the focus of the prior chapters of this book.

Alongside Lord Kelvin's work on mechanical strain were several other milestones that eventually led to the development of electrically based commercial monitoring and control systems. The observation by T. J. Seebeck in the early nineteenth century that dissimilar metals joined together in a closed loop could be made to cause electrical current to flow when subjected to different temperatures led to the invention of the thermocouple. Further research by Sir Humphrey Davy on the relationship between temperature and resistance led Sir William Siemens to the invention of the resistive temperature detector (RTD) in the late 1880s.

The principles of self-induction and mutual induction discovered by Joseph Henry and Michael Faraday, respectively, led Samuel Morse, in 1836, to the invention of the telegraph sounder and the electromagnetic relay. The relay became an important device in electrical control and is still used today. Further, Michael Faraday's invention of the dynamo led not only to the invention of the electric motor but to the invention of the magnetic flow meter where a liquid conductor induces a current in a surrounding electromagnet.

Despite these nineteenth-century advancements, it was not until the 1940s that electrically based commercial control and monitoring made significant progress. Technological advances in telemetry and communications systems during World War II were largely responsible. Developing solutions to address the adverse conditions associated with electrical signaling in an industrial environment was also necessary. Alternating current (AC) technology was used first to transmit temperature and pressure, and then later direct current (DC) became popular [2].

It was the late 1950s before an electrical standard was finally agreed upon and widely developed in the United States [9]. Like the 3- to 15-psi pneumatic standard, the 4- to 20-milliamp (mA) DC electrical standard has an inherent robust design. The absence of current, like the absence of air pressure, indicates a failure. Overcurrent is also a fault condition. In the 4- to 20-mA standard, a transmitter varies a DC current in proportion to a process variable. The current flows in a loop to a sensing resistor whose voltage is measured. Using Ohm's law and the scaling factors of the transmitter, the process variable is telemetered.

There are several variations to this design, and the standard can be used for monitoring and for control; however they are usually referred to collectively as *analog current-loop technology*. The 4- to 20-mA standard maintained by the ISA with their joint standard, ANSI/ISA-S50.1-1982 (R-1992) "Compatibility of Analog Signals for Electronic Industrial Process Instruments," is widely used today [10]. By one estimate, approximately 25% of new process control and monitoring installations are still using 4- to 20-mA technology today [11].

The advent of digital communications has impacted commercial control and monitoring in several ways. First, with the codevelopment of the digital computer, there has been a movement to use modern digital communications directly in the process. Serial digital communications is the most popular format, and there are over 60 such protocols in wide use today [12]. Industrial Ethernet has become the newest entry into the marketplace and it is estimated that "at least 5 percent of . . . manufacturing and process plants have Ethernet-based systems on the production floor . . ." [13].

Some digital communication techniques that can be found in the control and monitoring environment include simple differential serial communications techniques like RS-422. Other techniques include elaborate protocols with multiple physical layer variations, like Echelon's LonWorks.

There is also a large hybrid marketplace that supports the installation of newer communications protocols over the existing 4- to 20-mA infrastructures that have been in place for over 20 years. Protocols like the Highway Addressable Remote Transmitter (HART) protocol are very popular with existing facilities that want to avoid replacing large incumbent cable infrastructures that were installed to support sometimes hundreds of 4- to 20-mA applications. HART superimposes a modulated signal on the 4- to 20-mA current loop to convey multiple variables and control signals to and from field equipment.

Another popular and relatively new protocol that also takes advantage of the existing current-loop infrastructure, but that uses a completely digital signaling technique, removing the 4- to 20-mA signal, is Foundation Fieldbus. Like HART, this protocol is scalable and supported by many vendors. New types of sensing transmitters that communicate using these protocols are generally referred to as smart transmitters, in contrast to traditional current-loop transmitters, because they can do more than transmit a single process variable.

It should be noted that there are still a vast number of monitoring and control applications that utilize nonstandard proprietary techniques. Unique requirements, like those of the New York Subway CBTC project, mentioned above, can lead to the development of specialized communications techniques to perform monitoring and control. For example, the New York CBTC project will utilize a new wireless radio interface that is being developed specifically for the mission critical nature of that subway's train movement control. Similarly, self-contained, unique manufacturing and testing environments sometimes require the use of proprietary protocols to streamline the process. What is sometimes given up in communications overhead and utility is replaced with speed and efficiency.

18.2.1 Computers

Prior to the 1940s, control and monitoring in the industrial process environment was supervised on the plant floor where the sensors and actuators were located. If changes in production were needed, the plant operator made the adjustments on the control devices and watched the monitoring devices for the correct reading. If this change affected another process in another part of the plant, the two or more responsible operators had to coordinate their efforts. An example of a classical control process is shown in Figure 18.2 describing a distillery operation and displaying the numerous controllers inherent in such systems.

A gradual shift to central control-room supervision began in the 1940s. Technological developments that permitted the telemetering of the process information some distance away became feasible [2]. The 3- to 15-psi pneumatic standard was used to connect process sensors with a central control room. By centralizing most of the monitoring and even some of the control side of the process, the number of technicians required to operate the process plant was significantly reduced, monitoring became more efficient, safety was enhanced by removing the operator from continuous duty on the production floor, and, ultimately, profits were increased. The functional block diagram for centralized computer control is shown in Figure 18.3. The ability to monitor different processes simultaneously and react to changes faster

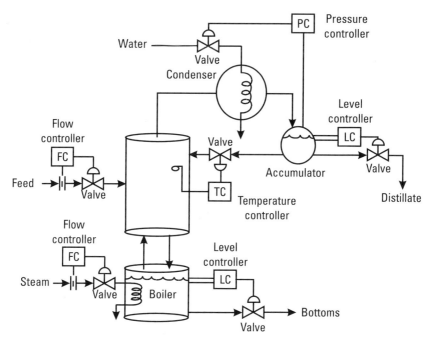

Figure 18.2 A classical distillery control process.

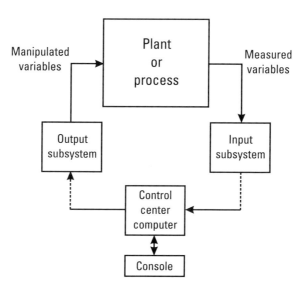

Figure 18.3 Centralized control block diagram.

meant increased efficiency and safety as well. The transition from 3- to 15-psi pneumatics to 4- to 20-mA electronics also improved central control room operations because, among other things, the space and maintenance required for electronic monitoring devices was less than that of pneumatic ones.

The invention of the computer dovetailed nicely into the central control-room configuration and by the early 1960s it was being used in process control and monitoring applications [14]. The miniaturization of the computer in the 1970s and 1980s led to improvements in central control-room processing, but more importantly, it led to the development of highly specialized industrial field computers that could carry out control functions locally. These programmable logic controllers (PLCs) are used in small applications, where they operate autonomously, or in large applications, where they share information with other PLCs and with the control center.

Strangely enough, the PLC has led back to the decentralization of process control. Distributed control systems (DCS), used in large facilities, are networks of PLCs operating in association with their respective process. The control room is still the center for plant supervision, but it relies on communications links between its DCS computers and PLCs to coordinate macro-level process administration. A block diagram for DCS is shown in Figure 18.4. The automated PLC of the late twentieth century has, in effect, replaced the human plant operator of the early twentieth century.

The most recent step in the evolution of computing in the monitoring and control environment has been to replace the highly specialized PLC with an actual microcomputer. Gordon Moore's accurate, and now famous, 1965 prediction that the growth of computer memory chip performance would continue at an exponential rate, doubling every 2 years, has created an environment that permits relatively high performance computers and digital signal processors to be used cost effectively right on the production floor. The proliferation of high-speed protocols such as Ethernet has contributed to this development as well.

18.2.2 Individual Circuits and Multidrop Circuits

Before the advent of smart transmitters, and looking back over 100 years, early sensors were called upon to sense and report only one piece of information, typically called a process variable (PV). A single sensor might report the process variable using a pneumatic or electronic signal and thus be called a "transducer." This transducer also conveyed one piece of information, although in a transformed state. When the pneumatic or electronic signal was

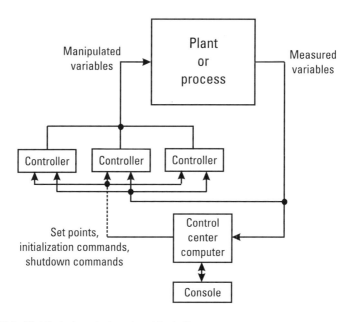

Figure 18.4 Distributed control system block diagram.

transmitted some distance away for reporting, the transducer was considered a "transmitter." It is the transmitter that is of particular interest to the industrial telemetry design engineer because it is the device that requires the communication of information some distance away.

The first electronic transmitters were similar to pneumatic transmitters in that they used a single analog current value to convey the process variable, much the same way as a pneumatic transmitter uses a single analog air pressure value. If the process had two variables, then there were two transmitters, each operating side by side with parallel communications circuits. Originally there was a one-to-one correspondence between the number of transmitters and the number of pneumatic tubes or wire-pairs. With the invention of the computer this relationship changed. If a computer was used to do the monitoring and control, then the analog pressure or voltage signal needed to become a digital signal at some point for interfacing with the computer. There was no reason that point could not be some distance from the computer, in the field, or at least at a strategically located place somewhere in the middle.

The use of computers and A/D signal converters meant it was possible to combine multiple analog transmitter signals from individual circuits, into a single digital circuit. Many of these devices are called multiplexers because the analog transmitter signals are sampled digitally and then commutated in time on the single digital circuit. These devices are often referred to as industrial signal conditioners and come in many different forms. A pneumatic valve-actuator multiplexer reverses the direction, providing control by using a digital electronic signal to set the 3- to 15-psi pressure on multiple pneumatic tubes that are each then connected to value actuators.

The most modern multiplexers are really digital "hubs," "switches," "bridges," or "routers," borrowing terms from computer networking, and they connect a single digital circuit from the computer (or computers) to multiple digital circuits, each of which extends as far as the transmitters itself. In this last case, the analog current-loop signal, or pneumatic signal, has been completely removed from the monitoring and control system. As might be expected, the topology for multiplexer networks is almost always a star topology. Protocols like HART, Foundation Fieldbus, and Industrial Ethernet can be configured to operate like this.

There is also another common type of network that provides efficiencies of scale over one-for-one transmitter-circuit configurations. The multidrop or bus-circuit forms a daisy chain between multiple transmitters and permits the most economical use of cable infrastructure. In some instances, a single copper wire pair can be used to communicate with dozens of transmitters spread over several miles. An example of this type of circuit is shown in Figure 18.5. There are several digital techniques that can utilize bus circuits, including RS-422, RS-485, and Modbus. Protocols like HART can also be installed by daisy-chaining transmitters together.

18.2.3 Telemetry Versus SCADA Versus Process Control

Remote monitoring came before remote control, and examples of early systems that used only remote monitoring, such as the railroad track circuit or the radiosonde, were therefore examples of telemetry systems in the strict sense of the word. In instances when a monitored process variable was used to then control that process, the loop was closed locally. The railroad industry, the petroleum and gas industries, and the power utility industry all developed uses for remote control, but in different ways and at different times in history.

During the 1960s, remote control capabilities were being developed in the petroleum, gas, and power utility industries and added to existing remote monitoring systems [14]. These systems continued to be called telemetry systems for another 10 years; then in the early 1970s the term "supervisory

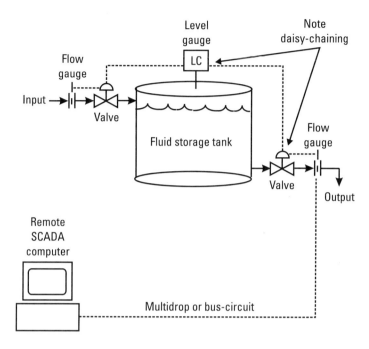

Figure 18.5 The multidrop or bus circuit.

control and data acquisition" (SCADA) replaced telemetry to more accurately describe the two-way nature of the monitoring and control systems.

The petroleum industry also uses the terms "process control" and "batch control" (as do many manufacturing industries), but these are generally limited to describing the control and monitoring of refinement processes and also particular events in the distribution process, such as tanker-truck loading. The term SCADA is used in describing the remote control and monitoring of well production and inventory control in both the gas and petroleum industries. It is also used universally to describe the remote control and monitoring of the transmission process in the gas, petroleum, and power utility industries.

Remote control in the railroad industry began much earlier than in these other three industries, and the railroads used a different name for it. It was on July 25, 1927, that the first remote control system for train traffic was cut in [15]. Centralized traffic control (CTC), as it has come to be known, was first used along 40 miles of right-of-way on a New York Central Railroad Subsidiary near Toledo. CTC uses track circuits to indicate track occupancy and relays or digital logic processors for control. The majority of large freight railroads and transit agencies, as well as Amtrak, use it today.

The distinction between "remote" and "local" control and monitoring is often blurred in many of these industries. There is often a combination of both local and remote capabilities. Many modern PID process control systems still "close the loop" locally using PLCs or embedded processors in the transmitters themselves but permit changes in scale factors, calibration, diagnostic probing, or process shutdown by remote control or locally through the PLC, or processor user interface.

Modern railroad CTC systems perform the actual control of train movement with relays or digital logic processors located near the interlockings and signals but allow a remote dispatcher to "request" the relays or processors to change a track switch position. The relays and processors will only comply if the request does not create an unsafe condition for approaching trains. Interestingly, a signal system maintainer with access to the relays or processors many times cannot locally request changes in track switch position. They must contact the dispatcher by radio or telephone and ask them to make the request! Is this remote control or local control, or both?

Similarly, many monitoring systems now present the system parameters locally and remotely. Many modern transmitters and PLCs have a digital display that shows one or several process variable values. Those same values are also communicated remotely for display on a computer screen. The same is true of some railroad CTC systems, where a local display may show the track switch positions and track occupancy circuit states for the nearby territory, and a remote display may show the same information to a dispatcher on a larger, more sophisticated computer display covering the entire railroad division.

18.3 Modern Industrial Applications

Telemetry, SCADA, and process control applications can be found in most any industry. There are several traditional applications and many unique ones. And the list grows daily. While a presentation of a list of all applications is impossible in the scope of this book, it is worthwhile to review some of the more prevalent applications to understand how communications helps to support remote control and monitoring.

18.3.1 Petroleum Industries

The oil and natural gas industries both have three distinct businesses within them, and the use of remote monitoring and control is somewhat different in each one. The three businesses can be roughly classified as production, transmission, and distribution. For both oil and natural gas, production starts at the well. Flow from the well is remotely monitored and sometimes remotely controlled through the use of valves. Flow meters, temperature meters, and even viscosity meters can be involved in the well-head monitoring process.

Digital radio communications with PLCs and remote terminal units (RTU), a less sophisticated PLC, located at the well head are often used to concentrate the information to an intermediate point where either a telephone line or satellite link brings the data to a central monitoring point. Well production can involve storage of the product before transmission to a customer. This storage is monitored and controlled using flow meters, level sensors, pressure sensors, and valves. It is important to note that in the United States, roughly half of the crude oil used is imported. Much of the imported crude is transported by ship, and these ships have large storage tanks that are monitored using similar level and flow meters.

In the oil industry, raw crude is refined to produce hundreds of different types of petroleum products. This chemical distillation of crude oil is conducted in refineries. Refineries are a form of manufacturing plant, and that topic will be covered under a separate application. Natural gas refinement by liquefaction can also be a complex process analogous to the manufacturing plant.

Transmission of natural gas and of oil products presents unique remote monitoring problems. Pipelines are monitored for leaks and corrosion. Oil and gas pump stations, natural gas compression stations, and cathodic protection systems are also remotely controlled and monitored. Pipeline storage facilities and terminals can be either locally or remotely monitored and controlled, or both. The distances between pump stations and between cathodic protection stations can be on the order of tens of miles. This makes them ideal candidates for remote monitoring and control.

The pipeline right-of-way is sometimes used to install the supporting communications infrastructure, copper, fiber, or radio. One pipeline company is famous for having exploited this infrastructure and has joined the communications network business. Williams Communications was once known only as the Williams pipeline company; then in the mid-1980s they began installing fiber-optic cable in their decommissioned natural gas pipelines. Today Williams is both a natural gas pipeline company and provider of communications network services to telephone companies, Internet service providers, and large corporations.

Distribution of natural gas is provided by local natural gas suppliers. After transmission, natural gas is stored near the distribution point, usually underground, in depleted former natural gas reservoirs. This storage technique requires compressors to force the natural gas underground and flow and pressure meters to monitor the storage reserve. Once the local supplier takes possession of the natural gas, it enters their local distribution pipeline. Again, monitoring and control is necessary. Here, though, the emphasis is also on providing accounting information.

The large number of customers supported by local natural gas distributors generally requires their SCADA computer systems to provide flow information in a format suitable for accounting. Recently, some distributors have started installing electronic meter-reading systems. Two techniques are generally being employed: telephone and radio. Telephone interconnection allows the electronic meter to call in its reading. Radiometer reading is either fixed or mobile. Fixed radiometer reading allows one central radio location to communicate with many radio-equipped meters. Mobile meter reading permits a data collection vehicle to pass nearby to each meter and receive the meter reading data.

Distribution of oil industry products like gasoline, diesel, and heating oil begins at pipeline terminals. These terminals store different types and grades of petroleum products. Some terminals store crude oil that is awaiting refinement, while others only handle distribution of refined products. Terminals are known as tank farms because their most recognizable features are the dozens of large product tanks. Larger terminals store millions of gallons of petroleum products. To fill their storage tanks, pipeline valves are opened at the appropriate time and a series of downstream valves and pumps divert the product to the appropriate tank. Pipeline-flow meters and tank-level gauges generate bulk product receipts for the transaction. If the source of the product is a tanker ship or railroad tanker car, then pumps retrieve the product and valves in the terminal divert it to the appropriate tank. Most of these valves, flow meters, and level gauges are remotely controlled and monitored from a central control room at the terminal.

At the other end of a distribution terminal, a loading rack is used to fill tank trucks from the storage tanks. Usually the appropriate additives and dyes are injected in the tank truck during the loading process. A tank truck may hold two or three different types of product comprising upwards of 7,500 gal. The loading process is controlled at the loading rack using flow meters and electronically controlled valves. Remote monitoring is sometimes used to generate receipts.

The petroleum industry uses pneumatic, analog, and digital control and monitoring equipment. The hazardous nature of the industry necessitates the use of pneumatic equipment and intrinsically safe electronic devices in many instances. However, wireless communications is also widely used throughout the three businesses of the industry.

18.3.2   Power Utility Industry

Like the petroleum industry, the power utility industry has three distinct branches: generation, transmission, and distribution. Unlike the petroleum industry, there is no efficient way to store electric power for future transmission or distribution. As a result, the demand and the generation have to be matched in real time. To do this, a complex control and monitoring system matches generator capacity with customer distribution needs and interconnects them to each other using a transmission network known as a grid.

The majority of generation uses either water or thermal sources to turn turbines connected to generators. In the United States, 95% of all electric power is supplied by large plants that generate at least 30 MW each [16]. Most of these facilities have multiple generators that are turned mechanically. Thermally driven turbines and water driven turbines are used to turn the generators. In the case of thermally driven turbines, coal-fired steam boilers predominate in the industry. To a lesser degree, nuclear, natural gas, and oil fuels are also used to boil water, generate steam, and turn turbines. Hydroelectric plants use static water pressure to force a turbine to turn. In the United States, roughly 10% of the electric power generated is generated by hydroelectric means.

All of these techniques require sophisticated control and monitoring systems to regulate the process of creating electricity. For fossil-fuel plants (coal, natural gas, oil) the added complication of maintaining a large fuel inventory is necessary. Generation plants operate in a unique fashion. They must be available to generate power 365 days a year; however they may be called upon to vary the amount of power they generate over a wide range and in a short period of time (sometimes adjusting generation several times a day). Such an operation requires real-time sophisticated SCADA systems.

The factors that dictate the amount of power generated are the customer demand, the market price per megawatt, and the price of fuel (in the case of fossil fuel plants). These factors are ultimately used to make control decisions that are entered into the SCADA system. Inventory control, boiler pressure and temperature, generator voltage, current, and frequency are just the obvious parameters that must be monitored and controlled. Multiply the task by the number of generators, add in safety systems and accounting requirements, and the sophisticated nature of electric utility generation SCADA systems begins to appear.

Except for the inability to store its product, transmission systems for electric power are surprisingly similar to that of petroleum products. In fact, in the United States, both of these transmission industries encompass interstate commerce and are jointly regulated by the Federal Energy Regulatory Commission (FERC). Electric transmission systems interconnect generation plants with local distribution substations, using a network, or grid, of transmission-line segments. Each segment is rated to carry a specific amount of power at a specific voltage. Transmission substations are used to interconnect different segments of transmission lines. These substations can include transformers and large circuit breakers.

They are monitored and controlled remotely using SCADA systems. RTUs at the substation are used to control individual circuit breakers and report alarms via the SCADA communications links to a centralized control center. To facilitate these links, the SCADA systems use radio frequency signals that are superimposed, not surprisingly, on the transmission lines themselves. Today, fiber-optic core transmission lines are becoming commonplace and replacing radio frequency-based SCADA communications systems.

The electric power transmission industry is a commodities-based market. On a daily basis, transmission companies negotiate the price of electric power with generation companies, local distributors, and other transmission companies. They must then monitor the megawatt hours of energy they buy, deliver, and wheel (transport for other companies), using their SCADA systems. The input and output voltage and current on transmission-line segments is monitored for accounting purposes, but also to ensure that the capacity of a line segment is not exceeded, and to ensure that a fault condition does not occur.

Local distribution substations are owned by the utility companies most familiar to consumers. These are the companies whose focus is selling small quantities of energy to hundreds of thousands of customers. Their business is customer-centric, and not surprisingly, some of the most advanced telemetry communication techniques in use today are found here. The days of the meter reader walking through neighborhoods once a month reading meters are quickly disappearing. New customer meters are being installed that include intelligent communication devices. Companies such as Schlumberger are providing products and services that tie together wireless cell-based communications techniques with meter-reading services. Other companies use telephone-line interfaces or mobile vehicles that use short-distance wireless interfaces to remotely read utility meters.

18.3.3 Railroad Transportation

As mentioned above, the railroad industry has a prominent and unique place in the history of remote monitoring and control. Today, Class 1 railroads, the largest railroads, have thousands of miles of track interconnecting large population centers, ports, and industrial suppliers and users. In general, these railroads are operated from centralized control centers. For example, the CSX railroad operates services on more than 20,000 miles of track in over 20 states and two countries. Their entire operation is controlled from their dispatch center in Jacksonville, Florida. Their track miles are divided into more than 75 segments that are controlled by computers called "codeline" computers located at the dispatch center. The track occupancy circuits and switch interlockings on each segment are monitored and controlled using individual communications circuits for each segment. Historically this circuit was physically connected to each switch-interlocking location and occupancy circuit using leased telephone lines and one of several simple proprietary DC codes.

Today, the protocols sent over these lines between the field processors and the computers are digital serial protocols operating over relatively low asynchronous modem speeds. In the case of CSX's 75 segments, approximately half of them are currently supported in this manner. The remaining segments utilize a relatively new technique involving what has become known as radio codeline. Telephone lines connect the codeline computers in Jacksonville to a few radio base stations strategically located along the segment of track. Using 900-MHz, 4,800-baud frequency-shift keying, the base stations send messages to field processors at switch interlockings. Control messages from the control computers in Jacksonville are relayed to these field processors, which then take the appropriate action, for example, throwing a switch. Indications and processor status information are radioed back to the computers for display on dispatcher computer terminals.

Railroads are taking advantage of other communications technologies to support remote monitoring and control as well. Global positioning satellite signals are being used for train location by some railroads, passive radio-frequency tagging is being used to track freight cars, and cellular digital packet data transceivers are being used in urban environments on some transit trains to relay health and welfare data about the onboard systems through the cellular telephone network. Further, the railroad caboose was made obsolete in part by an "end of train" device that can radio the air pressure in the braking system at the end of the train to a monitor in the engine.

18.3.4 Manufacturing

An example of a manufacturing process is shown in Figure 18.2 and is indicative of most manufacturing operations. The distillation process shown is commonly used in refineries, chemical plants, and food processing. There are several inputs to the process including feed flow rate, feed composition, steam rate, cooling water flow rate, and the reflux ratio. The outputs include distillate composition, distillate flow rate, bottoms composition, bottoms flow rate, and column pressure. The distillate output from the tower is often accumulated before final delivery to permit a backflow into the distillery for the purpose of controlling the purity. Typically, the temperature of the distillate boiling point in the column is inversely proportional to purity. Thus, an increased temperature can be used to increase the reflux ratio in the process.

Each of the controllers uses some combination of proportional, integral, or derivative control. The values of the gain terms for each controller can be set remotely if the controller has a remote communications interface. The controllers can also be sophisticated enough to permit a complete change of the type of control by remote reprogramming. For remote monitoring, if the level, pressure, and temperature sensors have remote communications capability, either analog (air pressure or electrical current) or digital, then their values can be relayed to a control center.

18.3.5 Municipal Water Supplies

Water supplies are controlled using traditional SCADA systems. RTUs are used to control pumps and monitor temperature, flow rates, and water storage levels. Many municipalities use radio modems to send data to and from the RTUs. The radio modems communicate with a centralized system controller or master terminal unit that is then interfaced with a monitoring and record-keeping computer. A small municipality with 10,000 customers may have a half-dozen wells and several storage tanks. The radio modems are typically licensed FSK modems operating at a low asynchronous modem speed. In a small system, a single frequency is used and each RTU is addressed, responding to polls from the system controller.

18.3.6   Fire-Life-Safety Systems

In large commercial or industrial facilities such as skyscrapers or factories, fire and smoke detection and prevention, environmental control (heating, cooling, and humidity), facilities control and monitoring (power, lighting, water, telecommunications, elevators, escalators, and signage), access control, and security monitoring are all necessary components of the building infrastructure.

The National Fire Protection Association (NFPA) publishes building codes for fire detection and protection systems. These regulations require facilities such as high-rise office buildings to have central control stations for fighting fires. Remote monitoring of fire and smoke detection devices and fire pumps are necessary as well as remote control and monitoring of elevators and automatic fire doors [17].

Vendors such as Johnson Controls and Siemens Building Technologies sell building automation systems and fire-life-safety systems that are highly integrated. These systems operate everything from automatic door locks to automatic fire extinguishing sequences. They provide graphic workstation interfaces and emergency fire command consoles. Large facilities may have several thousand telemetry and control points. These points are usually concentrated at an intermediate location containing a small controller or RTU. From the controller, the link to the central control station may be a simple serial link or, in newer installations, a network protocol such as LonWorks or TCP-IP over Ethernet.

18.3.7   Intelligent Transportation Systems

In 1991 the United States Congress legislated the creation of the Intelligent Transportation Systems (ITS) program as a component of the Department of Transportation (DOT). Since the federal government subsidizes the maintenance and construction of the highways and interstates in each state, and the DOT regulates interstate commerce on those roads, the ITS program at the DOT has an important contribution to make to transportation around the country. The focus of the ITS program is the use of modern computer and communications technologies to improve safety and efficiency. Traffic control and electronic toll collection are two of the most active areas of focus. The use of remote cameras and entrance ramp signals have been increasing in the past few years with some benefit to traffic mobility. The interstate success of automatic vehicle identification systems for electronic toll collections, for example, EZ-Pass, is due in part to the efforts of the federal ITS program.

Passenger rail service is benefiting from remote monitoring and control as well. Amtrak's new Acela passenger rail coaches use 18 computer systems connected over a local area network that, among other things, monitor differences in outside air temperature from one end of the car to the other. They also control the amount of tilt that the train should apply to provide a comfortable ride based on speed and centrifugal acceleration through a curve [18].

From the commercial side, companies like United Parcel Service and Federal Express are revolutionizing freight tracking by using wireless communications services to track package location. On a larger scale, services like Qualcomm's Omnitracs have been providing satellite-based truck tracking for freight transportation management as well as theft control for several years. There is also a relatively new field in ITS called telematics that is gaining in popularity. It permits limited real-time remote control and monitoring of automobile parameters like door locks and air-bag deployments. Services such as General Motors's OnStar and Mercedes-Benz's TeleAid are prominent in this new industry.

18.3.8 Telephone and Cable Network Monitoring

Telephone companies use the infrastructure they sell to also provide a means of monitoring and controlling their own network. Most medium- and long-distance telephone networks use synchronous optical network (SONET) equipment. An individual high-speed SONET connection operating at 2.4 Gbps can carry the equivalent of over 30,000 voice or Internet dial-up phone calls. The overriding protocol, SONET, has built-in alarm and status information about the network that can automatically inform the connecting network equipment of a failure and reroute communications around the problem. This is a sophisticated technique but it is widely used. In the central-office environment, where customer telephone connections are made, the devices that terminate customer circuits, called channel banks, support network-management interfaces that permit remote monitoring. Protocols such as the Simple Network Management Protocol (SNMP) are common in data communications equipment and are quickly becoming popular in voice communications equipment. Loop-back tests can now be performed on customer wiring, permitting telephone maintainers to instantly, and remotely, isolate a problem as being inside or outside the customer premises.

The cable television industry has been modifying its infrastructure to take advantage of fiber-optic communications. Many cable television systems are now a hybrid of fiber and copper cable networks, utilizing fiber to bring analog television channels from the cablehead to a neighborhood. Once at the neighborhood, a Hybrid-Fiber-Coax (HFC) device converts the analog signals from fiber to copper for distribution to individual homes. The cable industry has been slow to develop monitoring standards techniques for their HFC equipment, but recent moves by cable television service providers into the Internet and telephone service business has required an increased level of network management and monitoring. The cable modem industry anticipated this and developed products with a built-in standard that includes some remote management, the Data-Over-Cable Service Interface Specification (DOCSIS). Networking equipment manufacturers such as Cisco Systems now market cable modem network management systems that can manage and troubleshoot entire HFC networks using the DOCSIS protocol.

18.4 Industrial Communications Equipment

The devices used in industrial communications include the input and output devices used in the field as well as the control devices and computer systems that interface with them. This section presents each type of device, explaining the different variations and under what circumstances they are used. The communications techniques that interconnect them will be discussed in the following section.

18.4.1 Temperature Measuring Devices

Four of the most common types of temperature sensors used in temperature measuring devices are resistance temperature detectors (RTD), thermistors, thermocouples (TC), and silicon sensors. The first two types convert changes in temperature into changes in resistance in an electrical circuit. The second two types convert changes in temperature into changes in voltage in an electrical circuit. RTDs can be very stable and have a relatively linear response. Thermistors are very sensitive but usually nonlinear. Thermocouples have a large operating temperature range but are susceptible to electrical noise. Silicon sensors are very stable and linear.

Sometimes the temperature sensor is built directly into a local controller, but often it is attached to a temperature transmitter. The transmitter is used to convert the resistance or voltage measurement into a signal that can be transmitted to a local controller, remote controller, or remote monitoring computer. Commercial transmitters are available with a wide range of analog and digital interfaces including current loops, pneumatic signals, HART, and serial protocols like Modbus. Temperature sensor and transmitter combinations can be inexpensive, starting at less than $200.

18.4.2 Fluid and Gas Flow Measuring Devices

Selecting a method for measuring flow depends on several factors including accuracy, viscosity, density, turbidity, and even conductivity. The four most common types of flow measuring devices are differential pressure meters, positive displacement meters, velocity meters, and true-mass flow meters.

Of these four, only the positive displacement meter directly measures flow. Pistons or gears are commonly used in positive displacement meters to incrementally count "segments" of the liquid. They work well with viscous liquids.

Differential pressure (DP) flow meters are the most common type of flow meter. By introducing a physical device into the flow, such as a flat plate with a specific size hole drilled in it that changes the kinetic energy of the liquid, a differential pressure is developed. The pressure difference increases in proportion to the square of the flow. DP meters are relatively inexpensive and usually have no moving parts.

Velocity meters use mechanical, electromechanical, or ultrasonic methods to measure the velocity of the moving liquid. Measured through a known cross section of pipe, the flow is proportional to the velocity.

True-mass meters use heat loss from a sensor in the flow or Coriolis forces imparted on a vibrating tube by a mass of liquid. Mass meters can be expensive but are less susceptible to changes in temperature and viscosity.

Pressure sensors and transducers are available in the marketplace and are used in local control applications. They usually require an additional electronic circuit nearby to utilize the signal in a control loop. Another device common in industrial control is the pressure switch. Typically pneumatic or hydraulic, and usually mechanical, these devices include either a transistor switch or relay contact that closes at a specific percentage of full-scale of the pressure sensor in the switch.

Pressure transmitters, like temperature transmitters, commonly use either standard analog or digital protocols. Some new smart pressure transmitters are actually multivariable transmitters, for example, providing access to differential and absolute pressure measurements. Still other new smart transmitters incorporate limited controller functions. Fisher Rosemount, ABB Automation, Endress & Hauser, SMAR, and Siemens-Moore all manufacture smart pressure transmitters.

18.4.3 Fluid Level Measuring Devices

Level measurement is used in inventory control, process control, and custody transfers of fluid products. These devices measure the level in tanks that can contain hundreds of thousands of gallons of petroleum products. The market for these devices is large, and some companies sell highly specialized products for specific segments of the market. Technologies that are used to measure fluid level include pressure-at-depth devices (hydrostatics), radar and ultrasonic distance devices, capacitance devices, and buoyant force or float devices. In some segments of the market, devices need to be very rugged because they are exposed to combustibles. Usually, "intrinsically safe" and explosion-proof level gauges are legally required in petroleum tank-level measuring applications because of the hazards involved.

Hydrostatic tank gauge (HTG) systems use a combination of pressure sensors and sometimes temperature sensors, to measure level. In open tanks, the process uses the difference between atmospheric pressure and the pressure at a specific depth on the tank. The differential pressure is proportional to the height of the fluid in the tank. For a given density of a fluid, its volume and mass can be calculated. Temperature sensors are used in these systems to compensate for temperature-related expansion and compression of the fluid. This is important for accurate measurements in large product tanks that hold hundreds of thousands of gallons. Closed tanks use the differential pressure between the fluid and the vapor space on top. Often HTG systems connect the various auxiliary gauges to one smart gauge/transmitter that calculates the level and transmits it to a central control center.

Radar and ultrasonic level systems are mounted above the fluid and measure the distance to surface. They can be temperature-compensated as well and are used when the fluid surface is stable and free of agitation. They are particularly useful in applications where contact with the fluid is to be avoided, or where the fluid tank cannot be taken off-line to perform maintenance and repair of the level gauge.

Other gauging techniques include capacitance gauges, which use the dielectric properties of the fluid to change the capacitance of an electronic probe inserted in the substance. Liquid glass is often measured this way [19, p. 87]. Float gauges can be used at the air interface of a liquid (surface) or at the interfaces between different types of liquids. These types of gauges are often used in level-controlled switches or as stand-alone visual sight gauges, but they can be used in very sophisticated continuous-level transmitters as well. The magnetostrictive float sensor uses the Villari effect to create a return radio frequency pulse, similar to radar, from a magnet-equipped float that surrounds a round vertical waveguide [19, p. 85].

Most level gauges include temperature compensation and are sophisticated measuring devices. The use of transmitters in level gauging applications is common. This permits the relay of all of the telemetry parameters involved in the measurement process as well as the update of calculation parameters in the gauge to accommodate different fluids. Calibration and diagnostic procedures are frequently performed on these devices, and the use of transmitters permits this to be done remotely. As a result, smart transmitters with digital interfaces are commonly installed in modern applications.

18.4.4 Other Measuring Devices

Telemetry devices that measure other process variables are used in many areas of industrial processing and manufacturing. Example parameters are listed below:

- Position;
- Size (milling and manufacturing);
- Presence or absence of electrical current or voltage;
- pH and oxidation reduction potential (waste water and drinking water);
- Conductivity;
- Oxygen and combustible concentrations;
- Vapor opacity (environmental monitoring, e.g., smoke stacks).

Some of these devices are available as sensors, transducers, or transmitters. In others, simple electrical input devices detect the presence of high voltage or current and convert it into a standard and safe signal. Electromechanical and solid-state relays as well as optical isolator circuits can also be forms of transducers when used as input devices in a remote monitoring application. Still other forms of transducers may amplify a weak signal or regenerate or filter a noisy signal. And finally, A/D electrical signal sensors are commonly available on PLCs and RTUs for receiving electrical signals directly from transducer circuits.

18.4.5 Control Output Devices

Output devices perform an action and are the empowering devices on the control side of remote monitoring and control systems. The action is typically mechanical or electrical. They can exist either as stand-alone devices, receiving a singular signal, pneumatically or electrically, and performing a single action or as one of several such devices within a controller.

One of the most prevalent output devices is the valve actuator. Variations of this device can be controlled pneumatically, electrically, or electro-pneumatically. These devices are available as open-close actuators or as continuous-motion positioners. There are also remote control interface options available from most commercial vendors. Analog and digital communication standards are supported.

Relays are often used as output devices, interfacing a control system to an AC circuit and controlling, for example, the start/stop of a pump motor. A PLC, RTU, or computer usually provide the control signals that activate these relays. Similarly, optical isolators may be used to isolate a PLC, RTU, or computer control signal from the actual process.

18.4.6 Field Control Devices

RTUs and PLCs are the two popular types of field controllers that are considered off-the-shelf devices. The two devices have traditionally been used in different sectors of industries. RTUs have been applied to the petroleum industry, power utility industry, and the waste and water processing industries. PLCs have been used in manufacturing and in applications in which networks of relays were previously used to perform complex processing tasks.

RTU devices provide a concentrating, common interface for numerous remote control and monitoring signals. To accomplish this they convert remote signals carried over a singular communications link into multiple field-device-compatible signals. The field-device-compatible signals can be digital or analog. Simple RTUs are relatively inexpensive and small. The Bristol Babcock 3301 is an example of such a device. It measures 3 in^2 and can support only two or three analog signals or a dozen discrete signals. It cannot be programmed. The 3301 uses a simple digital interface to communicate with a master device via a modem or a short-distance bus.

In contrast, PLCs are more sophisticated than RTUs. They include a programmable memory space that can contain control functions. These control functions can react to signals coming from input field devices or from a remote central-control facility. The output signals from the PLC can be digital or analog. PLCs are available in a large range of configurations from small devices no larger than a Bristol Babcock 3301 RTU to rack-mounted shelf units that can process hundreds and even thousands of inputs and outputs, performing independent and dependent control functions.

Many PLC customers install PLCs with some form of remote communications. In such cases, the control loop parameters can be programmed remotely. Similarly, the controller can be reset, disabled, or overridden remotely. In fact, the entire control algorithm can be replaced and started over, remotely. It can also receive control inputs and outputs from other PLCs. The local inputs to a PLC can be monitored, usually in real time, over the communications link as well. This creates a powerful tool, which provides local control of a process and also remote monitoring and a remote oversight function.

The communications links available in the PLC marketplace includes standard digital interfaces like RS-232, RS-485, Profibus for decentralized peripherals (Profibus DP), and Ethernet. Many PLC manufacturers have developed protocols as well that are highly specialized, in particular for PLC to PLC interconnections, though they are now also used for remote monitoring and control. Allen-Bradley has developed one of the most popular protocols, Data Highway+, while Modicon developed Modbus and Honeywell developed Smart Distributed System (SDS). The more popular of these proprietary protocols are available on PLCs made by different manufacturers through licensing agreements with the inventing manufacturer.

18.4.7 Remote Control and Monitoring Computer Systems

Most manufacturers of smart transmitters, PLCs, RTUs, and large input/output interface modules also market host software for use on computers that can remotely control and monitor their devices. This software can also usually be used to configure the device and load parameters, or in the case of PLCs, load processing instructions. There are also third-party host software vendors that are focused on smart transmitters that use a particular industrial protocol, such as HART or Foundation Fieldbus.

At the high end of this commercial, off-the-shelf software market are powerful software tools that can be used to control and monitor entire refining, manufacturing, or processing facilities. Some of these tools can monitor and control large production or distribution networks in the petroleum industry as well. Since many large process control installations interface dozens and even hundreds of controllers and input-output stations over large multinode communications networks, the term DCS is often applied when viewing the entire system, including the host software.

Commercial, off-the-shelf DCS software is available from companies like Intellution and Wonderware. Remote monitoring and control applications in SCADA, where data gathering is the primary focus and where valve or circuit breaker actuation is more event driven than process driven, also can take advantage of these off-the-shelf packages. The computers that run this software are often networked to communications nodes on the control network that connect to PLCs or RTUs using standard local area network protocols like Ethernet. The software can interpret and generate commands to and from field devices from dozens of manufacturers. Ethernet is used to move the commands to and from convenient network node locations. From there the field device or controller protocol is often what is used in closing the link.

Custom software is still used in some large process control and SCADA applications. In some instances, the software is actually built from a suite of preexisting modules. Rockwell Automation and Honeywell Industrial Automation & Control both use their own hardware and software tools to build entire process control and SCADA systems for customers.

18.4.8   Interface Equipment

There is a general category of devices that serve the purpose of interfacing field instruments, such as transducers, smart transmitters, PLCs, RTUs, valves, or relays, with central or distributed control and monitoring computers and their local area networks. These interfacing devices come in many forms. They take on the various roles of data concentrators, signal conditioners, and front-end processors for protocol conversion, depending on the application and its size. Most of them have certain common features. They are all generally used to simplify the interface to the field and streamline the communications process. Some of the more simple devices are conceptually similar to RTUs, but they are considered more of an interface device without memory. The Opto22 SNAP-I/O and the Grayhill Open Line products use discrete plug-in modules to interface individual analog and digital lines to a master panel where dozens of control and monitoring lines can connect. These rack-mounted master panels can then communicate with process or SCADA computers using serial protocols like TCP/IP, Modbus, and Foundation Fieldbus. More sophisticated devices like the Safetran Wayside Communications Controller are highly specialized to receive both digital serial and analog direct current signals from different railroad-control field equipment, and convert them to a common protocol for communication to a railroad-control computer.

18.5   Conclusions

Understanding how the technologies of industrial telemetry came into being provides the design engineer with the reasons why certain applications utilize different types of telemetry and SCADA equipment. The actual devices that connect to the applications are numerous and sometimes tailored specifically for the required task. The computer has had a vast impact on this industry and has revolutionized the office side and control side of the industrial telemetry and SCADA application. What remains is to discuss how these various devices interconnect. There are several prominent and well-established communications techniques that complete the remote control and monitoring system. New techniques as well as techniques borrowed from other industries are also contributing to the industrial telemetry and SCADA landscape of the twenty-first century. Chapter 19 addresses commercial communications techniques, new and old, of interest to the industrial telemetry design engineer.

Problems

Problem 18.1

Write several paragraphs about telemetry systems. What does the term mean? How does it differ from just a telephone connection between two people?

Problem 18.2

Discuss the term *monitor* for a telemetry system and what it entails.

Problem 18.3

List five industrial applications where a telemetry system could be used to monitor the processes.

Problem 18.4

Discuss what *control* means for industrial applications. Give five examples.

Problem 18.5

How does monitoring differ from control? How do the two go together? List five applications that it would be useful to both monitor and control.

Problem 18.6

Discuss the terms *supervisory control* and *data acquisition.*

Problem 18.7

List five applications in which data acquisition would need to be used in telemetry due to a hostile environment.

Problem 18.8

Discuss how the pneumatic diaphragm-actuated valve performs both supervisory control and data acquisition. List five applications where this valve might be employed.

Problem 18.9

Discuss how in the early years of the Industrial Revolution humans were always in the supervisory-control and data-acquisition loop. What function did the humans play?

Problem 18.10

What is the main, if not the only, limiting factor for pneumatic valves?

Problem 18.11

What was one of the earliest electrical devices used for data acquisition and how did it work?

Problem 18.12

Give a brief description of how the electromagnetic relay works.

Problem 18.13

From reference [12], list five different protocols used in digital communications for control.

Problem 18.14

Discuss the current-loop 4- to 20-mA standard. What changes are being made on this technique to give it more capability?

Problem 18.15

Discuss the advantages of centralized monitoring.

Problem 18.16

Discuss PLCs and what functions they perform. How have they changed monitoring and control in the context of Problem 18.15?

Problem 18.17

Discuss DCSs. How did the PLCs lead to these distributed systems?

Problem 18.18

Discuss transmitters and their history, computers, and A/D converters and how the three devices now fit together. How have these devices changed data acquisition?

Problem 18.19

Discuss SCADA with respect to railroads.

Problem 18.20

Discuss SCADA with respect to well production.

Problem 18.21

What are pipeline right-of-ways used for other than pipe networks?

Problem 18.22

Discuss SCADA as it is used by the Power Utility Industry and how SCADA is used in manufacturing.

Problem 18.23

Discuss RTUs and PLCs and their importance.

Problem 18.24

Discuss telemetering systems and their importance in fluid-level measuring. Describe the devices used for this procedure.

Problem 18.25

List 10 process variables that are commonly measured.

References

[1]   Ashcroft, E. H., *Pressure Gauge*, U.S. Patent No. 9836, Boston, MA, July 12, 1853.

[2]   Feeley, J., et al., "The Early Years," *Controls Magazine*, Online Archives, December 1999, http://www.controlmag.com/, accessed January 3, 2001.

[3]   Westcott, L. H., *Steam Locomotives*, Waukesha, WI: Kalmbach Books, 1960, p. 97.

[4]   Johnson, D., "Pneumatic Control: Not Dead Yet," *Control Engineering*, Online Archives, July 1999, http://www.controleng.com/, accessed January 3, 2001.

[5]   Robinson, W., *Improvement in Electric Signaling Apparatus for Railroads*, U.S. Patent No. 130661, Brooklyn, NY, August 20, 1872.

[6]   Whalen, R., Railway Signal Engineering, personal communications with author, Danvers, MA, January 3, 2001.

[7]   Hollanden, B. R. (ed.), *Transactions in Measurement and Control, Vol 3*, Force-Related Measurements, Omega Handbook Series, Stamford CT: Putman Publishing Company and Omega Press, L.L.C., 1998, p. 15.

[8]   United States National Weather Service, Division of the National Oceanic and Atmospheric Administration, http://www.nws.noaa.gov/er/rah/education/eduit.html, accessed January 3, 2001.

[9]   Liptak, B., "Computer-Based Control in the Sixties," *Controls Magazine*, Online Archives, December 1999, http://www.controlmag.com/, accessed January 3, 2001.

[10]   "Compatibility of Analog Signals for Electronic Industrial Process Instruments," Instrumentation, Systems, and Automation Society, 67 Alexander Drive, Research Triangle Park, NC, 27709, Document No. ANSI/ISA-S50.1, 1982.

[11]   Helson, R., "Ten Years On, At Least Fifteen to Go," *The Hart Book*, Issue 10, 2000/2001, pp. 12–13.

[12]   Harrold, D., "Here We Go Again," *Control Engineering*, Online Archives, January 2000, http://www.controleng.com/, accessed January 17, 2001.

[13]   Kaplan, G., "Ethernet's Winning Ways," *IEEE Spectrum*, Vol. 38, No. 1, 2001, pp. 113–114.

[14]   Boyer, S. A., *SCADA Supervisory Control and Data Acquisition*, Research Triangle Park, NC: ISA, 1999, p. 22.

[15]   Frailey, F. W., "Tools That Turned the Tide: 'Here Comes a Non-Stop Meet!' " *Trains*, Vol. 60, No. 1, 2000, pp 43–48.

[16]   United States Department of Energy, "Energy Resources," United States Department of Energy Research and Development Portfolio, Vol. 1 of 4, February 2000, Washington, D.C., p. 106.

[17]   Code 5000 "NFPA Building Code," Draft Version, National Fire Protection Association, a Compilation of NFPA 101 "Life, Safety Code, and the EPCOT Building Code, p. 32-2.

[18]   Johnson, B., "A Coming-Out Party for Acela Express," *Trains*, Vol. 61, No. 2, 2001, p. 26.

[19]   Hollanden, B. R. (ed.), *Transactions in Measurement and Control, Vol. 4, Flow & Level*, Omega Handbook Series, Stamford CT: Putman Publishing Company and Omega Press, L.L.C., 1998, pp. 85, 87.

19

Commercial Communications Techniques for Industrial Telemetry
by Dr. Brian Kopp

The industrial telemetry and SCADA system is divided into two general subsystems: field equipment and office equipment. To connect these subsystems there are numerous communications techniques available to the industrial telemetry system designer. These techniques use copper cables, fiber-optic cables, and wireless infrastructures.

The future of industrial telemetry offers advances in the intelligence designed into field equipment and the communications techniques used to interface with office equipment. Most noticeably, the Internet and the concept of "browsing" are already demonstrating an impact on industrial monitoring and even control.

19.1 Learning Objectives

Upon completion of this chapter, the student should be familiar with the following aspects of industrial telemetry:

- Communications techniques used in industrial telemetry;
- Communications protocols used in industrial telemetry;
- The future of communications in industrial telemetry.

19.2 Communications Techniques

Both analog and digital communications techniques are still in use today. Of the analog techniques, both pneumatic and current-loop systems are relatively standardized. There are numerous digital communications techniques and most are standards-based. The use of wireless communications in SCADA is prevalent, and while there are no standards for air interfaces that have been developed for SCADA, some have been borrowed from other industries such as cellular communications. The computer communications industry has been a rich source of communications techniques for industrial control and monitoring. Wireless LANs, Ethernet, and RS-232 are all examples of such techniques. This section will review several of these communications methods focusing on their lower layers in the open systems interconnection (OSI) model.

19.2.1 The Analog 4- to 20-mA Current Loop

By the 1950s, analog current and voltage loops were being used to transmit monitoring and control signals from electrical sensors and controllers. After about 10 years, the 4- to 20-mA current loop became standard [1]. In 1961 the SP50 committee of the Instrument Society of America ratified its 4- to 20-mA standard and it is still used today [2, 3]. Two common configurations for current loops are shown in Figure 19.1. Only one transmitter can be on

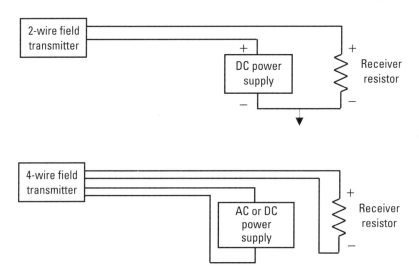

Figure 19.1 Current-loop configurations.

an individual current loop. In both configurations the field transmitters vary the loop current between 4 and 20 mA, which manifests itself as a deterministic voltage across a sensing resistor in the receiver. The sensing resistor is high precision and the ISA standard calls for a value of 250 ohms, ±0.25 ohms and a temperature coefficient of not over 0.01% per degree centigrade. The wire gauge of the loops is generally number 22 American wire gauge or larger. The cabling is low capacitance and shielded twisted pair.

The current loop varies a direct current and does so at a slow rate. By definition it is not therefore a high-bandwidth device and is incapable of sending large amounts of information. The ISA specification severely limits the transmitter response above 10 Hz as measured through a single-pole resistor-capacitor lowpass test filter under maximum signal conditions and maximum power supply noise conditions. This suggests that the current-loop transmitter cannot be used to measure variables that change substantially more than once a second.

19.2.2 The Recommended Standards RS-232, RS-422, and RS-485

There are three common standards that facilitate the transmission of digital, serial, binary information in industrial applications. Recommended standard 232 (RS-232), RS-422, and RS-485 are joint products of the Telecommunications Industry Association (TIA), the Electronic Industries Association (EIA), and the American National Standards Institute (ANSI) [4–6]. However, committees made up of industry professionals meeting through the TIA wrote the standards.

The current version of RS-232 is ANSI/TIA/EIA-232-F-1997 "Interface Between Data Terminal Equipment and Data Circuit-Terminating Equipment Employing Serial Binary Data Interchange." It is the sixth revision of this very popular standard, which was originally ratified in 1962 by the EIA to help standardize computer-to-modem communications. It supports synchronous and nonsynchronous serial point-to-point binary data communications with parallel hardware handshaking and clocking at data rates of up to 20 Kbps over an unspecified cable length. It does set generator output and receiver input voltage limits. Further, it specifies the combined receiver and cable capacitance, therefore indirectly setting the distance limit for a given selected cable type. The signaling is unbalanced.

All circuits are referenced to the "AB," or signal common circuit, which is electrically neutral and grounded within the attached communications equipment. A positive received voltage on any data circuit of at least 3V, referenced to AB, is a "space" or "zero," while a negative voltage less than −3V, referenced to AB, is a "mark" or "one." Note the inherent fault logic insofar as a zero-volts reading is not allowed on any circuit referenced to AB. The RS-232 configuration and an example waveform are shown in Figure 19.2.

While the standard specifies 25 circuits, it uses only two to send the actual serial data. Many implementations use only a subset of the 25 circuits, implementing as few as three circuits (transmitted data, received data, and signal common) for a simple nonsynchronous application. RS-232 is the de facto industry standard for interfacing computers with modems and so has become a vital protocol for interfacing many PLC and RTU products with DCS and SCADA computers.

The RS-422 standard and the RS-485 standard are both balanced signaling techniques. Neither standard discusses handshaking, but both address point-to-point transmitter-to-receiver configurations over balanced two-wire circuits. RS-485 also includes standards for combined transmitter-receiver (transceiver) devices that use the same balanced two-wire circuit to send and receive binary serial data between two or more users. These two standards are specified to signaling speeds of 10 Mbps, but as with RS-232, they do not directly specify cabling lengths.

The use of balanced two-wire circuits for communications significantly improves performance in noisy environments. In RS-422 and RS-485,

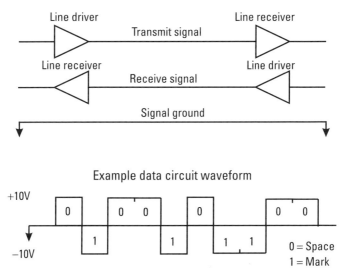

Figure 19.2 RS-232.

information is sent using the polarity of the difference in voltage between the two wires in the circuit. An additive noise signal will tend to affect both conductors of a balanced circuit equally, thus adding a voltage spike equally to both wires and not affecting the difference signal. This is called common-mode noise rejection. It is a prime reason why RS-422 and RS-485 are considered more robust than RS-232 in noisy environments and why they are preferred over longer cable runs. The voltage differences are in the absolute value range of between 0.2V and 12V. The two terminals on the receiver that interface with the two-wire circuit are arbitrarily designated as A and B. Positive voltage differences measured from the B terminal with respect to the A terminal are defined to be a "mark," or binary 1. RS-422 and RS-485 are shown in Figures 19.3 and 19.4 respectively.

RS-485 is also popular in industrial applications because only two wires are needed to communicate in both directions if transceivers are employed, as per the specification. It is used extensively by many small controllers and I/O devices and is also one of the underlying physical layers for several popular protocols, such as Modbus and Allen-Bradley's Data Highway 485 PLC protocol.

19.2.3 NRZ and Manchester Binary Signaling

With all of the recommended standards in Section 19.2.2, the signaling techniques are bipolar techniques because there are only two valid voltage

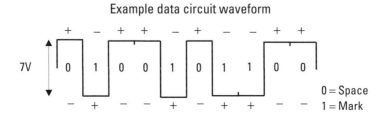

Figure 19.3 RS-422.

Data circuit configuration

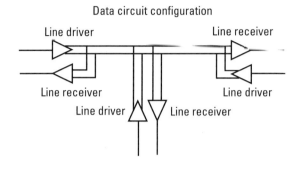

Example data circuit waveform

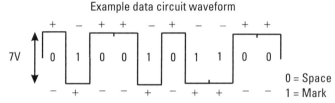

Figure 19.4 RS-485.

levels that are permitted for the transmission of information. The voltage levels are sustained for the entire time duration of the mark or space corresponding to a bit period. This is referred to as NRZ signaling. The name distinguishes this technique from return-to-zero signaling wherein the voltage level that conveys a mark or space is removed after an initial fraction of a bit period. NRZ signaling formats apply signaling voltage, that is, bit energy, for the entire bit period and therefore can have better error performance than return-to-zero formats.

The recommended standards in Section 19.2.2 are most often used in asynchronous communications protocols where the receiver must trigger on the leading edge of the first bit in a limited sequence of bits. Any timing difference between the transmitter and receiver clocks is cumulative with each bit, and this limits the number of bits that can be reliably transmitted without retriggering. The RS-232 standard does include two circuits for clock signals, one to accompany each of the two opposing data signals. When used together, the clock and data signal transmitted in each direction represent a synchronous transmission technique. The receiver receives the clock, synchronizes to it, and then uses it to determine the received bits. Such a technique can be used to more efficiently send data, but the cost is an increase in receiver complexity and the use of nearly twice as much cabling infrastructure.

A compromising technique that is popular in many communications and telemetry systems, including some networking standards like 10 BaseT Ethernet, is Manchester signaling. This technique is synchronous but embeds a clock in the data itself rather than carrying it on a separate circuit. The clock is sent by using unique waveforms for both a mark and space that include a transition midway between the start and end of the bit. The mark is sent by using a waveform that transitions from a low voltage to a high voltage at midbit. The space is sent by sending a waveform that transitions from a high voltage to a low voltage at midbit. The receiver must still synchronize to the embedded clock, but a separate communications circuit is not necessary. The drawback with Manchester signaling is that the addition of midbit transitions effectively doubles the required bandwidth. On copper twisted-pair cables commonly used in communications and telemetry systems (including 10BaseT Ethernet), the available bandwidth decreases with cable length. Therefore, at a given bit rate the maximum cable length is significantly less for Manchester signaling compared to that of NRZ signaling.

19.2.4   Wireline Modems

RS-232, RS-422, RS-485, NRZ, and Manchester signaling techniques are all forms of baseband signaling because the energy used to signal is contained in the frequency bandwidth immediately bordering 0 Hz or DC. When the signaling energy is translated to a bandwidth centered on another frequency besides 0 Hz the technique is said to employ modulation at the transmitter end. A corresponding demodulation at the receiver end returns the signaling energy to a bandwidth immediately bordering DC. Devices that perform modulation for the transmitter in one circuit and demodulation for the receiver in the opposing direction circuit are referred to as modems.

Modems are widely used in remote control and monitoring techniques both on wireline circuits and on wireless radio links. Wireline modems are used on private cabling infrastructures, leased telephone circuits, and sometimes dial-up telephone circuits. While the 56-Kbps modem is now widely used to interface personal computers to the Internet via dial-up circuits, this speed is not yet common in remote monitoring and control applications that use modems. Much slower speeds such as 1,200 bps, 4,800 bps, and 9,600 bps are common in contemporary industrial applications.

There are several different modulation techniques used by these modems, and their sophistication has grown with the maturity of the microprocessor and the development of equalizer techniques and, later, error correction and compression techniques. Although the Bell Telephone Company used its own standards for modems throughout the 1960s and 1970s, today, the premier modem standard's body is the International Telecommunications Union (ITU). The ITU began as the International Telegraph Union in 1865. It changed its name in 1932 and then became a specialized agency of the United Nations in 1947 [7]. The Telecommunications Standardization Sector of the ITU, designated ITU-T, was formed in 1993, replacing the ITU consultative committee on international telegraph and telephone issues, known as the CCITT. The early ITU modem standards were similar to the Bell Telephone standards, and there is some equipment interoperability. However, after divestiture of the Bell Telephone Company on January 1, 1984, the ITU essentially stood alone in developing new standards, and as of today the ITU "V.90" 56 Kbps modem standard is the fastest analog phone modem standard in use on ordinary audio phone circuits.

Modulation techniques in use in modem standards applicable to industrial applications use discrete analog symbols to convey digital bits of information. By keying discrete values for the amplitude, frequency, or phase of an analog carrier signal, individual digital bits, or groups of bits, can be sent through the analog channel. BFSK uses two different frequencies, conveyed, for a period identical to the bit period, to send a single bit. Differential QPSK and differential 8-PSK send 2 and 3 bits respectively by keying four or eight phase changes respectively. A differential phase change, relative to the previous symbol, is sent, rather than an absolute phase, because this simplifies demodulation. A more sophisticated technique that supports faster bit rates is differential 16-QAM. This technique simultaneously keys 16 different combinations of discrete amplitude and differential phase values to convey 4 bits of information.

The popular slower speed modem standards used in industrial applications include V.21, V.22, V.23, V.26, V.27, V.29, and V.32. The noteworthy features for each standard, including modulation techniques are presented in Table 19.1 [8]. Where appropriate, the corresponding old Bell Telephone standard is included for reference. It is important to note that a few Bell Telephone standards are still maintained by Telcordia Technologies, which was formerly known as Bell Communications Research, the administration and standards organization formed as part of the 1984 divestiture. As an example, Bell 202, the 1,200-bps BFSK modem standard, which is almost 40 years old, is still maintained by Telcordia because it is used to send automatic number identification information in the 4 seconds between the first and second rings in switched telephone networks. Unless otherwise noted, all the standards support switched or leased telephone circuits and

Table 19.1
ITU Wireline Modem Standards

Standard	Bit/Symbol Speed	Modulation	Notes
V.21	300 bps 300 symbols per second	BFSK	Ratified 1964.
V.22	1,200 bps 600 symbols per second	Differential QPSK	Ratified 1980. Similar to Bell 212.
V.22 *bis* (second edition)	2,400 bps 600 symbols per second	Differential 16-QAM	Ratified 1984.
V.23	1,200 bps 1,200 symbols per second	BFSK	Ratified 1964. Similar to Bell 202.
V.26	2,400 bps 1,200 symbols per second	Differential QPSK	4-wire leased operation only. Point-to-point or point-to-multipoint. Ratified 1968. Similar to Bell 201B.
V.26 *bis*	2,400 bps 1,200 symbols per second	Differential QPSK	Switched telephone operation only. Ratified 1972.
V.26 *ter* (Third Edition)	2,400 bps 1,200 symbols per second	Differential QPSK	Ratified 1984. Includes echo cancellation.
V.27	4,800 bps 1,600 symbols per second	Differential 8-PSK	Leased circuit operation only. Includes a manual equalizer. Ratified 1972. Similar to Bell 208A.
V.27 *bis*	4,800 bps 1,600 symbols per second	Differential 8-PSK	Leased circuit operation only. Includes an automatic equalizer for lower-quality leased lines. Ratified 1976.
V.27 *ter*	4,800 bps 1,600 symbols per second	Differential 8-PSK	Switched telephone operation only. Includes an automatic equalizer. Ratified 1976.
V.29	9,600 bps 2,400 symbols per second	Differential 16-QAM	4-wire leased operation only. Point-to-point operation only. Nonstandard QAM: 2 bits encoded as amplitude and 2 bits encoded as phase change. Ratified 1976. Similar to Bell 209.
V.32	9,600 bps 2,400 symbols per second	Differential 16-QAM or 32-QAM	32-QAM includes trellis-coded modulation for increased performance. Ratified 1984.

(*Source:* [8].)

simultaneous transmission in both directions. Further, only the fastest speed in each standard is listed in Table 19.1. Most of the ITU standards include slower speeds for alternative suboptimum configurations. These incorporated slower speeds are often implementations of the earlier standards. This also helps to make some modems backwards compatible.

19.2.5 Wireless Modems

In contrast to wireline modem standards, wireless modem standards for use in industrial applications have seen very little standardization. One somewhat successful effort in the field of wireless data communications standardization has been the recent development of the Institute of Electrical and Electronics Engineers (IEEE) 802.11 wireless local area network standard. The applicability of this infant standard to industrial applications has not yet been determined.

Until recently, wireless modems used in remote control and monitoring were operated only in licensed narrowband frequency channels that can also carry voice signals. In the United States, these channels are licensed to the operator by the Federal Communications Commission (FCC). Typical industrial operators can include petroleum industry production or transmission companies, power utility transmission companies, or municipalities operating a water works. The available radio frequency bands include 150 to 174 MHz, 220 to 240 MHz, 330 to 512 MHz, 400 to 420 MHz, 450 to 470 MHz, and 800 to 960 MHz. Not all frequencies within each band are available. Companies such as Microwave Data Systems, Johnson Dataradio, and Electronic Systems Technology all manufacture products for these bands. They support signaling rates of between 300 and 19,200 bps. Their output power is on the order of 2 to 4W and can offer ranges of operation on the order of 15 miles.

Almost without exception, these types of modems utilize frequency-shift keying. Depending on the band, the channel spacing that is usually utilized varies from 5 kHz to 25 kHz. The majority of higher-speed modems support 19,200 bps by using variations of frequency-shift keying that more efficiently utilize the available bandwidth. These variations are members of a family of modulation techniques called continuous-phase modulation (CPM). In its most simple form it appears as a BFSK waveform where the symbol transitions occur without an instantaneous change in phase. Further, the relationship between the keyed frequencies and the symbol rate is not arbitrary as it can be in ordinary BFSK.

In 1985, the FCC made three bands available for wideband, low-power communications using a particular type of modulation called spread spectrum modulation. Spread spectrum modems that operate in these bands can use up to 1W of transmitter power. High-gain directional antennas are also permitted in some bands. Further, these types of modems can support data rates on the order of 100 Kbps. The original allocated frequency bands were in the range of 900 MHz, 2.4 GHz, and 5.7G Hz. This particular allocation by the FCC was unique because it did not require users to license their operation. While their reliable range is generally less than that of the licensed frequency data modems mentioned above, the higher data rates and speed of implementation (no lead time for licensing) makes them attractive for remote monitoring and control.

19.2.6 Accessing the Channel Using Modems

Networks that utilize these wireline and wireless modems operate in either a point-to-point configuration or in a multipoint configuration. In the former configuration, communications can be relatively simple because there is no ambiguity about who is transmitting and who is receiving. However, in multipoint configuration, addresses are commonly used to identify who is transmitting and who should listen. In the very simple version of multipoint configurations one communication entity (usually the host computer in a remote monitoring or control system) is selected to be the master and all other communications entities are designated as slaves. In a round-robin fashion the master will sequentially transmit to each slave and then receive from them in a "poll-response, poll-response, poll-response, et cetera" transaction process. More sophisticated versions of multipoint modem networks can use carrier sense multiple access techniques where ad hoc communication is permitted by first listening to the channel to see if another communication entity is transmitting. If the channel is busy, the entity desiring to transmit waits until it is clear.

19.2.7 Commercial Terrestrial and Satellite Communications Services

There are several wireless mobile communications services available in the commercial marketplace that have been adapted to fixed (and mobile) remote monitoring and control applications. Most of the technologies used in these services present an interface to the remote monitoring and control system that is similar to that of the wireless modems discussed above.

The United States' paging and cellular telephone markets have both been adopted to remote control and monitoring. In particular, cellular digital packet data (CDPD) and narrowband personal communication services (PCS), also known as two-way paging, are currently being used in industrial data applications. CDPD operates as an overlay on the first generation analog cellular telephone network, transmitting at 19,200 bps using a form of CPM called Gaussian minimum-shift keying. Two-way paging was developed during the second generation of cellular communications in the 1990s. The most commonly used technique, developed jointly by Glenayre and Motorola, transmits outbound to paging units at rates of up to 25,600 bps and receives inbound from them at rates of up to 9,600 bps. On the outbound side, the maximum data rate is achieved in a 50-kHz channel spacing using four 4-FSK signals, each supporting 6,400 bps for an aggregate bit rate of 25,600 bps [9].

CDPD modems and two-way pagers have been adapted to remote monitoring and control applications where telephone lines are not practical or are inappropriate, and where wireless licensed or unlicensed modems are also not practical. Mobile applications in the transit industry are potential users, as are petroleum industry production and transmission applications. The applications can be quite novel, for instance in retrofitted irrigation control or security applications where communications cables cannot easily be installed.

Satellite-based remote monitoring and control is used to a lesser extent, due to high operational costs. However, the petroleum industry has made limited use of satellite-based communications in remote locations. There are a wide variety of services available. Higher-end services including high-speed data connections and dial-up telephone services are available from Inmarsat and Intelsat, although both these agencies have also recently targeted the remote monitoring and control markets by offering low-speed data services. There are also several service providers that have focused solely on low-speed, low-cost data services and these include Comtech Mobile Datacom and Orbcomm.

19.3 Industrial Communications Protocols

Analog current loops and pneumatic transmitters use standardized techniques to provide interoperability between different vendors' equipment. While these techniques are not protocols, they do perform the same function in establishing a set of rules dictating how to communicate information. Similarly, the recommended specifications RS-232, RS-422, and RS-485, as well as the ITU modem specifications are not by themselves protocols, but their rules dictate how binary information is sent over a particular type of circuit.

Protocols encompass not only these types of lower-layer specifications but also the specifications associated with the higher layers of the OSI model dealing with the transport and delivery of the communicated data. HART, Profibus, and Foundation Fieldbus are commonly used to interface to smart instruments while Data Highway+, Ethernet, and Modbus are used extensively to connect PLCs, RTUs, and interface electronics with host computer systems. There are many other protocols that could be included in a comprehensive list, but most have only a small share of the industrial communications market.

19.3.1 HART Protocol

Originally designed to leverage previous capital investments in copper cabling for 4- to 20-mA current loops in large facilities, this protocol overlays a low-power analog modem signal on the current-loop signal permitting both to coexist. The analog modem signal permits multiple field devices and as many as two master devices to communicate with each other at 1,200 bps.

The protocol is administered by a nonprofit organization called the HART Communications Foundation (HCF) that is comprised of member companies that sell HART-compatible products. The Rosemount Corporation (now Fisher-Rosemount) invented HART but transferred custodianship to the foundation to create an open environment in which other vendors could participate in its oversight and easily create conforming products.

The protocol defines different command structures that all use asynchronous bytes transmitted in packets. Command data can be different depending on the type of instrument and on the function of the command and so there are universal and common-practice commands as well as device-specific commands. HART has found strong acceptance in retrofits and additions at existing facilities but is also being installed in new facilities.

The HART protocol permits operation in hazardous environments by accommodating intrinsically safe, low-power implementations and by accommodating electronic isolation barriers. In fact, in low-power configurations HART transmitters can actually draw enough operating power from the 4-mA minimum current-loop signal level, and not disturb the attached analog current-loop device. While HART has a large share of the smart transmitter marketplace it is utilized more for its calibration, testing, and diagnostic capability than its data communications capability. However, the

HART protocol is well suited for transmitting the value of the primary variable, as well as other variables in the transmitter, and can achieve a poll-response packet transaction rate of better than twice per second in communicating such data.

19.3.2 Profibus

Profibus is a protocol that was sponsored by the German Federal Ministry for Research [10] and is popular in Europe. Like HART, the protocol is administered by an organization made up of member companies that participate in maintaining the specification. Currently, the most popular Profibus configuration supports communications between devices interconnected over multidrop RS-485 circuits. It is called Profibus DP, where DP stands for decentralized peripherals. Through the use of a logical token ring, multiple masters share access to the RS-485 multidrop network. The token-holding master can then communicate with passive field instruments using a poll-response transaction.

A maximum of 127 devices can be addressed in a Profibus network, with up to 32 of them being masters. A Profibus network can be made up of multiple RS-485 segments to overcome the electrical limitations of RS-485 that limit the number of devices supported on a segment. Profibus can support signaling speeds that take full advantage of the RS-485 protocol, achieving data rates on the order of 10 Mbps. An intrinsically safe lower-speed signaling specification, IEC1158-2, is also supported by Profibus, as is low-speed fiber optics. The IEC1158-2 standard is a lower-power technique that is similar to RS-485 in most other regards.

Communication takes place between masters or between masters and slaves using three operational states. In the "operate" state, user data is read from and sent to other devices from the token-holding master. In the "clear" state, user data can be read, but no outputs on devices can be set or modified. The third state is called "stop" and is a diagnostic and configuration state [11].

19.3.3 FOUNDATION Fieldbus

The Fieldbus Foundation was created in 1993 when two other standards organizations, WorldFIP North America and the Interoperable Systems Project merged [12]. The foundation is also a nonprofit organization, like HART, and administers its protocols through the use of member-company working groups.

There are two protocols that make up what is called FOUNDATION Fieldbus (FF). The first protocol uses the same physical layer as the intrinsically safe Profibus standard, IEC1158-2, and can support speeds of up to 31.25 Kbps over balanced, multidrop, power-providing cables. The second has just recently become available and supports 100-Mbps Ethernet speeds.

FOUNDATION Fieldbus is unique in that it has moved away from the master-slave packet transaction relationship in favor of a peer-to-peer relationship that empowers the field devices more so than in HART or Profibus. However, this does add complexity and expense to the cost of field transmitters with which HART and Profibus are not encumbered.

19.3.4 The PLC and RTU Protocols

There are several protocols that are commonly used to network PLCs and RTUs. They are not necessarily for interfacing to field devices. They are used to interface PLCs to other PLCs and to interface RTUs or PLCs with host computers. Most of these protocols were developed by manufacturers of RTUs and PLCs. The most popular PLC networking protocol is Allen Bradley's Data Highway+ protocol [13]. While Ethernet is ranked second, Modicon's Modbus protocol is third.

Data Highway+ is effectively a local area network protocol that permits PLCs to be interfaced with each other and computers. The protocol supports up to 64 devices per segment, or link, and can support up to 99 links in a single network. Cable networks that implement Data Highway+ can be trunked, or daisy-chained, but not hubbed, or laid out in a star pattern. A special type of twin-axial cable is used to link devices, and by using this special cable Data Highway+ can support 57.6 Kbps over a 10,000-ft trunk (and faster speeds in specific equipment configurations). Like Profibus, Data Highway+ uses a token-passing protocol to gain access to the communications network.

Modicon created Modbus in the late 1970s for use with RTUs, and it has since become popular with PLCs and input/output interface electronics. It is a master-slave protocol that is supported over RS-485 and RS-232 interfaces at speeds up to 57.6 Kbps. Multidrop segments can be created with RS-485 and each Modbus network can support up to 247 slaves.

19.4 The Future of Industrial Remote Monitoring and Control

The computer and information ages have impacted industrial remote control and monitoring just as they have most other technology-based markets.

Computers have been used to control large processing plants and refineries since the early 1960s, and the miniaturization of the computer has created a trend wherein computer intelligence is moving out into the plant or field from the central control center while maintaining an oversight presence in the center. Distributed control systems are more common in new facility designs since they take advantage of this local control and remote monitoring scenario. Further, the PLC, RTU, and many interfacing devices are becoming increasingly more sophisticated in terms of processing power, programmability and communications. This has enabled these devices to keep up with the increased demand for information from the field.

The same is true for field instruments themselves. The popularity of protocols like HART, Profibus, and Fieldbus is due in part because they increase accessibility to field telemetry and control points. The multivariable transmitter is a contemporary marketing term for this new breed of field instruments that do more than measure a single temperature or pressure. Remote calibration, testing, and diagnostics are also motivating factors that have helped the proliferation of these new protocols.

The increased demand for field information is also attracting the use of Ethernet in the connections between the control center and the field concentration points. These points can be PLC locations or large input/output signal conditioning nodes; however, they can also be intermediate computer systems that communicate with specific areas or process equipment within the facility using different protocols.

Some existing protocols are endeavoring to keep pace with the increased demand for field information as well. Both the HCF and Profibus International are currently working to improve the performance of their protocols. Profibus International is moving to adapt the Profibus protocol to Ethernet and the Transport Control Protocol/Internet Protocol (TCP/IP) stack, and the HCF is pursuing a faster version of their HART protocol that continues to maintain compatibility with 4- to 20-mA current loops.

At the host computer end there is an exciting development involving the Internet that is redefining the meaning of remote monitoring [14]. It is now possible to access the information available in the central monitoring computer systems using the World Wide Web. The information is generally collected using computer servers at the control center. With modifications to the server software and the addition of an Internet connection, it is possible to access information from remote locations inside or outside the plant environment, and indeed, all over the world. The addition of two-way paging technology and personal communications services technology means a user can have wireless access to the server information as well. As the technology associated with this enhanced level of data access matures, it will propagate out to the field instruments, eventually making information in the instrument itself available over a wide area network infrastructure, and where appropriate, the Internet.

Problems

Problem 19.1

List the ways office and field equipment might be connected in a communication system for data acquisition.

Problem 19.2

Using the two-wire system of Figure 19.1, discuss the operation of the 4–20 current loop and how it works.

Problem 19.3

What are several limitations of the 4–20 current loop? From [2, 3] discuss the 4–20 standards.

Problem 19.4

What is the industrial de facto standard for interfacing computers and modems?

Problem 19.5

Discuss what type signal is sent on a RS-232 line. What is its maximum bit rate? What is the maximum voltage to be expected on this line between ground and the receive line? What is the minimum threshold voltage to indicate a space?

Problem 19.6

Why does a balanced two-wire circuit between transmitter and receiver, such as the RS-422, tend to have noise immunity. And what is this noise immunity called?

Problem 19.7

What is the maximum range of voltage on an RS-422? What does a positive voltage designate?

Problem 19.8

From Chapter 18, draw a figure of a telemetry system for level control.

References

[1]   Liptak, B., "Computer-Based Control in the Sixties," *Controls Magazine,* Online Archives, December 1999, http://www.controlmag.com/, accessed Jan. 3, 2001.

[2]   Harrold, D., "Here We Go Again," *Control Engineering,* Online Archives, Jan. 2000, http://www.controleng.com/, accessed Jan. 17, 2001.

[3]   "Compatibility of Analog Signals for Electronic Industrial Process Instruments," Standard of the Instrumentation, Systems, and Automation Society (formerly the Instrument Society of America), Document No. ISA-S50.1-1982.

[4]   "Interface Between Data Terminal Equipment and Data Circuit-Terminating Equipment Employing Serial Binary Data Interchange," Telecommunications Industry Association, Document No. ANSI/EIA/TIA-232-F-1997, Oct. 1997.

[5]   "Electrical Characteristics of Balanced Voltage Digital Interface Circuits," Telecommunications Industry Association, Document No. ANSI/TIA/EIA/422-B-94, May 1994.

[6]   "Electrical Characteristics of Generators and Receivers for Use in Balanced Digital Multipoint Systems," Telecommunications Industry Association, Document No. ANSI/TIA/EIA-485-A-98, March 1998.

[7]   International Telecommunication Union, http://www.itu.int/aboutitu/history/history.html, accessed April 20, 2001.

[8]   "Data Communications over the Telephone Network, the Blue Book" Vol. VIII-Fascicle VIII.1, Series V Recommendations, The International Telecommunications Union, Geneva 1989.

[9]   Motorola, Inc., http://www.motorola.com/MIMS/MSPG/spin/library_files/wad.pdf, accessed April 20, 2001.

[10] Jordan, J. R., *Serial Networked Field Instrumentation,* West Sussex, England: Wiley, 1995, p. 154.

[11] Profibus International, http://www.profibus.org/, accessed April 20, 2001.

[12] Fieldbus Foundation, http://www.foundationfieldbus.org/, accessed April 20, 2001.

[13] "Monthly Product Focus," *Control Engineering,* Online Archives, May 1999, http://www.controleng.com/, accessed March 14, 2001.

[14] Hebert, D. "It's Alive!" *Control Magazine,* Vol. 14, No. 4, April 2001, pp. 38–48.

20

Industrial Security Applications
by Dr. Guillermo Rico[1]

In recent years, the many areas of the security industry have gone through an accelerated pace in technological advancements just to keep up to date with the demand for more efficient and sophisticated security systems. In particular, the area of physical security for protecting certain special facilities requires the use of very specialized electronic equipment.

When tight security is required in facilities with buildings scattered over several square miles, remote sensing and telemetry can play a very important role. For example, at a remote point on the boundary of a facility, a microphone is installed for detecting the approach of people or vehicles. However, to eliminate the possibility of a false alarm caused by vegetation and wind, a wind speed meter is installed nearby. Furthermore, to better assess the cause of the alarm, an operator could have direct visual contact by remotely manipulating a video camera at the site.

20.1 Learning Objectives

Upon completing this chapter, the reader should understand the following:

- Interior and exterior sensors;
- Sensor placement, interior or exterior;

---

1. Dr. Rico is an associate professor in the Department of Technology at New Mexico State University. His e-mail address is gurico@NMSU.edu.

- Sensor classification according to its application: passive or active, covert or visible, line of sight or terrain following, volumetric or line detection;
- Motion detectors;
- Alarm assessment.

20.2 Sensors

Sensing events and conditions for security purposes requires the use of electronic sensors that respond to different kinds of stimuli. Sensors are therefore classified in terms of different parameters, characteristics, and applications and can be classified into two categories: interior and exterior. Although most of the sensing functions apply to both, the physical construction of the sensors will obviously be very different. Within these two groups, sensors can be further classified as passive or active, covert or visible, line of sight or terrain following, volumetric or line detection, and mode of application [1].

20.2.1 Passive and Active Sensors

Passive sensors are those that react to forms of energy emitted by persons or objects as they enter into their operating zone. Motion detectors commonly seen in residences and many commercial installations are typically passive infrared motion sensors.

Active sensors generate signals that interact with the surrounding environment. They normally consist of a transmitting element that emits the signal, and a sensor acting as the receiving element. When the received signal is altered by the presence of an intruding person or object, an alarm is generated.

20.2.2 Covert and Visible Sensors

Depending on the application, sensors could be placed in plain view and obvious to people. This provides a deterring factor against intrusions. However, a visible sensor can also allow a decisive intruder to move cautiously around its protection zone to avoid detection. This action is called "spoofing" [1]. Covert sensors are those hidden from view or disguised into equipment or furniture.

20.2.3 Other Classifications

Volumetric sensors are those that detect when a person or object enters into a space or volume. Typical examples are passive infrared sensors and microwave sensors. Sensors that detect when a person or object crosses a threshold are referred to as line-detection sensors. Examples are fiber-optic, vibration, and active (emitter-detector) infrared.

Sensors that can only operate over a straight line are classified as line-of-sight detectors. On the other hand, terrain-following detectors are those that can be installed on flat or irregular surfaces. An active infrared sensor is a line-of-sight detector, whereas a seismic sensor would be classified as terrain following.

Sensors are also characterized in terms of their probability of detection and their propensity to falsely respond to extraneous stimuli. This effect is referred to as nuisance alarm rate. In simple terms, a nuisance is an alarm that is not caused by an intrusion.

An ideal sensor has a probability of detection equal to 1.0 and a nuisance alarm rate of zero. In the real world, however, all sensors interact with their environment, and they cannot discriminate between intrusions and other events in their detection zone. This is why it is extremely important to have an assessment system in place. The most common means for assessing an alarm is through strategically located video cameras.

20.3 Boundary Protection

In large facilities, it is common to see parallel fences installed around the perimeter of the facility. The distance between the two fences is typically 8 to 15m. The area between the fences serves as a buffer zone where detection and surveillance equipment can be installed away from people, animals, vehicles, and vegetation. Only hardware and its associated power and signal cables are allowed in this zone. This fence system offers significant benefits to the overall performance of the security system such as higher probability of detection and lower nuisance alarm rate. Additionally, it represents a visible deterrent factor and a double obstacle for intruders.

Boundary protection in facilities requires the use of very diverse types of sensors. Some are buried; others are associated with a fence, while others are freestanding.

20.3.1 Ported Coax System

This sensor may be used for detection of people walking through a protected area. This sensing technique employs two "ported" coax cables buried a few inches deep and maintaining about 3 ft of separation between the two. The term "ported" comes from the fact that the cables have a small area of the insulating jacket and shield removed every foot or so. Figure 20.1 shows a typical configuration of a ported coax pair. A transmitter drives one of the cables, while the other is connected to a receiver. A magnetic field with a finite volume is produced. When the magnetic field is distorted by the presence of an intruder, the signal change at the receiver end is used as a detection event, which in turn can activate an alarm or initiate some specific outcome. This type of detection is considered volumetric even though the basic purpose is to detect the crossing of a protected boundary, which could be seen as line detection.

20.3.2 Optical Cables

Optical cables can be used effectively in intrusion detection applications. The reflections of the light traveling through the optical fibers are affected by the shape of the cable. To detect intruders as they walk through a protected area, a mesh of optical cable can be buried a few centimeters below ground, normally covered by gravel. As the intruder steps over this area, the slight change of shape of the optical cable will be enough for a sensitive electronic instrument to detect the intrusion.

20.3.3 Microwave Sensors

Another very important remote sensing application is in bistatic and mono-static microwave intrusion sensors. With this technology, intrusion detection takes place when an intruder enters into the detection volume of the sensor. Bistatic microwave sensors employ two antennas, one transmitting and one receiving. Monostatic microwave sensors have both antennas in the same housing. These sensors are very appropriate for perimeter protection of large

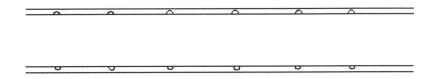

Figure 20.1 A ported coax sensor.

facilities due to their relatively large range. Bistatic sensors can operate up to about 200m between antennas and have a detection width of up to about 12m. Figure 20.2 shows the typical shape of the volume obtained from bistatic microwave sensors. The shorter the distance between antennas, the narrower the width of the volume will be. Intruders entering the protected zone cause a change in the received signal strength or phase, which in turn triggers an alarm or protection event.

Monostatic microwave sensors provide a detection zone with a teardrop shape, as shown in Figure 20.3, that can be adjusted from close range up to 120m [2]. The width of the zone can range from 1 to 7-m wide depending on the length selected. Both types of microwave sensors transmit signals in the X or K frequency band, or around 10 GHz. Detection in microwave sensors is based on the Doppler frequency shift between transmitted and received signals caused by an intruding person or object.

20.3.4 Magnetic Field Sensors

Magnetic field sensors respond to the presence of metallic objects moving near their magnetic field. These sensors employ loops of coiled insulated conductors buried in the ground. These sensors are widely used in automated traffic light control in cities. In intrusion detection, they are particularly useful for detecting intruders carrying metal tools and equipment.

20.3.5 Ultrasonic and Infrasonic Sensors

Ultrasonic sensors operate at frequencies somewhat above the human hearing frequency range (20 Hz to 20 kHz) while the infrasonic types operate at

Figure 20.2 A bistatic microwave sensor.

Figure 20.3 A monostatic microwave sensor.

frequencies below that range. Both operate under the same principle as microwave sensors (Doppler effect) but their protection zone is substantially smaller, typically a few meters. Their use should be limited to indoor applications because these sensors are very sensitive to air movement around their protection zone. In fact, the operation of infrasonic sensors is based on air movement caused by moving objects or persons that generates frequencies in the order of less than 1 Hz.

20.3.6 Video Motion Detectors

A video camera placed in a protected area and connected to a special electronic circuit can be used very effectively in motion-sensing applications. Even in dark locations such as storage rooms, night-vision cameras can offer an extra degree of security. The signal from a still video image due to the absence of moving objects or persons is a static signal. Upon the entry of an object or person within the angle covered by the camera, the video signal will immediately change, causing the detection.

The resulting outcome upon the activation of a video motion might depend on different factors. For example, during normal operating hours of the facility, an operator could be alerted so that he can assess whether an intrusion is taking place or is simply an authorized employee going into the protected area. During hours when nobody is supposed to be in a protected area, an audible alarm could be generated, followed perhaps by the recording of video images on videotape or hard-disk memory.

Remote sensing through wireless digital cameras can be implemented through the use of digital communications and computer processing. Recording can be activated by an intrusion, and even though the number of frames per second is normally set low to preserve memory, the quality of the picture is far superior to that obtained from low-speed VCRs that are typically used in most commercial applications.

20.3.7 Capacitance Proximity Sensors

These sensors are intended for detecting people getting close to large metal objects such as cabinets and desks. The metal object is isolated from ground by means of suitable insulating feet and becomes one of the plates of a capacitor with respect to ground. This capacitor is then covertly connected to an electronic circuit, such an oscillator, to produce a certain effect in the circuit. In the case of an oscillator, the oscillation frequency depends on the capacitance. When a person gets close to the protected metal object, the capacitance between the object and ground changes, and so does the oscillation frequency. This change is then used for triggering an alarm.

When isolating metal objects from ground represents an inconvenience, if appropriate, a large metal plate can be placed between sheetrock and nonconductive studs to effectively perform the capacitance function.

20.4 Alarm Assessment

As previously mentioned, alarms should be assessed by the security system operator on duty. Alarm assessment is important for two main reasons. First, it determines the cause of an alarm, including whether it was caused by an intrusion or it was assessed to be a nuisance; and second, it provides first-hand information to security personnel so that they can respond appropriately and accurately.

The most common means for assessing alarms is through closed-circuit television (CCTV) video cameras strategically located throughout the facility, mostly in and around sensitive areas. To better assess the cause of an alarm, designers can resort to pan-and-tilt actuators that can point a CCTV camera in any direction. Furthermore, most pan/tilt controllers can be ordered with an optional zoom control for the camera that allows viewing the target in more detail to better assess the alarm. When zoom control is desired, besides ordering the pan/tilt controller with zoom control capability, the lens must be specified when ordering with motor-driven zoom. It is also important to specify the lens with automatic iris control to relieve the operator from manually adjusting this function.

These controllers are normally hardwired to the pan/tilt unit by means of a multiconductor cable. However, in long-distance applications, wireless communication and control via remote sensing can be more effective than cable due to the significant wire resistance present in long cable runs.

20.4.1 Prioritized Assessment

Alternate forms of semiautomatic assessment are typically implemented through the use of the so-called dual-technology sensors. These sensors can respond to two or even three different stimuli, and they are normally used in *priority assessment* schemes. These schemes are designed to discard alarms with low probability of being actual intrusions based on the number of activated sensors. For example, a single sensor indicating intrusion in a 3-sensor detection system should probably be interpreted as a nuisance. On the opposite situation, if all three sensors get activated, an intrusion is almost certain, and quick response must be initiated.

Problems

Problem 20.1

Classify sensors into two main categories. Can the same sensor be used in both categories? If so, discuss how the sensor would differ in each case.

Problem 20.2

List the subclassifications of sensors.

Problem 20.3

(a) How does a passive sensor work? What are its advantages? (b) Name three specific passive sensors. List three places each could be used. (c) How does an active sensor work? What are its advantages? (d) Give three specific active sensors. Describe situations where each could be used.

Problem 20.4

Describe line-of-sight detectors and give a specific type.

Problem 20.5

Give a specific type of terrain-following detector.

Problem 20.6

Under what circumstances would a terrain-following detector be used? A line-of-sight dector?

Problem 20.7

Discuss nuisance alarms. How can these be minimized by placement of sensors?

Problem 20.8

How do sensors interact with the environment? Be specific by giving a sensor and its operation.

Problem 20.9

Discuss sensors with respect to boundary protection. Give five examples of when this would be important.

Problem 20.10

Discuss ported coax systems and how they work and when they would be used.

Problem 20.11

Discuss microwave sensors, both monostatic and bistatic. When are they generally used? What type of geometrical space do they protect?

Problem 20.12

Describe both ultrasonic and infrasonic sensors. What type of geometric space do they protect? When would they be used?

Problem 20.13

Describe video motion detectors. What are their advantages?

Problem 20.14

What is an assessment system? Give an example.

References

[1]   Garcia, M. L., "The Design and Evaluation of Physical Protection Systems," Butter-worth-Heinemann, Woburn, MA, 2001, pp. 67–75, 97–103, 115–123.

[2]   Racon Intrusion Detection Systems; http://www.racon.com.

Appendix A: IRIG Specifications

Table A.1
Proportional-Bandwidth Subcarrier Channels

Channel	Center Frequencies (Hz)	Lower Deviation Limit (Hz)	Upper Deviation Limit (Hz)	±7.5% Channels[1] Nominal Frequency Response N = 5 (Hz)	Nominal Rise Time N = 5 (ms)	Maximum Frequency Response[1,2] (Hz)	Minimum Rise Time[5] (ms)
1	400	370	430	6	58	30	11.7
2	560	518	602	8	44	42	8.33
3	730	675	785	11	32	55	6.40
4	960	888	1,032	14	25	72	4.86
5	1,300	1,202	1,398	20	18	98	3.60
6	1,700	1,572	1,828	25	14	128	2.74
7	2,300	2,127	2,473	35	10	173	2.03
8	3,000	2,775	3,225	45	7.8	225	1.56
9	3,900	3,607	4,193	59	6.0	293	1.20
10	5,400	4,995	5,805	81	4.3	405	0.864
11	7,350	6,799	7,901	110	3.2	551	0.635
12	10,500	9,712	11,288	160	2.2	788	0.444
13	14,500	13,412	15,588	220	1.6	1,088	0.322
14	22,000	20,350	23,650	330	1.1	1,650	0.212
15	30,000	27,750	32,250	450	0.78	2,250	0.156
16	40,000	37,000	43,000	600	0.58	3,000	0.117
17	52,500	48,562	56,438	790	0.44	3,938	0.089
18	70,000	64,250	75,250	1,050	0.33	5,250	0.057
19	93,000	86,025	99,975	1,395	0.25	6,975	0.050
20	124,000	114,700	133,300	1,860	0.19	9,300	0.038
21	165,000	152,624	177,375	2,475	0.14	12,375	0.029
22	225,000	208,125	241,875	3,375	0.10	16,875	0.021
23	300,000	277,500	322,500	4,500	0.08	22,500	0.016
24	400,000	370,000	430,000	6,000	0.06	30,000	0.012
25	560,000	518,000	602,000	8,400	0.04	42,000	0.008

Table A.1 (continued)

Channel	Center Frequencies (Hz)	Lower Deviation Limit (Hz)	Upper Deviation Limit (Hz)	±15% Channels[2,3] Nominal Frequency Response $N = 5$ (Hz)	Nominal Rise Time $N = 5$ (ms)	Maximum Frequency Response[1,2] (Hz)	Minimum Rise Time[5] (ms)
A	22,000	18,700	25,300	660	0.53	3,300	0.108
B	30,000	25,500	34,500	900	0.39	4,500	0.078
C	40,000	34,000	46,000	1,200	0.29	6,000	0.058
D	52,500	44,625	60,375	1,575	0.22	7,875	0.044
E	70,000	59,500	80,500	2,100	0.17	10,500	0.033
F	93,000	79,050	106,950	2,790	0.13	13,950	0.025
G	124,000	105,400	142,600	3,720	0.09	18,600	0.018
H	165,000	140,250	189,750	4,950	0.07	24,750	0.014
I	225,000	191,250	258,750	6,750	0.05	33,750	0.010
J	300,000	255,000	345,000	9,000	0.04	45,000	0.008
K	400,000	340,000	460,000	12,000	0.03	60,000	0.006
L	560,000	476,000	644,000	16,800	0.02	84,000	0.004
M	700,000	595,000	805,000	21,000	0.017	105,000	0.0033
N	930,000	790,500	1,069,500	27,900	0.013	139,500	0.0025

Table A.1 (continued)

Channel	Center Frequencies (Hz)	Lower Deviation Limit (Hz)	Upper Deviation Limit (Hz)	± 30% Channels[4] Nominal Frequency Response N = 5 (Hz)	Nominal Rise Time N = 5 (ms)	Maximum Frequency Response[1,2] (Hz)	Minimum Rise Time[5] (ms)
AA	22,000	15,400	28,600	1,320	0.265	6,600	0.053
BB	30,000	21,000	39,000	1,800	0.194	9,000	0.038
CC	40,000	28,000	52,000	2,400	0.146	12,000	0.029
DD	52,500	36,750	68,250	3,150	0.111	13,750	0.022
EE	70,000	49,000	91,000	4,200	0.083	21,000	0.016
FF	93,000	65,100	120,900	5,580	0.063	27,900	0.012
GG	124,000	86,800	161,200	7,400	0.047	37,200	0.009
HH	165,000	115,500	214,500	9,900	0.035	49,500	0.007
II	225,000	157,500	292,500	13,500	0.026	67,500	0.005
JJ	300,000	210,000	390,000	18,000	0.019	90,000	0.004
KK	400,000	280,000	520,000	24,000	0.015	120,000	0.003
LL	525,000	367,500	682,500	31,500	0.011	157,500	0.0022
MM	700,000	490,000	910,000	42,000	0.008	210,000	0.0016
NN	930,000	650,000	1,209,000	55,800	0.006	279,000	0.0012

1. Rounded off to nearest Hz.
2. The indicated maximum data frequency response and minimum rise time is based upon the maximum theoretical response that can be obtained in a bandwidth between the upper and lower frequency limits specified for the channels.
3. Channels A through N may be used by omitting adjacent lettered and numbered channels. Channels 13 and A may be used together with some increase in adjacent channel interference.
4. Channel AA through NN may be used by omitting every four adjacent double lettered and lettered channels and every three adjacent numbered channels. Channels AA through NN may be used by omitting every three adjacent double lettered and lettered channels and every two adjacent numbered channels with some increase in adjacent channel interference.
5. Correct channel separation of adjacent channels can be determined by comparing the higher frequency deviation of the lower center frequency channel and the lower frequency deviation of the higher center frequency channel. No overlap should occur; otherwise a channel must be deleted.

Table A.2
IRIG Constant Bandwidth Subcarrier Channels[1]

A Chans $f_d = \pm2$ kHz, Nom Freq Res = 0.4 kHz, Max Freq Res = 2 kHz		B Chans $f_d = \pm4$ kHz, Nom Freq Res = 0.8 kHz, Max Freq Res = 4 kHz		C Chans $f_d = \pm8$ kHz, Nom Freq Res = 1.5 kHz, Max Freq Res = 8 kHz		D Chans $f_d = \pm16$ kHz, Nom Freq Res = 3.7 kHz, Max Freq Res = 16 kHz		E Chans $f_d = \pm32$ kHz, Nom Freq Res = 6.4 kHz, Max Freq Res = 32 kHz	
Channel	Center Freq. (kHz)	Channel	Center Freq. (kHz)	Channel	Center Freq. (kHz)	Channel	Center Freq. (kHz)	Channel	Center Freq. (kHz)
1A	16								
2A	24								
3A	32	3B	32	3C	32				
4A	40								
5A	48	5B	48						
6A	56								
7A	64	7B	64	7C	64	7D	64		
8A	72								
9A	80	9B	80						
10A	88								
11A	96	11B	96	11C	96				
12A	104								
13A	112	13B	112						
14A	120								
15A	128	15B	128	15C	128	15D	128	15E	128
16A	136								
17A	144	17B	144						
18A	152								
19A	160	19B	160	19C	160				
20A	168								
21A	176	21B	176						
		23B	192	23C	192	23D	192		
				27C	224				
				31C	256	31D	256	31E	256
				35C	288				
				39C	320	39D	320		
				43C	352				
				47C	384	47D	384	47E	384
						55D	448		
						63D	512	63E	512
						71D	575		
						79D	640	79E	640
						87D	704		
						95D	768	95E	768
								111E	896

1. The indicated maximum frequency is based upon the maximum theoretical response that can be obtained in a bandwidth between deviation limits specified for the channel. The maximum frequency response represents a modulation index or deviation ratio of 1 (N = 1).

Table A.3

IRIG Constant Bandwidth Subcarrier Channels with the New Nomenclature

F CHANNELS Deviation Limits = ±64 kHz Nominal Frequency Response = 12.8 kHz Maximum Frequency Response = 64 kHz Center Frequency (kHz)	G CHANNELS Deviation Limits = ±128 kHz Nominal Frequency Response = 25.6 kHz Maximum Frequency Response = 128 kHz Center Frequency (kHz)
<u>256</u>	<u>512</u>
<u>512</u>	<u>1,024</u>
<u>768</u>	1,536
<u>1,024</u>	2,048
1,280	2,560
1,536	3,072
1,792	3,584
2,048	
2,304	
2,560	
2,816	
3,072	
3,328	
3,584	
3,840	

Note that only those center frequencies underlined may be supported at a range. The nomenclature for the constant bandwidth channels has been changed for all channels as indicated above for the F and G channels. In the new nomenclature, channel designation shall be the channel frequency in kilohertz and the channel letter indicating deviation limits. For example, the designation by 64A would indicate a channel with f_c = 64 kHz and deviation limits = ±2 kHz. The previous designation for this channel was 7A. The designation by 64C would indicate a channel with f_c = 64 kHz but with deviation limits = ±8 kHz. The previous designation for this channel was 7C. There is an additional higher frequency channel, H, which is not necessarily supported by the ranges. For further information on these channels, see IRIG 106-93, [1], Chapter 1.

Appendix B: Frame Synchronization Words

Pattern Length	Patterns							
7	101	100	0					
8	101	110	00					
9	101	110	000					
10	110	111	000	0				
11	101	101	110	00				
12	110	101	100	000				
13	111	010	110	000	0			
14	111	001	101	000	00			
15	111	011	001	010	000			
16	111	010	111	001	000	0		
17	111	100	110	101	000	00		
18	111	100	110	101	000	000		
19	111	110	011	001	010	000	0	
20	111	011	011	110	001	000	00	
21	111	011	101	001	011	000	000	
22	111	100	110	110	101	000	000	0
23	111	101	011	100	110	100	000	00
24	111	110	101	111	001	100	100	000

For additional synchronization words up to a pattern length of 33, see IRIG 106-93, pp. c2, reference [1], Chapter 1.

Glossary

Telemetry Section

A	Zero-to-peak amplitude of the transmitted signal
A_1	1
A_2	f_{dc2}/f_{dc1}
A_3	f_{dc3}/f_{dc1}
A_n	f_{dcn}/f_{dc1}
A/D	Analog-to-digital converter
APK	Amplitude and phase keying
AWGN	Additive white Gaussian noise
B	Baud rate—number of digital signal symbols per second
B_{bpi}	Bandwidth of ith subcarrier bandpass filter
B_c	$B_{IF} = B_t$ (at the receiver) approximately
B_c	Carrier IF bandwidth, also B_{IF}
$b(t)$	The complex baseband envelope of a *spread spectrum signal* $s(t)$
B_e	Source baud rate—number of symbols per second generated by the digital source
B_{IF}	Bandwidth of the carrier IF
B_t	Transmission bandwidth
B_v	Bandwidth of the video filter of the receiver
B_{99}	The bandwidth containing 99% of the unfiltered transmitted power
C	Channel capacity in bps

$c(t)$	Spreading code generator signal
CDMA	Code-division multiple-access
chip rate	$1/T_t$—the rate of the spreading code generator signal
DEPSK	Differentially encoded phase-shift keying
DS/BPSK	Direct sequence BPSK spread spectrum
DS/QPSK	Direct sequence QPSK spread spectrum
DSSS	Direct sequence spread spectrum—modulation by $c(t)$ causes spectrum spreading
D	f_d/f_m = mod index or deviation ratio (FM/FM)
$D = \Delta f / f_m$	Mod index (PCM/FM)
D_{ci}	Mod index of the carrier and the ith subcarrier
D_{si}	Mod index or deviation ratio of subcarrier and message in the ith channel
E	Number of levels the encoder digital input signal amplitude may assume
$e(t)$	Encoder input digital signal
E_b	Energy per bit of the received signal in joules
E_b/N_o	Received energy per bit to noise spectral density ratio
EFQPSK	Enhanced Feher-patented quadrature phase-shift keying encoder *rate* defined as $1/N$
E_s	Transmitted energy per symbol
E_s/N_o	Received energy per symbol-to-noise spectral density ratio
Δf	Peak frequency deviation (PCM/FM) of the carrier
f_{3db}	Corner frequency of the postdetection lowpass filter and for PCM/FM the bandwidth of the video filter of the receiver whose output goes to the bit synchronizer
f	Frequency of the transmitted signal
FHSS	Frequency-hop spread spectrum—rapidly changing carrier frequency
FQPSK	Feher-patented quadrature phase-shift keying
f_c	Carrier frequency
f_d	Peak deviation of the carrier by a single tone
f_{dc1}	The deviation of the carrier by the highest frequency subcarrier
f_{dc2}	The deviation of the carrier by the next highest frequency subcarrier
f_{dcn}	The deviation of the carrier by the lowest frequency subcarrier
f_{dci}	Deviation of the carrier by the ith subcarrier

f_{dcpi}	$\sqrt{B_c}\,f_{dci}$ = preliminary deviation of the carrier by ith subcarrier
f_{dn}	RMS of the deviation of the carrier by the subcarriers
f_{dnc}	RMS deviation of the carrier by the PCM/FM + FM/FM composite modulation
f_{dp}	Peak deviation of the carrier by the subcarriers
f_{dsi}	Deviation of the ith subcarrier by the ith message
f_m	Frequency of the single tone or highest modulating frequency of single message and if the 3-dB corner frequency of the post detection filter is set equal to f_m, these two are equal
f_{mi}	Maximum frequency of the ith message = message bandwidth ($f_{3\mathrm{dB}}$ is often set equal to this frequency)
f_{s1}	Highest-frequency subcarrier
f_{s2}	Next-highest-frequency subcarrier
f_{sn}	Lowest-frequency subcarrier
f_{sh}	Highest-frequency subcarrier
f_{si}	Center frequency of the ith subcarrier
f_{sL}	Lowest-frequency subcarrier in a PCM/FM + FM/FM system
k_d	Constant of FM discriminator; units V/Hz
GMSK	Gaussian minimum-shift keying
I	Information rate—number of bps of a digital signal = baud rate($\log_2 E$)
I_e	Source bit rate—number of bps generated by the digital source
I_e / B_{tnn}	A metric ratio that represents the *bandwidth efficiency* (bps/Hz)
IJF-QPSK	Interference and jitter free quadrature phase-shift keying
I_k	Polarity of the in-phase baseband signal
M	Number of levels the encoder digital output signal amplitude may assume
$m(t)$	Encoder digital output signal, also the modulating signal
M-ary ASK	M-ary amplitude-shift keying
M-ary FSK	M-ary frequency-shift keying
M-ary PSK	M-ary phase-shift keying
MSK	Minimum-shift keying
$m^2(t)$	Average value of the modulation power
N	Number of bits of information required to represent one symbol of the signal

N	Number of encoder output symbols generated by each encoder input symbol
N_o	Single-sided noise spectral density in watts/hertz
OQPSK	Offset quadrature phase-shift keying
P_b	Probability of a received bit being detected in error (equivalent to BER)
PCM	Pulse code modulation
QAM	Quadrature amplitude modulation
Q_k	Polarity of quadrature baseband signal
QPSK	Quadrature phase-shift keying
$R = R_b$	Data rate in bpss
$r(t)$	Received signal
r_{ij}	Output of correlator i when $s_j(t)$ is transmitted
$s(t)$	Transmitter (modulator) output time signal
SS	Spread spectrum
$[S/N]_c$	Carrier-to-noise ratio in the carrier IF
$[S/N]_i$	$[S/N]_c$ by definition
$[S/N]_o$	Signal-to-noise ratio in the output lowpass filter
$[S/N]_{oi}$	Output signal to-noise for ith subcarrier channel
$[S/N]_{opcm}$	Signal-to-noise out in the PCM channel or the signal-to-noise of the bit synchronizer
T_b	Bit duration
T_c	Chip duration or chip interval (as contrasted with the bit interval T)
T_h	Chip duration—duration (period) of the hopping signal $c(t)$
T_s	Symbol duration (period)
V_p	Analog signal zero-to-peak voltage
V_{pp}	Analog signal peak-to-peak voltage
Φ	Phase of the transmitted signal
fast-frequency hopping	FHSS in which several frequency hops occur for one symbol
orthogonal signaling	A signaling set in which the cross-correlation parameters are zero
power efficiency	E_b/N_o ratio required for a given BER
signal space diagram	A plot of the baseband signal $q(t)$ as a function of $i(t)$
slow-frequency hopping	FHSS in which several symbols are sent on each frequency hop

spectral efficiency	I_e/B_{tnn}—number of bps sent per Hz of channel bandwidth
symbol error rate	Rate at which symbols are received in error (equivalent to P_s)
vector diagram	Same as a signal space diagram

Antenna and Link Analysis Section

$\overline{A}$	Magnetic vector potential
A_r, A_θ, A_ϕ	Spherical components of magnetic vector potential
A_{rx}, A_y, A_z	Rectangular components of magnetic vector potential
A_e	Effective area
A_{er}	Receive antenna effective area
A_p	Physical area
A_{path}	Path attenuation in dB
AR	Axial ratio
AR_{dB}	Axial ratio in dB
A_{rc}	Receive cable attenuation in dB
A_{tc}	Transmit cable attenuation in dB
B	Bandwidth
B_v	Video bandwidth
c	Speed of light
C	Carrier power
C/N	Carrier to noise ratio
D_{dish}	Diameter of dish antenna
dim	Nominal dimension of circularly polarized microstrip patch
d_x	Feed point offset along x direction
d_y	Feed point offset along y direction
D	Directivity
D_{dB}	Directivity in dB
$\overline{E}$	Electric field
E_b/N_o	Bit energy to noise spectral density ratio
E_θ, E_ϕ	Transverse components of electric field
EIRP	Effective isotropic radiated power
E_{rad}	Radiation efficiency
f	Frequency in Hz
F	Noise figure (or noise factor)
F_{cas}	Noise figure (or noise factor) of cascade
F_{dB}	Noise figure in dB

$F(\theta, \phi)$	Normalized field pattern
F_{rec}	Receiver noise figure (or noise factor)
FSL	Free space loss
F_{sys}	Noise figure (or noise factor) of system
g_a	Element pattern
G	Gain
G_{dB}	Gain in dB
G_{pol}	Gain to pol polariation
G_r	Receive antenna gain
G_t	Transmit antenna gain
G/T	Gain to temperature ratio
H	Height of microstrip patch or horn antenna
$\overline{H}$	Magnetic field
H_θ, H_ϕ	Transverse components of magnetic field
$HPBW$	Half-power beamwidth
I	Electric current
I_0	Electric current amplitude
k	Boltzmann's constant
L	Antenna length
L_{atm}	Atmospheric loss factor
L_{cor}	Solar calibration technique correction factor
L_{fsl}	Free space loss factor
L_{mpl}	Multipath loss factor
L_{path}	Path loss factor
L_{rc}	Receive cable loss factor
L_{tc}	Transmit cable loss factor
$Margin$	Link margin in dB
MML_t	Transmit mismatch loss in dB
MML_{tot}	Mismatch loss in dB
MPL	Multipath loss in dB
N	Noise power
N_a	Number of radiating elements in wrap-around antenna
N_{in}	Input noise power
N_{out}	Output noise power
N_{ref}	Reference noise power
p	Polarization mismatch factor
p_{pol}	Polarization mismatch factor between transmit antenna and receive antenna of pol polarization
$P(\theta, \phi)$	Normalized power pattern

$P_{dB}(\theta, \phi)$	Normalized power pattern in dB
P_{ant}	Power delivered to transmit antenna
PC_{pol}	Percent spherical coverage to *pol* polarization
P_{cold}	Received power when antenna is pointed at cold sky
P_{hot}	Received power when antenna is pointed at the Sun
P_{line}	Power delivered to receive transmission line
P_N	Noise power
P_r	Available power at terminals of receive antenna
P_{rec}	Power delivered to receiver
P_{rad}	Power radiated by transmit antenna
P_r^M	Power at the receive antenna terminals in the presence of multipath
P_s	Source power
P_t	Available transmit power
PML	Polarization mismatch in dB
q_r	Receive impedance mismatch factor
q_t	Transmit impedance mismatch factor
q_{tot}	Total impedance mismatch factor
r_{ff}	Far field distance
$\hat{r}$	Unit vector in direction of far-field point
$\bar{r}_i'$	Vector to ith element of wrap-around antenna
R	Range
R_d	Distance along direct path
R_r	Distance along reflected path
R_{ant}	Antenna resistance
R_{rad}	Radiation resistance
R_Ω	Ohmic resistance
$\bar{S}_{ave}$	Average Poynting vector
SF	Solar flux
S_{in}	Input signal power
S_{out}	Output signal power
S_{11}	Reflection scattering parameter
T_A	Antenna temperature
T_b	Brightness temperature
T_{cas}	Equivalent noise temperature of cascade
T_e	Equivalent noise temperature
T_o	Reference noise temperature
T_p	Physical temperature
T_{rec}	Equivalent noise temperature of receiver

T_{sun}	Equivalent system noise temperature when antenna is pointed at the Sun
T_{sys}	Equivalent noise temperature of system
$U(\theta, \phi)$	Radiation intensity
U_{max}	Maximum radiation intensity
V_N	Noise voltage
$VSWR$	Voltage standing wave ratio
V_{tot}	Total voltage
W	Width of microstrip patch or horn antenna
X_{ant}	Antenna reactance
Y_{inx}	Microstrip patch total input admittance
Y_{inx1}	Microstrip patch input admittance along x due to slot 1
Y_{inx2}	Microstrip patch input admittance along x due to slot 2
Y_{iny}	Microstrip patch total input admittance
Y_{iny1}	Microstrip patch input admittance along y due to slot 1
Y_{iny2}	Microstrip patch input admittance along y due to slot 2
Y_0^x	Characteristic admittance along x direction
Y_0^y	Characteristic admittance along y direction
Y_{slot}^x	Microstrip patch slot admittance along x direction
Y_{slot}^y	Microstrip patch slot admittance along y direction
Z_{ant}	Antenna impedance
Z_0	System characteristic impedance
β_o	Free space phase coefficient
β_x	Phase coefficient along x direction of microstrip patch
β_y	Phase coefficient along y direction of microstrip patch
ϵ_{ap}	Aperture efficiency
ϵ_{reff}^x	Effective relative permittivity along x direction
ϵ_{reff}^y	Effective relative permittivity along y direction
(ϵ, τ)	Wave polarization state parameters
(ϵ_r, τ_r)	Receive antenna polarization state parameters
ϵ_r	Relative permittivity of microwave substrate
(γ, δ)	Wave polarization state parameters (alternate form)
(γ_r, δ_r)	Receive antenna polarization state parameters (alternate form)
Γ_g	Ground field reflection coefficient
Γ_r	Receive antenna complex voltage reflection coefficient
Γ_t	Transmit antenna complex voltage reflection coefficient
λ_o	Free space wavelength
μ_o	Magnetic permeability of free space

η_0	Intrinsic impedance of free space
ω	Radian frequency
Ω_A	Beam solid angle

About the Authors

Dr. Frank Carden is a professor in the Department of Electrical and Computer Engineering at New Mexico State University (NMSU), and he occupied the Frank Carden Telemetry and Telecommunications chair before retiring in 1994. Dr. Carden has been the principal investigator on research contracts and grants from NASA, White Sands Missile Range, Sandia National Laboratory, Lockheed, and the National Science Foundation. Dr. Carden has published and presented more than 100 papers and articles in the communications systems area in the United States and abroad. He has taught academic courses and short courses in the design of telemetering systems at the International Telemetering Conference for a number of years for NASA, and at industrial sites, and has also taught similar courses in Indonesia and Australia. Dr. Carden has worked for NASA in the area of space communication and in the telemetry industry, and has been a consultant to White Sands Missile Range and the telemetry industry in the communications systems area. He was head of the Department of Electrical and Computer Engineering at NMSU from 1968 to 1987. He received his B.S. in electrical engineering from Lamar University in Beaumont, Texas, in 1959; his M.S. in electrical engineering from Oklahoma State University in 1960; and his Ph.D. from Oklahoma State University in 1965. Dr. Carden currently teaches short courses in PCM/FM. His e-mail address is cardenb@zianet.com.

Dr. Robert Henry is currently a professor and department head of electrical and computer engineering at the University of Louisiana at Lafayette, where he has been a faculty member for 25 years. Previously, he served 2 years on the faculty of the Electrical Engineering Department at Northern

Arizona University. Dr. Henry has published and presented over 60 papers and articles in the computer control and communications systems areas. Dr. Henry was a senior engineer for the Philco Ford Corporation, where he conducted telecommunications research for NASA at the Johnson Space Center in Houston. He has presented several industrial short courses and serves as a consultant for the telecommunications industry. He is a registered professional engineer in the state of Louisiana and is recognized as a forensic engineering expert. Dr. Henry received his B.S. and M.S., both in electrical engineering, from the University of Southwestern Louisiana in 1967 and 1969, respectively. He received his Ph.D. in electrical engineering from NMSU in 1974. His e-mail address is henry@louisiana.edu.

Dr. Russell Jedlicka is an assistant professor in the Klipsch School of Electrical and Computer Engineering at NMSU. Prior to joining the faculty, he was branch manager of electromagnetic systems at the Physical Science Laboratory and also taught courses in electromagnetic theory, microwave engineering, and antenna design for 16 years. At the Physical Science Laboratory he was involved in telemetry antenna design and measurements for sounding rockets and spacecraft. Dr. Jedlicka has been principal investigator on research contracts and grants from NASA, the Army Research Laboratory, Sandia National Laboratories, and the National Science Foundation. He has presented papers at numerous technical conferences and given short courses on telemetry antenna design. He received his B.S. in electrical engineering from the University of Kansas in 1977 and his M.S. and Ph.D. from NMSU in 1979 and 1995, respectively. His e-mail address is rjedlick@nmsu.edu.

Dr. Brian Kopp, consultant and vice president of network design for Clifton, Weiss & Associates, Inc., of Philadelphia, has worked in communications for 15 years. From 1988 to 1994 at NMSU, his doctoral research for NASA focused on creating and studying bandwidth-efficient digital satellite communication signals. An entrepreneur as well, in the mid-1990s, Dr. Kopp cofounded Wireless Scientific, Inc., which manufactured a product line of 20 wireless data radio transceivers for industrial communication applications. Since selling the company in 1997, Dr. Kopp has worked as a consultant in communication technology, designing and overseeing the construction of both wired and wireless data and voice communications networks. He advises state agencies and communication protocol organizations on technical and regulatory issues. Dr. Kopp received his Ph.D. in 1994, his M.S. in electrical engineering in 1990, and his B.S. in electrical engineering in 1988 from NMSU. He can be reached via e-mail at bkopp@cliftonweiss.com.

Dr. Guillermo Rico, an associate professor in the Engineering Technology Department at NMSU, is interested in and teaches courses in analog electronics, networks, digital signal processing, and security technology, and works as a consultant in these technical specialties. He has published papers in these areas and has coauthored a book, *Electronic Devices and Circuits,* 5th edition. Dr. Rico received his B.S. in electrical engineering from the Technological Institute of Chihuahua in 1974, his M.S.E.E. from NMSU in 1978, and his Ph.D. from NMSU in 1989. He can be reached at gurico@NMSU.edu.

Index

575

Videoconferencing and Videotelephony: Technology and Standards, Second Edition, Richard Schaphorst

Visual Telephony, Edward A. Daly and Kathleen J. Hansell

Wide-Area Data Network Performance Engineering, Robert G. Cole and Ravi Ramaswamy

Winning Telco Customers Using Marketing Databases, Rob Mattison

World-Class Telecommunications Service Development, Ellen P. Ward

For further information on these and other Artech House titles, including previously considered out-of-print books now available through our In-Print-Forever® (IPF®) program, contact:

Artech House	Artech House
685 Canton Street	46 Gillingham Street
Norwood, MA 02062	London SW1V 1AH UK
Phone: 781-769-9750	Phone: +44 (0)20 7596-8750
Fax: 781-769-6334	Fax: +44 (0)20 7630-0166
e-mail: artech@artechhouse.com	e-mail: artech-uk@artechhouse.com

Find us on the World Wide Web at:
www.artechhouse.com